Second Edition

Scientific Examination of Questioned Documents

Second Edition

Scientific Examination of Questioned Documents

Edited by

Jan Seaman Kelly
Brian S. Lindblom

CRC Taylor & Francis
Taylor & Francis Group

Boca Raton London New York

A CRC title, part of the Taylor & Francis imprint, a member of the Taylor & Francis Group, the academic division of T&F Informa plc.

Published in 2006 by
CRC Press
Taylor & Francis Group
6000 Broken Sound Parkway NW, Suite 300
Boca Raton, FL 33487-2742

Printed in the United States of America on acid-free paper
10 9 8 7 6 5 4 3 2 1

International Standard Book Number-10: 0-8493-2044-5 (Hardcover)
International Standard Book Number-13: 978-0-8493-2044-6 (Hardcover)
Library of Congress Card Number 2005054142

Library of Congress Cataloging-in-Publication Data

Scientific examination of questioned documents / edited by Jan Seaman Kelly and Brian S. Lindblom.--
2nd ed.
p. cm.
Includes bibliographical references and index.
ISBN 0-8493-2044-5 (alk. paper)
1. Writing--Identification. 2. Forgery. I. Kelly, Jan Seaman. II. Lindblom, Brian S.

HV8074.S356 2006
363.25'65--dc22 2005054142

informa
Taylor & Francis Group
is the Academic Division of Informa plc.

Visit the Taylor & Francis Web site at
http://www.taylorandfrancis.com

and the CRC Press Web site at
http://www.crcpress.com

About the Editors/Authors

Jan Seaman Kelly is employed as a forensic document examiner with the Las Vegas Metropolitan Police Department Forensic Laboratory. She received her formal training in forensic document examination from George Lewis and Susan Morton at the U.S. Postal Crime Laboratory in San Bruno, California. Kelly became a diplomate in 1993 after successfully completing the certification testing process with the American Board of Forensic Document Examiners (ABFDE).

Accomplishments in the field of forensic document examination include one term as president of the ABFDE, from 2001 to 2003, and service as ABFDE's secretary from 1999 to 2001. She was a director on the board from 1995 through 1999. Kelly is a fellow with the Questioned Document Section of the American Academy of Forensic Sciences, and a regular member of the American Society of Questioned Document Examiners.

In 2004, she received the Ordway Hilton Award presented by the Questioned Document Section of the American Academy of Forensic Sciences. In September 2004, the Rotary Clubs of Southern Nevada named her safety officer of the year. ABFDE presented the Charles C. Scott Award to Kelly in 2002 for her defense of the forensic document profession as a member of the Daubert Group.

Kelly is the author of *Forensic Examination of Rubber Stamps*, released by Charles C Thomas Publisher in 2002. In 1994, the ABFDE published her 62-page monograph, *Significant Dates of Typing Methods*. She has published several articles in professional journals such as the *Journal of Forensic Sciences* and the *Journal of the American Society of Questioned Document Examiners*.

Brian S. Lindblom graduated from the University of Manitoba with a B.A. in sociology. He received his training in forensic document examination at the Royal Canadian Mounted Police Forensic Laboratory (Ottawa) in the early 1980s, where he worked until 1987. Since that time, Lindblom has been in private practice, with offices in Ottawa and Toronto. His firm, Document Examination Consultants, Inc., is an international forensic consulting company that accepts document cases from government and private sectors.

Lindblom is a member of the Canadian Society of Forensic Science, Forensic Science Society (England), American Society of Questioned Document Examiners, and American Academy of Forensic Sciences. He was certified by the American Board of Forensic Document Examiners in 1990 and served as a director between 1995 and 1998 and again from January 2004 until August 2005. He has a diploma in forensic document examination from the Forensic Science Society (England).

Lindblom has presented many papers and published several articles in professional journals. He was invited to write chapters on document examination for two scientific texts, *The Litigator's Guide to Expert Witnesses* (1997) and *Forensic Evidence in Canada* (1999). He was a guest lecturer for several years in the forensic science courses at the

University of Ottawa Law School and University of Toronto School of Continuing Education. Lindblom has also been an instructor and workshop coordinator for seminars held by the American Board of Forensic Document Examiners, the Southeastern Association of Forensic Document Examiners, and the lecture series hosted by the Federated Press on Workplace Investigations.

One of Lindblom's proudest achievements is his involvement in the development of Write-On©: from conceptualizing this computer program to finding a programmer to create the software; to assisting with development and testing; through to introducing the final product to the forensic document community.

Contributors

Bonnie L. Beal received a bachelor of science degree in chemistry with an emphasis in criminalistics from Metropolitan State College in Denver, Colorado. In 1998, she began her training under the tutelage of Frederick H. Panhorst at Indiana State Police (ISP). Since 2003, she has been a forensic document examiner, assigned to the Memphis, Tennessee laboratory of the U.S. Postal Inspection Service. Beal is a diplomate of the American Board of Forensic Document Examiners. She is also an associate member of the Questioned Document Section of the American Academy of Forensic Sciences and a regular member of the American Society of Questioned Document Examiners.

William J. Flynn holds a bachelor of science degree in computer information systems from the University of Phoenix. He began his training in the field of forensic document examination at the Philadelphia Police Crime Laboratory. In 1972 he accepted a position with the Arizona State Department of Public Safety (DPS) Crime Laboratory and retired as the chief of the Questioned Document Unit in 1991. In 1983 Flynn founded Affiliated Forensic Laboratory, Inc., a private laboratory that continues to perform forensic document examinations in both civil and criminal cases. He is a member of the American Society of Questioned Document Examiners, the American Academy of Forensic Sciences, and the Southwestern Association of Forensic Document Examiners, which he cofounded in 1981. He is a diplomate of the American Board of Forensic Document Examiners and served as this organization's president from 1989 to 1991.

Susan L. Fortunato has a bachelor of arts degree from Gustavus Adolphus College in St. Peter, Minnesota. She has been employed with the Forensic Services Division of the U.S. Secret Service for the past 17 years. Fortunato was trained and certified by the U.S. Secret Service as a qualified document chemist. Her responsibilities include determining the authenticity of documents and their age by analyzing the paper, printing, writing ink, inkjet ink, toner, and other trace components. As assistant chief of the Forensic Automation Branch, she is responsible for the connectivity, use, and proliferation of its offices' numerous databases, including access to the International Automated Fingerprint Identification System, the Forensic Information System for Handwriting, and the Questioned Identification Document database. Fortunato is an associate member of the American Academy of Forensic Sciences, a member of the Mid-Atlantic Association of Forensic Scientists, and a member of the American Society for Testing and Materials International.

Robert Gervais holds an electronic engineering technologist diploma from AMJ Electronics College in Ottawa, Ontario. In 1984 he began a 16-year computer graphics career in the awards and recognition industry. Over that time he mastered a wide variety of graphics software programs, such as Macromedia FreeHand® and Adobe Photoshop® and gained experience in the use of scanning and imaging devices as well as numerous output machines, including laser and inkjet printers. In 2000, Gervais changed his career direction and joined Document Examination Consultants, Inc., as a forensic technician. Under the direction of forensic document examiner Brian S. Lindblom, he has completed training in the development of indented impressions and works extensively in the design and preparation of demonstrative charts. Gervais has been involved in research projects and the customization of software applications as they relate to forensic document examination and has presented the findings of this research at forensic conferences.

Frank Hicks is a forensic document examiner with Rile & Hicks located in Long Beach, California. After graduating from the University of Mississippi with a bachelor of science degree in forensic science, Hicks received his professional training in the Questioned Document Section of the Georgia Crime Laboratory in Atlanta. He then returned to Mississippi and started the Questioned Document Section at the Mississippi Crime Laboratory in Jackson. After 25 years at that laboratory, he retired from government service and entered into full-time private practice with Rile & Hicks. Hicks is board certified by, and past president of, the American Board of Forensic Document Examiners (1999 to 2001) and a fellow in the Questioned Document Section of the American Academy of Forensic Sciences. He is also a member of the American Society of Questioned Document Examiners, the Southeastern Association of Forensic Document Examiners, and the Southwestern Association of Forensic Document Examiners.

Ordway Hilton (1913–1998). Sections of this book include contributions by Hilton that appeared in the revised edition (1982).

Kirsten Jackson earned a bachelor of arts degree in psychology from the University of Virginia and a master of science degree from George Washington University. She has been a forensic document examiner for 19 years and is currently employed with the U.S. Postal Inspection Service Forensic Laboratory in Dulles, Virginia. Prior to her current position, she was employed with the Internal Revenue Service, Internal Security, Forensic Laboratory. She received her training with the Virginia Division of Forensic Science during her 7-year employment with that agency. Jackson is a member of the Questioned Document Section of the American Academy of Forensic Sciences, the American Society of Questioned Document Examiners, and the American Society for Testing Materials International. The American Board of Forensic Document Examiners presented the Charles C. Scott Award to Jackson in 2002, in recognition of her work in defending the profession as a member of the Daubert Group. Jackson is a diplomate of the American Board of Forensic Document Examiners and is currently serving as a director on this organization's board of directors.

Mary W. Kelly holds a bachelor of science degree in genetics from Ohio State University and a Juris Doctor from Cleveland–Marshal College of Law. She began her training in 1983 under the tutelage of Dr. Philip Bouffard. Kelly was employed as a forensic document examiner for 22 years with the Cleveland Police Forensic Laboratory. Currently, she works part-time for the Lake County Crime Laboratory. She is a member of the American Society of Questioned Document Examiners, the American Academy of Forensic Sciences, and the Midwestern Association of Forensic Scientists. Kelly is a diplomate of the American Board of Forensic Document Examiners. She also served as president of this organization from 1995 to 1997.

Carl R. McClary received his bachelor of arts degree in psychology from the University of South Carolina in Columbia. He obtained his training in forensic document examination while employed at the South Carolina Law Enforcement Agency in Columbia. Since 1996, McClary has been a forensic document examiner with the Bureau of Alcohol, Tobacco, Firearms, and Explosives (ATF) Forensic Science Laboratory in Atlanta. McClary is chair of the E30.02 Questioned Documents subcommittee of the American Society for Testing and Materials International® (ASTM), a standards development organization. He is also a member of the American Academy of Forensic Sciences and the American Society of Questioned Document Examiners, as well as a member and past president of the Southeastern Association of Forensic Document Examiners. McClary is also a diplomate of the American Board of Forensic Document Examiners. He received the Ordway Hilton award in 2006, presented by the Questioned Document Section of the American Academy of Forensic Sciencies.

Susan E. Morton received her bachelor of arts degree from Agnes Scott College in Decatur, Georgia. Her formal training in forensic document examination consisted of two 1-year apprenticeships, one with John McCarthy of the Florida Department of Law Enforcement and the second with James Kelly at the Georgia State Crime Laboratory. As a forensic document examiner, she has held positions with the following law enforcement organizations: Georgia State Crime Laboratory (1971 to 1976), U.S. Postal Inspection Service (1976 to 1998), and San Francisco Police Criminalistics Laboratory (1998 to present). Morton is a fellow of the Questioned Document Section of the American Academy of Forensic Sciences and a regular member of the American Society of Questioned Document Examiners (ASQDE). She is a diplomate of the American Board of Forensic Document Examiners (1978 to present) and is the 1999 co-recipient of the John Hale Award from the Southwestern Association of Forensic Document Examiners. She is also the 1991 recipient of the Ordway Hilton Award presented by the Questioned Document Section of the American Academy of Forensic Sciences.

Dan C. Purdy obtained a bachelor of science degree in mathematics and physics from the University of British Columbia. He joined the Royal Canadian Mounted Police in 1969 and was trained to examine questioned documents at its Vancouver Laboratory. In 1979, he transferred to Ottawa, where he later supervised the Document Section of the Central Forensic Laboratory. In 1989, he was appointed chief scientist of document examination and served in that capacity until 1999, when he left the RCMP to establish Forensic Document Examination Services, a private consulting company in Ottawa, Canada. Purdy was awarded a diploma in forensic document examination by the Forensic Science Society in 1986. He is also certified by the American Board of Forensic Document Examiners and served as its president from 1993 to 1995. He is a member of the Forensic Science Society, the American Academy of Forensic Science, and the American Society of Questioned Document Examiners, and a fellow of the Canadian Society of Forensic Science.

Howard C. Rile, Jr., earned a bachelor of science degree in chemistry fom California State University, Los Angeles. Following an apprenticeship with the firm of Harris & Harris in

Los Angeles, he has been engaged as a document examiner, working first for the Los Angeles County Auditor's office and then the Colorado Bureau of Investigation before entering private practice in 1983. Rile is currently (2004 to 2006) the vice president of the American Society of Questioned Document Examiners. He is certified by and has previously served as a director, vice president, and president of the American Board of Forensic Document Examiners. Rile is a full-time forensic document examiner in private practice with the firm of Rile & Hicks in Long Beach, California.

Farrell C. Shiver holds a bachelor of science degree in law enforcement from Jacksonville State University, Alabama, and a master of science degree in criminal justice from Troy State University, Alabama. He is currently a forensic document examiner in private practice in Woodstock, Georgia. Shiver has been in the field for 16 years. He was trained in the U.S. Army Criminal Investigation Laboratory, where he also held positions as the chief and training officer of the Questioned Document Division. He is a member of the American Society of Questioned Document Examiners, the American Academy of Forensic Sciences, and the Southeastern Association of Forensic Document Examiners, and he is a diplomate of the American Board of Forensic Document Examiners.

Tom Vastrick earned his bachelor of science degree in forensic science from California State University at Sacramento. Currently, he is in private practice with offices in Memphis, Tennessee, and Orlando, Florida. He has been a practitioner of forensic document examination for 28 years. Vastrick received his training from the U.S. Postal Inspection Service in Washington, D.C. While employed with the Service, he was assigned to the laboratories in Washington, D.C., and Memphis, Tennessee. He is member of the American Society of Questioned Document Examiners, the Questioned Document Section of the American Academy of Forensic Sciences, and the Southeastern Association of Forensic Document Examiners. Vastrick is also a diplomate of the American Board of Forensic Document Examiners.

Preface

The second edition of *Scientific Examination of Questioned Documents* serves two purposes: first, to inform the layperson about the field of forensic document examination, the accepted methodologies, and the qualifications of a properly trained forensic document examiner (FDE); and second, this book is offered to the forensic document examination community as a reference for trainee document examiners and journeymen alike. This edition adds to an already considerable number of texts and reference materials in the profession.

In addition to the two editors, thirteen authors have made this compilation possible. Each chapter is written by a contributor with considerable expertise in a specific area. The editors felt that by compiling the works of several FDEs whose practices emphasize one or more aspects of our field, this extensively revised and updated edition would be a comprehensive book reflecting post-1982 forensic document examination, including contemporary examination methodologies.

Throughout the process of overseeing the contributors' work, and in writing our own chapters, the editors were cognizant of Ordway Hilton's leadership[1] and the passion that he exhibited for this profession. He has authored numerous articles, published two books,[1,2] and was instrumental in establishing the Questioned Documents Section in the American Academy of Forensic Sciences.

In the preface of the revised edition, Ordway Hilton wrote:

> Now another 20 years has elapsed since an up-to-date treatise has appeared. During that period, many new instruments and materials for preparing documents have been introduced, including porous tip and roller pens, single element and electronic typewriters, and dry-process copiers. Recognizing and identifying their work has presented challenges to the progressive document examiner. In response to these new problems, there have been significant modifications in technique. At the same time, workers in the field have developed other new methods, such as thin-layer chromatography, infra-red luminescence, and examination with diachronic filters, to answer older questions more completely and accurately. In a number of respects, document examination today is more advanced than it was, or had to be, in the 1950s.[1]

The advancements in establishing this profession as a science continue through today. The explosion of modern technology has influenced every facet of our lives, from introducing new avenues of written communication to improvements in ink and ergonomic design of writing instruments. The introduction of desktop computers and printers has

[1] Ordway Hilton (1913–1998) authored the first edition (1956) of this book.

impacted on the nature of examinations and methodologies that are used. Computers and accompanying software have made it possible to insert text or signatures onto a document, making it challenging to detect the alteration on a copy. A formally trained FDE is equipped to identify the insertion by applying accepted methodologies. This treatise explores the complexities of the various aspects of the forensic document profession and the variety of documents and cases submitted for analysis.

As a result of how much the world has changed since the 1982 version of this book, the reader will note that outdated subjects, such as mimeographs and office duplicating machines, have been excluded. Other sections, such as the chapter on typewriters, are more abbreviated given their limited impact in today's business world. Nevertheless, they do constitute an up-to-date summary. This book focuses on contemporary office equipment, such as modern-day digital photocopiers, fax machines, and computer printers.

Ordway Hilton chose to curtail the discussion regarding methodology because his book was not supposed to be a training manual, but rather its purpose was to educate attorneys, investigators, and others interested in the forensic discipline. Although this book covers some of the subject areas in considerably more detail, neither is it intended to be a sole-source training manual. It is, however, intended to be a valuable reference tool for summarizing up-to-date and generally accepted methodologies for the examination of handwriting, rubber stamps, photocopies, facsimiles, and computer-generated documents.

The editors, contributors, and the American Board of Forensic Document Examiners (ABFDE) offer this treatise as a reference tool to further the knowledge of practitioners and to educate the non-forensic scientist as to the foundations of forensic document examination.

References

1. Hilton, O., *Scientific Examination of Questioned Documents, Revised Edition,* Elsevier Science Publishing Co., New York, 1982.
2. Hilton, O., *Detecting and Deciphering Erased Pencil Writing*, Charles C Thomas Publisher, Springfield, IL, 1991.

Acknowledgments

The editors thank each author for his or her contribution to the second edition of the *Scientific Examination of Questioned Documents*. Each contributor accepted a tremendous responsibility in his or her agreement to write one or more chapters. It is difficult to find forensic document examiners (FDEs) who have the courage, vision, and willingness to donate personal time to take on such a task. We are grateful to the following individuals for demonstrating their commitment: Bonnie Beal, William (Bill) Flynn, Sue Fortunato, Robert Gervais, Frank Hicks, Kirsten Jackson, Mary Kelly, Carl McClary, Susan Morton, Dan Purdy, Howard Rile, Farrell Shiver, and Thomas Vastrick.

Our appreciation goes out to the board of directors of the American Board of Forensic Document Examiners (ABFDE) for their financial and moral support in the preparation of this book. ABFDE is an independent certification body for forensic document examiners. This organization has always supported the document examination community by offering quality workshops and supporting the Daubert Group, who assists those facing a Daubert Hearing or critic's challenge. Those who comprise the board of directors are individuals who have their eyes cast toward the future of the forensic document examination profession. We applaud the board's insight in seeing this worthy project through to its fruition.

The editors offer a special acknowledgment to Grant Sperry for his encouragement and support. During the early phases of our involvement as editors, many decisions had to be made and Grant was there for guidance and direction on many issues that confronted us.

We also recognize the countless contributions Rob Gervais of Document Examination Consultants, Inc., in Ottawa made in preparing the manuscript for publication, from file management to the construction and formatting of numerous illustrations in this text.

A number of other individuals provided literature references, illustrations, and other information used in chapters of this book. Our thanks to Phil Bouffard, Donald Coffey, Lloyd Cunningham, Nanette Davis, John Harris, Hans Gideon, Anne Greeson, William Jungbluth, Paul Llewellyn, Robert Radley, Karen Smith, Andy Szymanski, Peter Tytell, and Robert Whritenour.

Brian's partner, Nancy Dobson, devoted many hours of careful proofreading and editing. We very much appreciate her dedication to this project, down to the most minute detail. Brian expresses his thanks for her support and patience throughout the writing and editing of the text.

Jan's husband, David, prepared the illustrations in Chapters 4, 10, and 11, and the ABFDE seal that appears on the cover of this version. His graphic arts ability provided quality illustrations for the book.

On a personal note, appreciation goes to the family of Jan Seaman Kelly: David, Katie, and Becca. A project of this magnitude can, at times, be stressful and all consuming. Thank you for your patience, love, and support.

Contents

Section I

Introduction

Organization and Content of Book

1

BRIAN S. LINDBLOM
JAN SEAMAN KELLY

The *Scientific Examination of Questioned Documents* is a comprehensive reference text taking the reader from a description of what document examination is and the training required to practice in this field, through the array of forensic examinations conducted, and finally to the giving of evidence in court. The focus has been placed on contemporary office equipment and examination procedures that are used in the day-to-day analysis of questioned documents.

In past decades experts in the area of analyzing documents were generally referred to as *examiners* of *questioned documents* or *questioned document examiner.* These titles have given way to the more contemporary term *forensic document examiner* (FDE), or simply *document examiner.* For purposes of this book, the latter two contemporary titles will be employed.

At various points in this treatise reference will be made to attorneys. This generic term is used to describe those in the practice of law who represent a party in the litigation process or who act as legal counsel. Their titles vary depending on the jurisdictions in which they practice; however, for purposes of this book the title *attorney* will be used throughout.

Section I informs the reader about the profession of document examination: what constitutes a document, the variety of materials submitted for examination, the scope of examinations, the requirement for comparison standards, the importance and uses of reference collections, and the scientific equipment found in a forensic laboratory. An outline of the training requirements for an FDE is documented in Chapter 3. The need for a structured apprenticeship incorporating reading assignments, essays, practical tests, workshops, peer review, study tours, and mock trials is emphasized. It is only through an intensive full-time program under the tutelage of a senior document examiner that the trainee can gain the necessary grounding in forensic document examination. Membership in forensic societies and attendance at meetings and training seminars ensures that the FDE remains current and can attend to a diversity of document examination questions. Certification by the American Board of Forensic Document Examiners

(ABFDE) or qualification for a diploma in document examination from the Forensic Science Society demonstrates that the FDE has subjected himself or herself to peer group testing and is competent.

In Section II, "The Document Case," Chapter 4 focuses on the appropriate care and handling of documents prior to and during analysis. Incidental markings, folding, or other damage to the materials can hinder the examination process and even, in the most extreme situation, result in their exclusion from court. A list of dos and don'ts in the handling of documents is included in Chapter 4. The next chapter, aimed at the FDE, deals with the need to make work notes and various methods of note taking, including the use of photography, computer graphics, and other software. To effectively present evidence in court, the FDE must be able to track his or her methods and observations. Failure to make detailed, precise, and organized notes can lead to confusion and errors when reviewing the file at a later date.

In Section III, "Science, Handwriting Examination, and the Courts," a comprehensive review of issues regarding the admissibility of handwriting evidence in U.S. courts is provided. Chapter 7 summarizes the guidelines for handwriting examinations and conclusions. Much of the information contained therein is taken from the American Society for Testing and Materials International® (ASTM) standards, which are recognized by the forensic document examination community.

In 1993 the U.S. Supreme Court issued a decision in *Daubert v. Merrill Dow Pharmaceuticals*[1] that outlined five criteria for the court to use in its function as a gatekeeper to determine if a proffered expertise is a science. Through a *Daubert* hearing, the courts evaluate the validity and reliability of scientific methodologies. In some cases testimony in a given field has been limited or excluded entirely. Forensic document examination (particularly the subject of handwriting comparison) was one of the first areas to undergo *Daubert* hearings. Initially the discipline did not fare well at these challenges because a forensic expert had never before been called upon to prove that his or her discipline was a science. Under *Frye*[2], all that was required was demonstration that the expert had knowledge beyond that of the trier of fact. The community responded by uniting to catalog the established elements of this 100+-year-old discipline, including examination procedures, emphasizing higher education and continuing education opportunities, and enumerating the many peer review publications in this discipline, so as to prove to the court that the profession is a science and meets the criteria outlined in *Daubert*.

Research establishing the individuality of and variation in a person's handwriting has been carried out at the State University of New York. These studies have gone a long way in proving long-standing axioms of forensic handwriting comparison and have helped to counter claims by critics of the profession that there is no scientific basis for this examination process.

Since 1991 the forensic document examination discipline has subjected itself to extensive testing to determine if practitioners outperform laypersons. Dr. Moshe Kam, a professor at Drexel University, Philadelphia, devised and continues to administer these tests. This section summarizes the results of Dr. Kam's testing and its impact on meeting *Daubert* criteria.

[1]*Daubert v. Merrill Dow Pharmaceuticals*, 509 U.S. 579 (1993).
[2]*Frye v. United States*, 293 F. 1013, D.C. Cir. (1923).

Section IV, "Handwriting Examination and Comparison," is devoted to the very important topics of hand printing, handwriting, and numeral and signature examinations. The number of pages dedicated to this section is evidence of their significance. Handwriting comparison represents the greatest proportion of the FDE's casework and is the facet that most often leads to court testimony. The authors emphasize the need to consider all characteristics of the handwriting at issue and to exercise logic and good reasoning in rendering conclusions. Several interesting and challenging case examples are presented to demonstrate the complexity of the subject matter. The examples illustrate that simply adding up the similarities and differences does not necessarily lead to an appropriate finding. Rather, taking into consideration factors such as the writer's age, health, signature styles, and other external factors are of paramount importance. The strength of any opinion is, in part, dependent on the quality of the samples provided — a subject dealt with in the final chapter of this section.

In the chapters in Section IV, the terms *character* and *letter* are synonymous. In many countries signatures are symbolic in nature and are not composed of individual letterforms, but rather symbolic forms or ideograms. Nevertheless, similar methodology is applied in the examination of these forms irrespective of their shapes and intended meaning.

Non-handwriting analyses are covered in Sections V and VI. The subjects of typography, computer printers, facsimile machines, photocopiers, and rubber stamps, among others, are explored in considerable detail, given their prominence in daily business activities. The most substantial changes and advances in the past 20 years are those pertaining to the creation of electronic files and their output. Increasingly, casework involves questions about the source, generation, and authenticity of faxed, copied, and printed documents. Many of today's business forms and related documents are the result of a merger of graphic arts printing and computer imaging technologies. The FDE must have a good grasp of the processes involved to properly examine the diversity of questioned material presented. An entire chapter of this section addresses the very intricate subject of typography — its progression from printing presses and manual typesetting to customized personal fonts. Also explained are type classification, measuring systems and devices, and the interplay between software programs and the output devices.

The chapter on typewriters is more abbreviated than the version found in the revised edition. This reflects its limited impact in today's business world. Nevertheless, Chapter 15 provides a historical review of significant advances in its development, as well as how to distinguish one machine class from another. Contemporary information, such as the use of the Haas Atlas, is also included.

The acceptance of non-original documents in both the private and public sectors has resulted in a proliferation of copied and faxed paperwork. Authenticating documents of this nature poses many challenges, from determining the copy generation to establishing what make and model of machine transmitted the questioned fax to identifying a particular printer as the source of an anonymous communication. The advent of multifunction machines that serve as copiers, scanners, printers, and faxes has compounded these challenges. In fact, the simple question "Is this an original or a copy?" can be a perplexing problem.

Advances in the application of modern technologies to stamp making have been considerable in the past 10 years. Chapter 20 chronicles the evolution of stamp making and the characteristics associated with each manufacturing process. The examination methodology and the criteria for reaching appropriate conclusions are outlined in detail.

Section VI deals with modern press printing, paper, and carbonless paper examinations. The characteristics of the various printing processes are described, as is the history and manufacturing process of paper. The destructive and non-destructive analytical techniques used to identify paper are discussed in detail.

The detection of indented handwriting impressions has gained prominence in recent years through its application in many high-profile cases. This examination method is considered in Chapter 26 and again in Chapter 29, where recent developments in indentation sequencing and impression enhancement have broadened its value in detecting backdated and altered documents.

A suspicion that a document has been changed since its original creation is frequently the basis for submission to a forensic laboratory. Section VIII looks at erasures, additions, obliterations, and substitutions on both handwritten and electronically generated material. Instruments and measuring devices used to answer questions about a document's integrity are discussed.

Many inquiries center on the age of a document. A vast array of examination techniques and reference materials can come into play in answering questions about how old a given entry or the entire document is. A multi-faceted approach is often necessary to discover instances of backdating or recent fabrication. Investigations can involve the study of paper, pre-printed forms, ink, signature evolution, indentation analysis, and staple hole examinations. These and other analytical techniques are detailed in Section IX.

Section X, "Digital Photography and Enhancement," has a significant role in the forensic document examination laboratory. Graphic software programs are used in the preparation of sophisticated court charts and as an aid to the examination process. Faint indented impression results or obscured writings can be revealed and enhanced using a number of software programs. Scanners can help to differentiate inks, and digital cameras are capable of capturing minute details in a handwritten entry or electronically generated document.

The guidelines discussed in Section XI, "ASTM Guidelines for Forensic Document Examination," have evolved through a lengthy peer review process prior to their approval for publication. The guidelines describe well-established forensic techniques in use for many years. They continue to undergo assessment, testing, and validation to reflect advances in technology. However, the basic principles of examination methodology will remain the same: the comparison of questioned to known material is a methodical, multistep process. The order of initial steps may differ, but the entire examination process is the same one used by the forensic document examination community. By following the published ASTM guidelines, the FDE is sure to progress through his or her examination in a methodical, systematic fashion.

The document case is not always concluded with the issuance of a report. Some cases find their way to court or to a hearing before a judge, jury, or arbitrator. While conducting the examination, the FDE must be cognizant that he or she may be called upon to articulate his or her findings before a judicial body. It is imperative that the FDE has an understanding of courtroom procedure, trial preparation, the construction of demonstrative charts to illustrate findings, and the effective delivery of court testimony. Each of these topics is thoroughly discussed in Section XII.

In Chapter 34 a number of demonstrative charts illustrate the array of examinations that can be effectively supported visually. The latest in software graphics utilities have been employed to create these charts. A properly prepared illustration speaks for itself and should leave the court feeling that it would have come to the same conclusion as the FDE.

The importance of being prepared for a *Daubert* hearing or critic challenge in the U.S. cannot be stressed enough. In Chapter 36 the reader will find a point-by-point guide to preparing for court. As part of his or her evidence, the FDE must make the court aware of, among other things, the research conducted in a given subject area, peer review publications, peer group testing, ASTM guidelines, and acceptance of the method at issue by the forensic document examination community.

Chapter 37 has been dedicated to discussing untrained document expert witnesses. Even though their backgrounds vary, all of these witnesses lack the necessary formal training to apply accepted examination methodologies. The document examiner has an obligation to assist the attorney or investigator by whom he or she has been retained to appreciate the difference between the two groups. Disclosing deficiencies in the skills and qualifications of the untrained is of paramount importance in getting the court to recognize the difference between the qualified FDE and the untrained witness.

The introduction of new office equipment, scientific instrumentation, and examination procedures make forensic document examination a dynamic field. Nevertheless, the basic philosophies and principles outlined are the foundation of this varied and challenging profession. This book should provide the reader with a wealth of knowledge that is now current and will stand the test of time.

What Is Forensic Document Examination?

2

BRIAN S. LINDBLOM

Contents

2.1 The Document

Documents feature prominently in the financial, legal, business, social, and personal affairs of most people. But is a document merely a sheet of paper bearing handwriting or computer-generated text? It is not, at least not as it relates to the science of forensic document examination. In the broadest sense, a document is any material containing marks, symbols, or signs that convey meaning or a message. While the great majority of documents are produced on paper either by electronic/mechanical means or by hand with pen or pencil, they can be produced with other instruments on a variety of surfaces. Writing in lipstick on a mirror; labels on CDs, packages, or boxes; graffiti on walls; and stamp impressions visible only under ultraviolet light are but a few examples of documents beyond the narrower definition. Among this diversity is a small group that comes to be referred to as *questioned or disputed documents* — those suspected of being fraudulent or whose source or history is disputed.

A questioned document may have been prepared with any of the numerous materials available. Sometimes the very material used brings discredit or raises suspicion. Even though the elements are entirely in keeping with the document's history and purpose, its authenticity may nevertheless be contested. Some of these documents involve fraud, forgery, counterfeiting, impersonation, blackmail, threats, murder, or a host of other crimes and offenses. The issues may impact on a person's wealth, property, possessions, or reputation. Thus, the truth about the integrity and authenticity of the documents assumes great importance.

2.2 Forensic Document Examiner

The profession of an FDE or examiner of questioned documents grew out of the courts' need for assistance in interpreting evidence relating to the preparation and subsequent treatment of documents. It is actually a pure forensic science in that it developed within the legal system rather than as an extension of other professions, such as medicine, dentistry, chemistry, or engineering.

For several decades FDEs have specialized in the study and investigation of documents to determine the facts about their preparation and history. Many inquiries involve handwriting, and consequently, those consulted are often referred to as handwriting examiners or experts. This is a rather restrictive label and fails to properly reflect the FDE's scope of expertise. Not only must these experts be able to identify handwriting and printed matter, but they must also be able to recognize non-genuineness; to analyze inks, papers, and other substances that comprise documents; to reveal additions and substitutions; and to restore or decipher erased and obliterated writing. When records produced by complex modern office machines are suspected of having been manipulated, FDEs may be among the first to be consulted.

The FDE must be more than a technician; he or she must be a scientist, for the methods at his or her disposal are those of applied science. Examination must be thorough, accurate, and entirely objective. The task is not to verify pre-conceived ideas of those who submitted the case, but to establish the facts of the document's preparation and history through a study of its identifying details and the comparison of its elements with those of known specimens.

The FDE's work does not end with the discovery of the identifying details. He or she must properly interpret them and, through logical reasoning, arrive at a correct conclusion regarding the problem at hand. After forming an opinion, a report must be prepared in a manner that allows a layperson, be it a judge, juror, or other interested party, to comprehend and appreciate its substance.

2.3 Standards

Many common document investigations require comparison of the disputed material with specimens from known sources. These known, authentic specimens, commonly referred to as *standards*, must be carefully selected so as to be truly representative. They, in fact, should be a cross section of the genuine or authentic material from the particular source. Unless the standards fulfill this condition, they cannot be considered adequate or representative.

In the examination of questioned documents with known specimens, it is necessary to compare like items. This rule holds true whether the issue is the analysis of inks, papers, handwriting, computer-generated text, or any other part of a document. For example, handwriting or typewriting examinations are more apt to lead to meaningful and definitive findings if the known and disputed materials are of a similar class. In contrast, little can be determined from a comparison of handwriting with hand printing or of signatures with anonymous notes.

2.4 Reference Collections

A somewhat special category of standards consists of the material compiled in reference collections. The FDE gathers together and classifies source material such as typewriting, facsimile Transmit Terminal Identifiers (TTIs), watermarks, etc. By constantly enlarging and bringing these essential reference collections up to date, FDEs have at their disposal the necessary material with which to answer many basic questions.

2.5 Scientific Equipment

Progress in document examination has resulted in no small measure from the application of scientific instruments and the development of specialized forensic equipment. The incredible strides made in several high-technology fields have aided the FDE with his or her inquiries. The digital camera and computer scanner are two such advances. Their ability to record fine detail at a high resolution makes them invaluable in a forensic laboratory. Their versatility and portability have greatly assisted the FDE when there is a need for an on-site examination.

Two types of microscopes are utilized to analyze documents. The first, the stereoscopic binocular microscope, gives a three-dimensional magnification and is relied upon daily by examiners to uncover any number of features and important minutia. The comparison microscope, which is specially constructed to facilitate side-by-side study of microscopic details, is also a valuable tool for certain types of examination. Modern versions of both have special lighting features and lens systems to facilitate a variety of analyses and to allow for microphotography. Macro-scopes and different styles of hand magnifiers are also routinely used.

Controlled illumination is very important to an FDE's work, and several light sources are available, including daylight, incandescent, fluorescent, and fiber optic. Some are built in, while others, such as the fiber-optic light source, can stand alone or be connected to a macro- or microscope through the use of special adapters and fixtures. Not only is visible light from these special sources employed, but also irradiation can be used to excite the document in the ultraviolet and infrared portions of the spectrum. This is achieved through the use of special light boxes for ultraviolet and imaging systems that convert invisible infrared reactions into a visible image. The document can be analyzed under both infrared reflectance and luminescence conditions. The optical filtering systems are now often connected directly to a computer to allow for further image enhancement and printing functions. Several software programs that assist in improving faint or obliterated images go hand in hand with the imaging systems.

A number of measuring devices and templates have been devised to examine the alignment and spacing characteristics of typed and computer-generated text. Some are produced on glass or plastic surfaces, while others can be created and employed in graphic software programs.

Electrostatic detection devices (EDD) were designed, and have been progressively upgraded, to aid in the discovery of indented writing impressions and accidental markings. Indentations are troughs created by pressure from the writing instrument through several sheets of paper. They can be visualized through the use of electrostatic equipment. The importance of these machines has grown increasingly in the investigation of medical charts, anonymous letters, diaries, agendas, and police notebooks.

All these tools, together with chemical testing, especially by means of thin-layer chromatography (TLC), mass spectrometry (MS), and gas chromatography (GC), have made possible a marked and steady progress in document investigation.

2.6 Scientific Examination of Documents

In general, analyses are conducted to determine by whom or what device a document was created, what changes have occurred since its original production, and if it is as old as it purports to be. The range of examinations include handwriting, hand printing, and signature comparisons; detection of alterations, deletions, and substitutions; mechanical and electronic impression comparisons; detection of counterfeits; decipherment of obliterated text; restoration of water-soaked, charred, stained, or torn documents; ink and paper comparison; sequence of entry determinations; detection and analysis of indented and other accidental markings; and document dating. A vast array of items are submitted for examination, such as wills; checks; credit card receipts; business contracts; mortgages, deeds of land, transfers, and leases; insurance applications; agreements; accounting ledgers and financial records; anonymous and threatening letters; tests and examinations; medical charts; diaries, agendas, and appointment books; and legal files. Any combination of examination techniques may be required on such documents depending on the questions raised about their integrity and method of production.

All document problems fall into two basic groups, one being those that require known specimens from an individual source to reach conclusions as to their authenticity. The other group is investigated either by a study of the questioned document alone or by comparison to reference collections. In subsequent chapters of this book, the reader will come to appreciate the full scope and limitations of each area of document examination.

Forensic examinations are directed toward the discovery of those elements that become a document's identifying attributes or characteristics. Actually, all the elements help to identify it, but the most unusual have the greatest importance. Since any single characteristic, even the most individual one, might be found in documents from another source, no single element can by itself be the sole basis for identification. There are very likely many identifying characteristics that are common to a larger group of documents from different sources — such features are termed *class characteristics*. While it is obvious that these class characteristics alone cannot identify the handwriting or typing on a document, they must nevertheless be considered and evaluated. The final identification or elimination, therefore, is based upon not one or two features, but the combination of all characteristics, both class and individual.

To ensure an objective scientific procedure, the examiner must seek out both similarities and differences. He or she must distinguish between actual differences and variations.

Everyone's writing will exhibit some degree of variation, be it within a series of signatures or several pages of handwriting. The same holds true for the output of any computer printer or other office machine. The examiner must be completely satisfied that the standards used in a comparison are representative of the person's handwriting, or that the machine-produced specimens reflect all of its defects and identifying characteristics, so that unusual variations are not misinterpreted as distinguishing characteristics. In this way the FDE is able to correctly interpret the observed features. The more unsatisfactory or incomplete the specimens, the more difficult it is to distinguish whether a particular feature is truly a difference vs. an unrepresented variation.

In the preceding paragraphs, reference has been made to variation. It is yet another important factor that must be considered because no repeated act is always accomplished with exactly the same results, regardless of whether it is produced by a machine or human effort. For instance, an individual's handwriting is made up of a complexity of habitual patterns that are repeated within a typical range of variation around the master patterns; the individuality of checkwriters is influenced to some extent by the variation in their moving parts, and an ink formula will show batch-to-batch differences in ingredient proportions. In the case of handwriting, a proper consideration of natural variation is essential in order to distinguish between genuine and non-genuine writing. Regardless of the problem type, variation is ever present and must be accurately evaluated. It is as much a basic part of the identification process as each identifying characteristic itself.

2.7 Non-Identity of Source

To establish that the known and disputed materials are not from the same source requires that there are some significant differences between them, that is, fundamental identifying characteristics that do not occur in the same way in both sets of documents. These characteristics will indicate that the handwriting is by a different person, or the typed text is from a different computer printer, or the paper is from a different stock. Significant differences in class characteristics alone are generally sufficient to eliminate the person from whom or the machine from which the samples are derived. Although there may be agreement in class characteristics, the presence of fundamental differences in individual features leads to an opinion of different origins. It is a basic axiom of identification that even a limited number of basic differences, in the face of numerous strong similarities, are controlling and accurately establish non-identity.

2.8 The Opinion

The collection of observed physical characteristics is considered and evaluated in reaching an opinion. The FDE should be able to demonstrate the reasons for the conclusions in a manner that makes the evidence clear to all concerned. Ideally, if findings are presented properly, the demonstration will lead the one reviewing the report or listening to the testimony to the same conclusion.

Although it may have been inferred that all conclusions resulting from the examination of a document are definitive, this is certainly not the case. Some questions, either because of their inherent nature or because of the material available, can only be answered with a qualified opinion. Obviously, these results are less desirable for both the document examiner and the parties involved. However, this does not alter the fact that qualified opinions are appropriate in some instances. In fact, there are certain types of problems that do not permit definitive findings, except perhaps in rare instances. The reader will find specific examples of these problems in subsequent chapters. The examiner who claims to employ methods that never lead to an indefinite or qualified conclusion is one whose opinions should be considered with caution, if not skepticism.

2.9 Basis of Effective Court Presentation

The FDE is often called upon to give testimony concerning his or her opinion. Effective presentation of the documentary evidence at trial is a matter that requires coordinated effort between the FDE and the attorney. The attorney must have a basic understanding of the theories, methodologies, observations, and findings. The document examiner must be familiar with the protocol for giving evidence. The skilled attorney brings a background of legal training and courtroom experience, while the qualified FDE supplies the technical knowledge supplemented by extensive experience. Effective delivery of the evidence can only be achieved through pretrial conferences between the FDE and the attorney. These discussions are proper not only from an ethical point of view, but also in a practical way. Both parties, thus, are better able to aid justice by the most effective presentation of the facts, thereby saving valuable court time through efficient and orderly testimony. Technical, scientific testimony can be difficult for the court and jury to understand. Therefore, it must be simply and directly presented so that its basic elements and conclusions are entirely clear. This is in part achieved through the use of illustrations and demonstrative charts that highlight the important aspects of the FDE's findings.

A Forensic Document Examiner's Training

3

BRIAN S. LINDBLOM

Contents

Training standards for all of the forensic sciences have come under greater scrutiny in the past decade. Recent legal decisions and increased exposure in the media have boosted forensic awareness and encouraged courts to more vigorously explore an expert's qualifications. Consequently, the training acquired by a forensic expert has become a source of interest, and FDEs are no exception.

Document examiners bring varied academic backgrounds to their profession, including undergraduate or graduate degrees in chemistry, physics, computer science, psychology, sociology, law, and criminal justice. Many FDEs have also earned master's degrees in forensic science from universities in countries such as the U.S., Scotland, and Australia. Like many other forensic sciences, document examination training is primarily acquired through full-time in-house apprenticeships or understudies. The few universities that offer academic study in the forensic sciences recognize that training is insufficient without exposure to actual casework under the tutelage of a qualified document examiner. For this reason, forensic document examination does not lend itself solely to correspondence courses or self-study.[1] Rather, a structured, comprehensive apprenticeship is mandatory. Typically, a full-time understudy is completed over a 2- to 3-year period.

A comprehensive training program in forensic document examination includes the study of areas such as handwriting/printing and signature comparison; forgery detection; mechanical and electronic impression comparisons (e.g., computer printers, marking devices, and checkwriters); the detection of altered and fabricated documents; indentation detection and analysis; ink and paper examination; counterfeit document detection; printing processes; restoration of damaged, water-soaked, and charred documents; dating methods; and physical matching techniques. The trainee must become competent in the operation of stereoscopic microscopes, infrared and ultraviolet imaging systems, electrostatic indented impression detection equipment, various measuring grids, enhancement systems, and photographic and scanning techniques.

A substantial portion of the FDE's training is directed to handwriting/printing and signature comparison. For many FDEs, this facet represents a significant percentage of his or her caseload. One critical aspect of handwriting examination is the ability to recognize subtle differences and similarities in shapes found in writing. For this reason, tests to identify any problems with form and color blindness are often administered to candidates at the outset of their training.

Many topics are covered in the study and identification of handwriting/printing and signatures: the historical development of alphabets and writing systems; natural variation in writing; individual writing habits; methods of disguise and simulation; influences of illness, old age, and other external factors; evaluating similarities and differences; and rendering conclusions.

Many training programs feature an extensive syllabus of books, presentations, and published articles that are required reading. Essays, practical tests, oral and written exams, tours of manufacturing facilities (e.g., pens, papers, security printers), and mock trials are generally a part of the apprenticeship. During the training period, the student is also exposed to proficiency tests and supervised examination of numerous real cases that allow the trainee to gain firsthand experience in the realities and variety of actual casework.

Certain individual laboratories and systems have chosen to divide their document sections into two groups; for example, one focusing on handwriting examination and the other on alterations, counterfeits, mechanical and electronic impression comparisons, and restoration of damaged documents.

Some FDEs go on to acquire specialized training and experience in particular areas; for instance, ink analysis and dating, typography, counterfeits, facsimiles, or computer printers. In fact, some of these specialists have contributed chapters in this book.

Peer group testing is available through a number of organizations. Three of the most notable are the American Society of Questioned Document Examiners (ASQDE), the American Board of Forensic Document Examiners (ABFDE), and the Forensic Science Society (FSS). The ABFDE offers individual certification of an FDE through written and practical testing, followed by an oral presentation. The ASQDE is an international forensic document organization that requires testing for membership or waives testing if a prospective member is ABFDE certified. The FSS offers a diploma program.

As with most professional endeavors, continuing education plays an important role in the ongoing training of an FDE. Changes in office machine technologies, such as computer printers and facsimile machines, have been the impetus for the development of numerous training workshops hosted by forensic societies, academic institutes, vocational colleges, and product manufacturers.[2] Additional knowledge is gained through regular attendance at forensic conferences and by review of published literature in the field. Forensic document examination is clearly not a static expertise, and a continuous accumulation of new information is essential to ensure that the FDE maintains competency.

No part of the training involves the study of graphology or grapho-analysis. This is because graphology attempts to distinguish personality traits by assessment of a writing sample, whereas forensic document examination involves the comparison of writing to determine authorship. While the subject matter in both cases is writing, the objectives are quite different. Unfortunately, a number of graphologists have, over the years, ventured into document authentication, in spite of the fact that they have no formal training in the field of forensic document examination.

The FDE who is incapable of presenting forensic evidence will be ineffective as an expert witness no matter how well educated, trained, and experienced he or she is. The FDE must learn how to objectively answer questions posed by attorneys and to clearly and concisely present technical findings to laypersons. Learning to prepare illustrative charts and other forms of demonstrative evidence goes hand in hand with effective testimony. Time and effort also are required to gain familiarity with the relevant court systems.

Training standards have become a crucial factor, not only for forensic document examination, but for all forensic sciences. A comprehensive, effective training program in forensic document examination must consist of a well-documented program of study and analysis of numerous actual and simulated cases in a controlled laboratory setting, supplemented with attendance at forensic meetings and access to forensic journals. Such full-time, comprehensive apprenticeships are essential.

References

1. American Board of Forensic Document Examiners, *Profile of a Forensic Document Examiner*, American Board of Forensic Document Examiners, Houston, n.d.
2. Frieman, M.J. and Berenblut, M.L., Eds., *The Litigator's Guide to Expert Witnesses*, Canada Law Book, Aurora, Ontario, 1997, p. 163.

Section II

The Document Case

Care and Handling of Documents

4

JAN SEAMAN KELLY

Contents

The improper or careless handling of a disputed document* can compromise the chain of custody or the examination. The legal arena after *Daubert* mandates proper chain of custody of the evidence, i.e., the questioned text and the known exemplars. In most cases, protocol for handling document evidence is mandated by the FDE's employer. Law enforcement agencies have policy manuals listing their procedures on the collecting, recording, and handling of evidence. Because the FDE usually is not one of the original crime scene responders or the investigating detective, he or she is only responsible for the proper care and handling of evidence once it comes into the document examiner's possession. Controlling the actions of those individuals handling the evidence prior to its submission is outside the scope of the forensic scientist's responsibility.

Forged checks, car titles, deeds, accounting records, and wills are just a few of the documents that will have been handled by numerous individuals in the normal course of business. Common problems, such as stamping with a rubber stamp over handwritten text, punching holes, stapling, and folding, occur during business transactions. The process of proper handling of document evidence begins when a document is first recognized as having some level of significance in a crime or other litigation or civil inquiry.

*Disputed documents are not limited to text on paper. Walls, mirrors, glass, and corpses are examples of other substrates that may be encountered, any of which may contain written or printed information, signatures, drawings, etc. For the purposes of this chapter, the term *document* is inclusive of all media containing information in any visual form.

4.1 Procedures in Handling Documents

Common sense is the first of two golden rules in the care and handling of evidence. Proper collection and handling procedures help maintain integrity of the evidence. Improper handling of the evidence can limit the examination results. Any alteration can seriously compromise the examination by preventing certain non-destructive testing to determine indented writing or ink differentiation. As a final note on this rule, prior to the examination, the FDE should be made aware of additional forensic testing that will be done upon completion of his or her examination. This information will help to ensure the evidence is not compromised, thereby interfering with or preventing a forensic examination by other disciplines.

The second golden rule: the document examination must take place before the evidence is examined by other forensic disciplines. For example, processing the document for latent prints prevents the FDE from examining it for indented writing. Latent print processing may contaminate or damage the ink. This type of damage can prevent the FDE from conducting a comparison of the disputed text to the suspected writer's known exemplar. In some instances, the document must be subjected to an examination involving destructive testing. Thin-layer chromatography (TLC), for example, requires removal of an area of ink and paper (media) for determination of the ink sample's chemical makeup. Therefore, to ensure a comprehensive document examination, the FDE has to have first priority in examining the evidence.

The procedures for handling document evidence discussed in this chapter are recommendations. Their application is determined, in part, by whether the case is criminal or civil. In criminal cases, law enforcement agencies usually have policy manuals outlining procedures for the collection and storage of evidence. Those policies tend to be generic, while the procedures discussed below address the specific actions that should be taken in the handling, marking, and packaging of evidence. In civil cases, a few of the recommendations may not be considered significant. For example, it may not be important to the litigators or the courts whether or not protective gloves were worn by any of the individuals handling the documents. Cutting to the chase, individuals outside the control of the document examiner determine application of these recommendations. Educating his or her clients as to how the evidence should be handled and making sure he or she follows the appropriate recommendations ensures the integrity of the evidence.

Even though the FDE is only responsible for the evidence once it is in his or her possession, educating those who have access to the evidence might lessen the occurrence of careless handling. Instruction on the care and handling of documents can be given to those who attend continuing education seminars sponsored by law enforcement agencies, bar associations, and law schools. When necessary, the FDE can provide individual instruction on a per case basis.

Because the actions of those who are the first to handle and collect the document can affect the examination, recommended procedures for them warrant discussion. The detective or investigator must take great care to preserve the documents. Wearing of gloves is recommended to ensure that latent prints from crime scene responders are not added to the disputed document. Careless handling can also cause cross-contamination that directly affects the interpretation of the evidence. For example, retrieving a suicide note from the decedent's desk and laying it on top of a bathroom counter can cause cross-contamination of the note.

The investigator should record the location of each document in a separate log or by writing this information on the evidence bag. If recorded on the evidence package, this information must be written prior to the document's insertion to prevent the introduction of writing indentations.

A document will retain its evidentiary value by observing the following guidelines:

Don't:

1. Use staples, pins, or clips.
2. Make perforations.
3. Create new folds.
4. Write on or otherwise mark the documents. If labeling and initialing is necessary, then make identification marks in blank spaces.
5. Write on the evidence envelope if it contains the document.
6. Touch the document with a pencil, pen, or other pointed instrument.
7. Glue the pieces of a torn or fragmented document onto a background. Submit them "as is" in a sealed evidence package.
8. Make erasures.
9. Make any tracings.
10. Cut or tear a document. Also, do not attempt to repair a cut.
11. Underscore, highlight, or circle words.
12. Carry disputed documents loosely in pockets.
13. Allow other people to handle or tamper with questioned writings.
14. Allow the suspect to handle or see the disputed document.
15. Allow anyone other than a properly trained forensic scientist to make chemical or other tests.

Do:

1. Use envelopes for storage and transportation. The envelopes should be large enough to hold the document without folding. If the document is to be examined for fingerprints or indented writing, wear gloves or hold it tightly between the sides of the fingers, but only long enough to place it in the evidence envelope. Be sure to label the envelope for fingerprinting.
2. Photocopy, photograph, or scan the documents as required — the process will not damage them in any way. *Do not* use the autofeed option on a photocopier.
3. Protect documents from excessive heat, light, or dampness.
4. Prepare a complete history of each document showing date, place, and from whom it was obtained.
5. Handle documents as little as possible.
6. Remember: The document evidence must be examined first before it is subjected to additional forensic examinations. Be sure to state on the evidence package or in the request to the FDE that additional forensic testing is desired.

The use of envelopes for storage and transportation of documents requires further discussion. The use of paper or lift-free plastic sleeves, page protectors, or plastic bags for evidence storage can be a personal choice or one mandated by an employer's policy. Storing a document in an improper plastic sleeve can be disastrous. A chemical reaction occurs when machine-generated text comprised of toner comes into contact with plastic. The toner-based text can be damaged as a result of toner adhering onto the plastic when the two materials are separated. If a plastic sheet protector is necessary, lift-free toner-safe sleeves should be used. The lift-free feature prevents the transfer of machine-generated text onto the plastic.

4.2 Photographic Evidence

A number of the previous recommendations are applicable only to the evidence that can be transported to the FDE. Occasionally, the disputed text will be on an object that cannot be easily sent to the FDE's laboratory. This type of case may include writing on corpses, pipes, beams, doors, and walls. Instead of original evidence being submitted, the FDE is provided with photographs. Since the best evidence will be a photograph, the best-case scenario is for the FDE to go directly to the crime scene to photograph and record the questioned text. If this is not possible and the opportunity presents itself, instruction on obtaining photographs of the text should be given. Part of the instruction should include informing the photographer that his or her camera position should be parallel to the text (Figure 4.1a and b). Photographs taken at odd angles are of no benefit in an examination. At least one photograph capturing the entire text (if very large, several shots are needed to achieve the panoramic view) and a close-up of each letter or set of letters will allow the FDE to view the text as a whole and by individual letterforms (Figure 4.2a–c). The person photographing the disputed text should record the date, time, and location of the subject matter. The photograph also should contain a ruler or scale. This will provide documentation of the actual image size. He or she also should use a camera suitable for the task. Polaroid or low-end point-and-shoot cameras are unlikely to record the detail needed for a document examination. Whether taken by the FDE or by other individuals, photographs should be handled the same as the original evidence.

4.3 Charred Documents

The handling of charred documents requires extraordinary care. Even though examinations of charred documents seldom occur, the FDE and investigator should have a basic understanding of how to collect and transport this delicate evidence to the forensic laboratory.

Charred papers vary in physical composition and are burned under many different conditions. The composition of the original paper has much to do with the fragility of the charred fragments. To avoid unnecessary breakage, it is advisable to treat all charred documents as though they were of the most fragile type.[1] The collection technique should be well thought out and not a rash decision. Some of the burnt documents may be pliable, while others will fall apart at the slightest touch. Charred documents found in a safe or storage box should not be removed from the container. Due to their fragile state, it is recommended the evidence be transported to the FDE as soon as possible. Prior to transporting, cotton wool or other cushioning material should be packed between the documents and the sides of the container.[1] Loose documents require an additional step to limit or prevent breakage. If stacked, they should never be separated into individual pieces. Instead, it is best to use a spatula to slightly lift the delicate evidence up and slip a pie tin or similar piece of thin metal or stiff cardboard underneath. In some cases, a large pan

(a)

(b)

Figure 4.1 Two illustrations of tagging (graffiti writing) on objects that cannot be submitted to the forensic laboratory for examination. (a) Photograph of a tagger's moniker on a block wall. (b) Photograph of a truck that was damaged by multiple writings of a tagger's moniker and his crew name. In both photographs, the entire text is photographed with the camera parallel to the writing. The image also includes as much of the object as possible that bears the graffiti. Having a photograph that captures all of the writing and surrounding area not only records its position in relation to its surrounding environment, but also demonstrates the size of the graffiti on the damaged property.

can be used if the mass of charred documents can be moved safely without causing further damage. The pan is then placed inside a sturdy box and packing material is used to hold it securely in place. Flat-bladed tweezers are recommended to gently place (exchange) single pieces in a flat box or pan to protect them from damage.[1]

4.4 Summary

This chapter has discussed the care and handling of documents. The dos and don'ts in marking and packaging of document evidence have been enumerated. As stated earlier,

(a)

(b)

Figure 4.2 (a) Photograph of a tagger's moniker on an electrical box. This photograph records the complete name. (b and c) Close-up photographs of the letters. These photographs aid the FDE in his or her examination of the writing, as they reveal the minute detail that cannot be observed in (a). Panoramic and close-up photographs are recommended for questioned text that is on an object that cannot be submitted to the forensic laboratory for examination.

(c)

Figure 4.2 *Continued.*

the recommendations listed provide a guideline that every investigator and FDE can follow. Preservation of the evidence through careful handling helps ensure that the truth, as revealed by the document evidence, can be discovered.

References

1. Doud, D., Decipherment of charred documents, *J. Criminal Law Criminol.*, 38, 812–826, 1948.

Work Notes

5

JAN SEAMAN KELLY

Contents

Work notes, also called case notes, are the written historical record of the FDE's activity as it relates to a specific case. Any format is acceptable if it achieves its primary objectives of stating or describing the methodology used in the examination(s) conducted, assists the FDE in identifying the documents submitted for examination, and states the results of all of the examinations conducted in the case. The text in the work notes can be handwritten or machine generated. The FDE may use pre-printed forms containing blank boxes or lines to aid in the recording of specific information. Great latitude is given to the formatting of work notes as long as they provide the document examiner's accurate record of the case and examination history.

5.1 Format of Work Notes

Formatting of work notes is influenced quite heavily by whether the FDE is a private practitioner or employed by an agency. If an accredited American Society of Crime Laboratory Directors Lab (ASCLD-LAB) is the FDE's employer, the agency's policy manual will set out the guidelines. As a rule, the format chosen is the minimum an FDE has to do on all casework. For example, if the policy manual states the document examiner will make photocopies of the documents and use a pen to mark the similarities and dissimilarities observed between the questioned and known texts, every case has to have this method of demonstrating these characteristics. If, in addition to marking photocopies, the FDE chooses to hand draw some of the letterforms, this is acceptable, as it exceeds the minimum requirements. Agency laboratories that are not accredited may or may not have policy manuals detailing the content of the FDE's work notes. In cases such as this, the formatting may reflect the one chosen by the FDE and not his or her employer.

Private practitioners are freer to choose their own method; he or she has the luxury in formatting case notes tailor-made for the case. Since private practitioners experience a wider range of document cases than those worked by government FDEs, this freedom provides the flexibility to style the work notes to fit the type of case that has to be examined.

5.2 Creating a Case File

The initial step in creating work notes is to prepare a case file that may be identified by either the client's name or a unique numerical identifier. The original request, whether it is an agency's laboratory request form or a letter from the submitting party requesting the examination, is placed in the file. In the case file, the FDE notes the date the evidence was received and makes an itemized listing of the documents. Written correspondence, meetings, e-mails, or telephone conversations should be noted and dated in the work notes. Some private practitioners have retainer contracts their clients sign once the case is accepted; this paperwork is also maintained. If the document examiner uses abbreviations, a legend should be included. This further promotes the objective of communicating to a third party (reader of the notes) the steps taken and the observations of the FDE.

The FDE usually chooses a format that works best in refreshing his or her memory as to the evidence submitted and the examinations conducted. For example, in a handwriting or signature case, some document examiners hand draw the questioned and known text as their method to visually capture the minute detail in each character of the writing. Their work notes may include a description of the significant similarities or differences observed between the questioned and known texts. Other FDEs prefer to make copies of the documents and note the similarities and differences using different-colored markers, pen inks, or pencils. Other examiners scan the documents and use computer and graphics software such as Adobe Photoshop® or CorelDraw® to electronically notate the characteristics used as the basis for their conclusion.

5.3 Documenting the Submitted Evidence

Preparation of work notes includes the reproduction of the submitted evidence. Using a scanner and computer, photocopying, and photographing each document are contemporary reproduction methods. If destructive testing is involved in any of the examinations, it is extremely important to record the evidence as it was originally submitted prior to such testing. If scanning or photocopying is the reproduction method chosen, do not use the autofeed option on the machines. Documents can be damaged if a malfunction occurs in the autofeed mechanism, causing the document to jam. Also, the mechanics of autofeeding can add marks during the grabbing and moving of the sheet of paper through the copying or scanning process.

A copy of the document reflecting its original size should be made. Copy settings for any enlargements or reductions to any portion of the document should be recorded in the work notes. If photographed, at least one photo of each document should be 1:1. A ruler should be in the photo as a means to determine the degree of enlargement or reduction.

In some situations, it may be desirable to record settings such as f-stop, lens size, film or digital media, etc.

Digital enhancement is a useful tool for documents bearing obliterated writing, faint impressions on a latent lift (indented writing) sheet, or faint writing on dark media. The steps taken to enhance the text should be recorded clearly enough so that another FDE could follow the procedure and achieve the same image quality. If hard copies are made, each copy should state what step was taken for that particular image and be placed in the file. It would also be wise to save the images on media such as a CD and store it in the case file.

Software programs such as Write-On© and Access® catalog individual letters, letter combinations, or words. These programs are tools FDEs use in their examinations of extended writings and cases involving a large number of questioned documents. They are quite beneficial in assisting the document examiner in knowing the frequency of occurrence of certain characteristics (Figure 5.1a and b). Both programs require the documents to be scanned into the computer. The scanning of the documents and the information retrieved as a result of using the software should be noted in the work notes.

Indented writing and suspected alterations or obliterations of visible writing require specialized instruments in order to conduct the examination. Use of infrared/infrared luminescence (IR/IRL) and imaging systems allows the FDE to document what he or she observes on a monitor screen. The results of the examination can be recorded by using the printer attached to either of these instruments. If the system is not connected to a printer, a photograph can be taken of the image on the monitor (Figure 5.2). Another method to record the results of IR/IRL is to videotape the images that appear on the monitor. Examinations involving instruments, including stereomicroscopes and even hand magnifiers, should be noted in the work notes. A description and the results of any control tests should be recorded.

The examination using an electrostatic detection device (EDD) produces a latent lift sheet that might or might not contain evidence of indented writing. Latent lifts containing indented writing are saved and stored according to the written policy of a government agency or the policy of the private practitioner. If the original lift is not stored in the case file, a reproduction copy of the latent sheet clearly showing the indented text should be made and added to the work notes. A notation as to the location of the original EDD lift should be made.

The date, control test results, and results of the examination involving instrumentation should be recorded in the work notes. As of 2004, according to ASCLD-LAB policy, it is not mandatory for a government agency or private practitioner to save negative-result EDD latent lifts, i.e., no discernible indentations observed. A number of government organizations do not require maintaining negative lifts because of the cost of materials and storage. In this circumstance, the FDE notes that the EDD examination was negative and the lift was not saved. Private practitioners are more likely to keep hard copies of all examination tests in order to demonstrate that they conducted the test. However, this is not a hard and fast rule, as it is the FDE's personal choice.

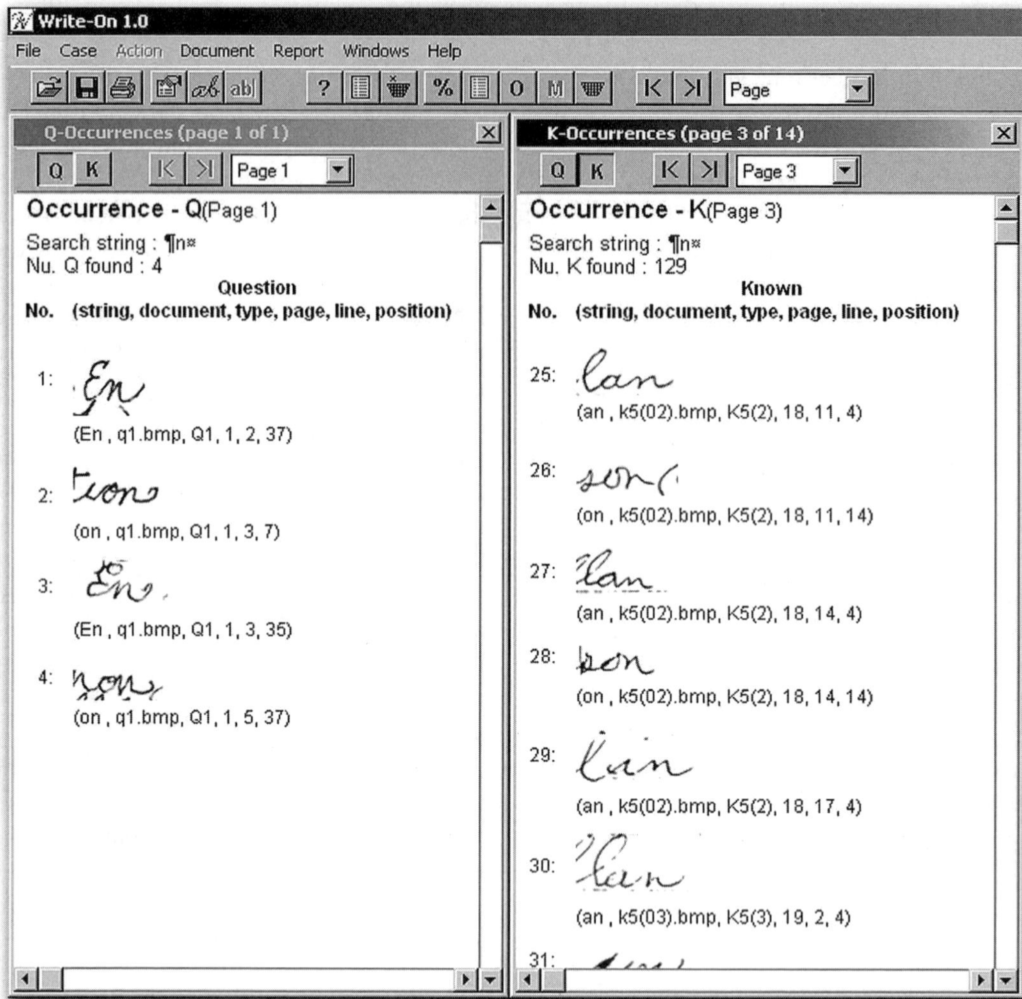

(a)

Figure 5.1 (a) A Write-On© occurrence chart showing the instances where "n" appears in the terminal position. As can be seen from the "Q found" and "K found," there are four questioned n's and 129 knowns. (b) The upper table shows the search statistics for queries made in Write-On© Handwriting Comparison Software. The data include the search number, word or letter combination searched for, occurrences in the questioned and known material, the date the search was conducted, and the documents on which the characters appear. In the second table, a word index is found listing the word or letter combination searched for, the number of occurrences, and the location where each word or letter combination appears on the documents searched.

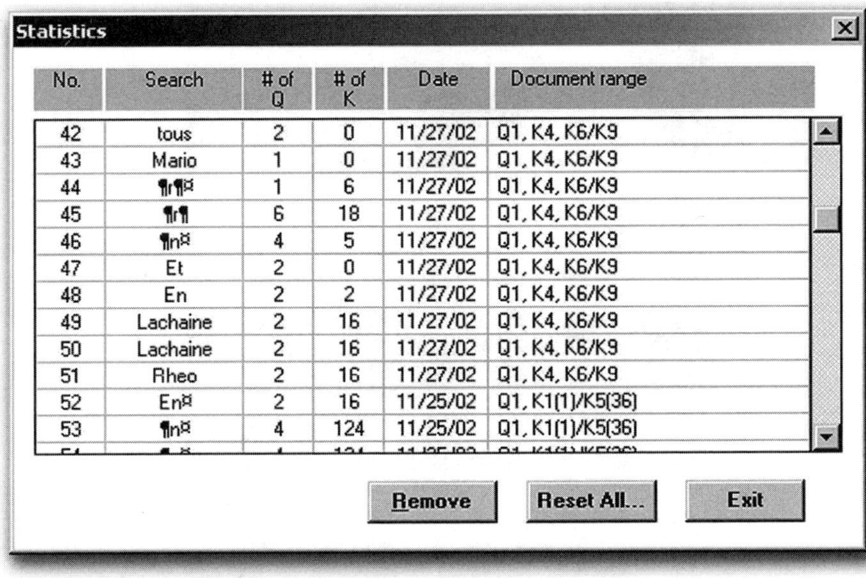

Statistics

No.	Search	# of Q	# of K	Date	Document range
42	tous	2	0	11/27/02	Q1, K4, K6/K9
43	Mario	1	0	11/27/02	Q1, K4, K6/K9
44	¶r¶¤	1	6	11/27/02	Q1, K4, K6/K9
45	¶r¶	6	18	11/27/02	Q1, K4, K6/K9
46	¶n¤	4	5	11/27/02	Q1, K4, K6/K9
47	Et	2	0	11/27/02	Q1, K4, K6/K9
48	En	2	2	11/27/02	Q1, K4, K6/K9
49	Lachaine	2	16	11/27/02	Q1, K4, K6/K9
50	Lachaine	2	16	11/27/02	Q1, K4, K6/K9
51	Rheo	2	16	11/27/02	Q1, K4, K6/K9
52	En¤	2	16	11/25/02	Q1, K1(1)/K5(36)
53	¶n¤	4	124	11/25/02	Q1, K1(1)/K5(36)

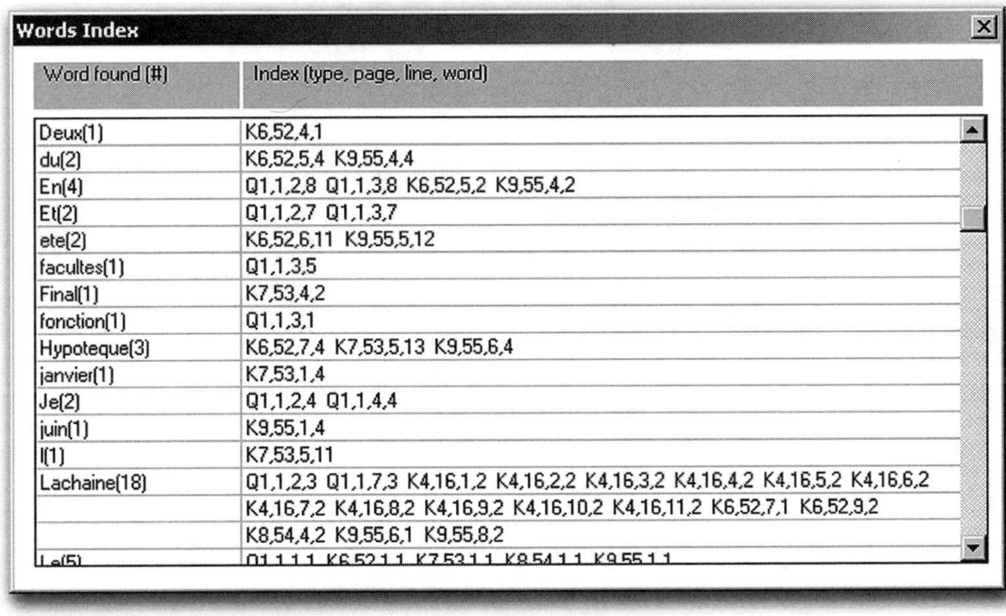

Words Index

Word found (#)	Index (type, page, line, word)
Deux(1)	K6,52,4,1
du(2)	K6,52,5,4 K9,55,4,4
En(4)	Q1,1,2,8 Q1,1,3,8 K6,52,5,2 K9,55,4,2
Et(2)	Q1,1,2,7 Q1,1,3,7
ete(2)	K6,52,6,11 K9,55,5,12
facultes(1)	Q1,1,3,5
Final(1)	K7,53,4,2
fonction(1)	Q1,1,3,1
Hypoteque(3)	K6,52,7,4 K7,53,5,13 K9,55,6,4
janvier(1)	K7,53,1,4
Je(2)	Q1,1,2,4 Q1,1,4,4
juin(1)	K9,55,1,4
l(1)	K7,53,5,11
Lachaine(18)	Q1,1,2,3 Q1,1,7,3 K4,16,1,2 K4,16,2,2 K4,16,3,2 K4,16,4,2 K4,16,5,2 K4,16,6,2
	K4,16,7,2 K4,16,8,2 K4,16,9,2 K4,16,10,2 K4,16,11,2 K6,52,7,1 K6,52,9,2
	K8,54,4,2 K9,55,6,1 K9,55,8,2

(b)

Figure 5.1 *Continued.*

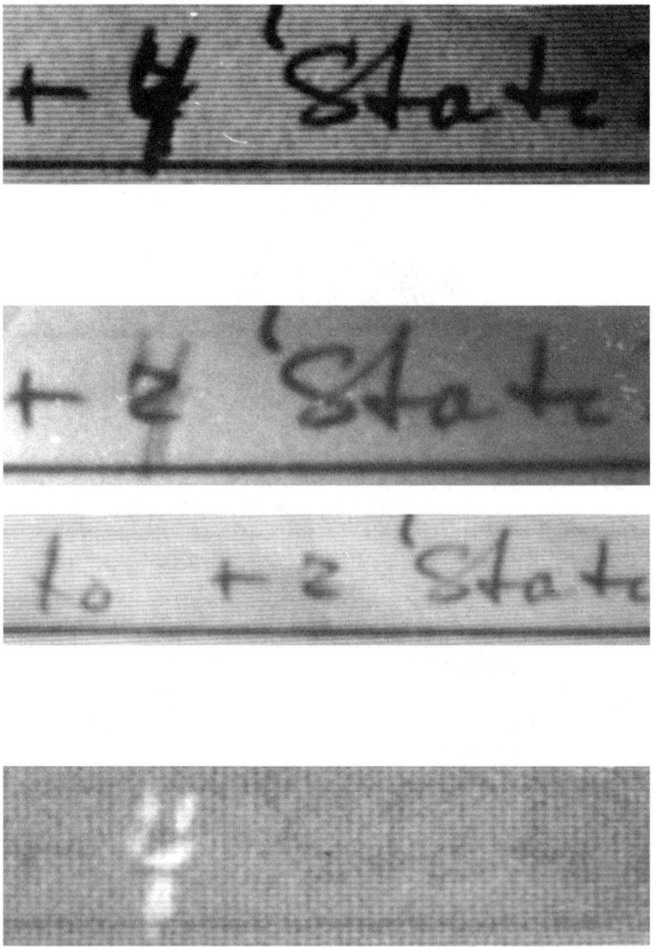

Figure 5.2 A handwritten entry examined under infrared excitation. The top window shows the unfiltered entry. The next two images show the reaction of the numeral 4 under infrared reflectance at different wavelengths. The bottom window shows the reaction of the 4 relative to the remainder of the entry under infrared luminescence.

5.4 Conclusion

The case file may be in the form of hard copies or may be electronically stored. It is comprised of the work notes, which are a historical recording of the FDE's activities related to the case and should include the following: copies of all of the submitted evidence (both questioned and known) used in the examination, any correspondence between the submitting party (attorney, investigator, the court, etc.) and the FDE, description of methodology used in the examination(s), and the issued report. All of this information should reflect the date of occurrence of each activity. The work notes should be in a format to allow another document examiner to duplicate the actions of the originating FDE in order to arrive at the same conclusion. In lieu of a discovery subpoena, copies of the case file may be given to opposing counsel. Thorough case notes will not only assist the FDE, but also may encourage the opposition to seek a settlement if the work notes indicate the document evidence is quite damaging to his or her client's case.

Section III

Science, Handwriting Examination, and the Courts

The *Daubert* Era

6

KIRSTEN JACKSON

Contents

The 1990s were a decade that significantly and forever transformed the relationship between forensic document examination and U.S. courts. The cumulative decisions by the U.S. Supreme Court in *Daubert*, *Joiner*, and *Kumho Tire* (the *Daubert* trilogy) decidedly changed the way that all expert testimony is presented and considered. Dr. Moshe Kam, a professor at Drexel University, began a series of controlled studies that continue to be crucial in the admissibility of the forensic examination of handwriting. And the discipline itself, under enormous pressure to comply with changing legal requirements, would demonstrate its ability to respond and strengthen as a profession in the era of *Daubert*.

6.1 *Daubert* Chronology

Since 1923, the majority of U.S. courts regularly applied the *Frye*[1] (general acceptance) test to admit or exclude scientific evidence. As an established expertise, forensic document examination, more specifically, forensic handwriting examination, was not contested under *Frye v. United States*. In fact, forensic handwriting evidence was instrumental in the conviction of Bruno Hauptmann in the highly publicized Lindbergh kidnapping trial in 1925.

When the Federal Rules of Evidence were adopted in 1975, many federal and state courts abandoned the *Frye* test in favor of the rules' relevancy standard, which provided less stringent admissibility requirements for scientific evidence. In essence, the rules allowed courts to admit evidence as long as it was more helpful than harmful to the jury.

An increasing number of circuit courts, however, were split over which admissibility standard to apply.

In 1993 the Supreme Court resolved the issue of dual admissibility standards in its review of *Daubert v. Merrell Dow Pharmaceuticals, Inc.*[2] by unanimously concluding that the rigid *Frye* test was now superseded by the more liberal Federal Rules of Evidence for determining admissibility of scientific expert evidence in federal courts.

The court's decision was significant. Scientific expert testimony would now be evaluated under a standard envisioned as less restrictive than *Frye* and admitted if it were found to be "scientific, technical or other specialized knowledge that will assist the trier of fact." Any concerns that the more lenient relevancy standard would admit questionable evidence were countered by the court when it reiterated that testimony may be challenged through "cross-examination, presentation of contrary evidence, and careful instruction on the burden of proof."[3]

In *Daubert* the court also appointed trial judges as gatekeepers to describe their increased responsibility in ensuring that reliable expert testimony was allowed to be admitted for trial. As guidance, the court suggested several criteria that judges could consider when evaluating a proffered expertise, including testing of basic premises, peer review, error rates, and general acceptance.

In *General Electric v. Joiner* (1997)[4] and *Kumho Tire Co. v. Carmichael* (1999),[5] the Supreme Court resolved some standing issues remaining from its *Daubert* ruling. In *Joiner* the court strengthened trial court decisions on expert admissibility by allowing appellate courts to reverse such decisions only if they were found to be based on flagrant error, such as unsupported facts, or an error of law. In *Kumho Tire* the court concluded that the judge's duty as gatekeeper extended to *all* expert opinion testimony, including expertise based on skill, observation, or experience. The court emphasized that trial courts had great latitude not only in determining the reliability of an expertise, but also in whether to apply the *Daubert* criteria at all.

6.2 *Daubert* Decisions and Forensic Document Examination

The first post-*Daubert* challenge to forensic handwriting admissibility occurred in the case of *United States v. Starzecpyzel*.[6] In considering the *Daubert* factors, the judge decided that forensic handwriting examination was practical rather than scientific, and as a result *Daubert* did not apply. Under Rule 702, however, the judge ruled that the expertise was admissible, but only as a "technical skill."

The *Starzecpyzel* decision stunned the forensic document community. Never before had one of the oldest forensic disciplines been described simply as a technical skill. Nor had a judicial opinion so sternly pointed out the dearth of published research, standards, and articulated methodology. *Starzecpyzel* heralded the need for change. What may have been sufficient for admissibility pre-*Daubert* was now clearly not enough.

Fortunately, such change had already begun. During the *Starzecpyzel* hearing, a small, but significant, published study was introduced into court for the first time. The study, conducted by Dr. Moshe Kam, a professor of engineering and computer science, took an unprecedented look at how FDEs and laypersons compare in their ability to correctly identify handwriting. The modest size of the study limited its significance for the hearing, but the results indicated that an expertise in forensic handwriting identification did exist.

6.3 Kam Proficiency Studies

Dr. Moshe Kam's initial role in the proficiency testing of FDEs was almost incidental. Contracted by the FBI in 1991 to design computer software for scanned handwriting images, Dr. Kam and his team searched the published literature to better understand the field of forensic document examination. They were surprised to find an absence of controlled studies and were highly skeptical of published statistics based on the flawed proficiency tests that did exist. Dr. Kam recognized the need for controlled studies to address whether document examiners possessed a skill that was absent in laypersons. With permission from the FBI, Dr. Kam conducted a pilot study involving 7 FDEs with the FBI and 10 graduate students (laypersons). Each participant was given 86 original handwritten documents and asked to sort the documents into piles so that each pile consisted of the writing of only one writer. Errors occurred if documents by two different writers were in the same pile or if documents by the same writer were in different piles.

In evaluating the results, Dr. Kam found the difference in performance between the FDEs and laypersons to be "striking." Although the sample sizes of the study were admittedly modest, the results indicated that an expertise in forensic handwriting identification did in fact exist. The study, entitled "Proficiency of Professional Document Examiners in Writer Identification," was published in 1993 in the *Journal of Forensic Sciences*.[7]

Dr. Kam published a second, large-scale study in 1997.[8] More than 100 FDEs and a control group of 41 graduate students (laypersons) were tested separately using the same documents. Each test taker was given 6 questioned and 24 known documents and asked to determine which were written by the same hand.

The results were significant. The study found that FDEs incorrectly matched 6.5% of the handwritten documents, while laypersons incorrectly matched 38.3%. In other words, laypersons were *six times* more likely to identify the wrong writer than FDEs. Interestingly, laypersons and document examiners correctly identified handwriting at the same rate, but this was because the laypersons freely identified so many of the handwriting samples that they not only made many correct matches, but also made many incorrect matches.

Dr. Kam's large-scale study undeniably proved that FDEs possess a skill in handwriting examination that laypersons lack. The difference in abilities provided assurance that FDEs are far more reliable in reaching correct conclusions than laypersons. It also equipped FDEs with a critical piece of information necessary for a *Daubert* inquiry, a known or potential error rate.

A third study published by Dr. Kam in 1998[9] addressed criticisms of the first two studies, namely, that the monetary rewards were not sufficient to motivate laypersons to perform as well as the professionals. The new study results showed, however, that despite the variable and sometimes lucrative rewards for correct answers, the error rate of laypersons remained significantly higher than that for FDEs in identifying handwriting.

In 2001 Dr. Kam published a fourth study[10] that compared the skills of FDEs with those of laypersons in determining whether a given signature was genuine or non-genuine (simulation). Again, Dr. Kam found a significant difference in the error rates between the groups. The error rate for FDEs concluding a genuine signature non-genuine was 7.05%, and a non-genuine signature genuine, 0.49%. In contrast, the laypersons' error rate in calling a genuine signature non-genuine was 26.1%, and a non-genuine signature genuine, 6.47%. In other words, laypersons erroneously called genuine signatures non-genuine 13 times more often than FDEs, and non-genuine signatures genuine 4 times more often.

6.4 Srihari's Individuality of Handwriting

Meanwhile, a professor at the State University of New York (SUNY) was in the process of developing a machine method to measure handwriting features with the intent to objectively support a basic premise of forensic handwriting examination: that no two writers share the same combination of handwriting features. Dr. Sagur Srihari of the SUNY Center of Excellence for Document Analysis and Recognition (CEDAR) had already invested over 20 years with the U.S. Postal Service to assist in the development of software that would recognize handwritten addresses on envelopes and convert them to bar codes. Dr. Srihari further refined the CEDAR software to extract and measure handwriting features to determine whether two documents were produced by the same or different writers. In a study involving over 1500 writers, the CEDAR software was able to correctly determine whether two writings were by the same or different writers with a 95% accuracy rate. This study was published in 2002.[11] The CEDAR studies continue today, as the database of handwriting samples is expanded and additional machine-extractable handwriting features are developed.

6.5 *United States v. Prime*

The turning point for forensic document examination occurred in a *Daubert* hearing in March 2002. *United States v. Prime*[12] was the culmination of intense efforts by the discipline to build a strong, structured response to *Daubert* inquiries. In just a few short years, the forensic document examination profession had developed and published standards of examination and had accumulated an effective array of relevant research articles.

The *Prime* ruling was a thorough, well-deliberated decision. The judge presented a succinct retrospective of *Daubert* and its progeny. He emphasized that the *Daubert* and *Kumho Tire* decisions were intended to exclude unreliable *novel* evidence, not established, time-honored techniques, such as forensic handwriting examination. In his final analysis, the judge concluded that the expertise satisfied all of the *Daubert* factors and, therefore, was reliable and admissible.

Key to the successful outcome of the *Prime* hearing was the extensive preparation by both the FDE and the attorney who proffered the expertise. Each provided substantial documentation demonstrating how the forensic document examination expertise satisfied each of the *Daubert* factors. The basic theory of handwriting individuality was supported by ongoing research involving computer measurement of handwriting features, as well as several published studies involving document examiners' ability to consistently distinguish the handwriting of identical twins. The studies of Dr. Moshe Kam not only provided the profession's error rate, but also convinced the judge that the Kam results superseded any purported statistical data previously based on the flawed proficiency tests conducted in the 1980s. Perhaps most compelling was the judge's observation that if forensic handwriting testimony was excluded based on an overly rigid application of *Daubert*, then a criminal defendant would be prevented from presenting contradictory, exculpatory evidence in his or her defense.

The 9th Circuit Court affirmed the *Prime* decision on appeal in 2004 and conspicuously noted the fact that all six circuits that had previously addressed the admissibility issue of forensic handwriting examination had also affirmed its reliability.

6.6 The Daubert Group

Prior to the *Daubert* decision, the most serious challenge a forensic scientist would face in court involved the questioning of one's credentials. Specific questions regarding the use of accepted methodology, validation of a procedure, peer review, general acceptance in the scientific community, and the error rate were rarely asked. In the post-*Daubert* era, however, such previously arcane areas of questioning were becoming increasingly commonplace, and expert witnesses were not always prepared with answers. Nor were attorneys sufficiently knowledgeable to argue for the admissibility of a proffered expertise.

United States v. Starzecpyzel was a wakeup call for the forensic document examination community and became the impetus for the profession to understand the impact of the *Daubert* trilogy and prepare FDEs to meet impending *Daubert* challenges. The American Board of Forensic Document Examiners (ABFDE) recognized the critical need for an organized response, and the Daubert Group, consisting of three FDEs, was formed. The group's primary functions are to track federal and state *Daubert* decisions and provide FDEs with condensed responses to each of the *Daubert* criteria. The effectiveness of the group is undisputed; it has assisted FDEs and attorneys in over 30 *Daubert* challenges, all of them successful.

Since the seminal *Daubert* decision in 1993, the importance of establishing the credentials of expert witnesses has acquired even greater importance in both U.S. federal and state courts. Although many state courts, and some countries, continue to apply a *Frye* type of standard for expert testimony, more and more state courts are replacing *Frye* with *Daubert*. And it may only be a matter of time before *Daubert* migrates beyond U.S. borders to international courts of law. The evolution of *Daubert* in U.S. courts and the organized response by the forensic document examination community may provide invaluable guidance to those experts who may eventually face the same challenges in their own *Daubert* era.

References

1. U.S. Court of Appeals, *Frye v. United States*, 54 App. D.C. 46, 47, 293 F. 1013, 1014, 1923.
2. U.S. Supreme Court ruling, *Daubert v. Merrell Dow Pharmaceuticals*, 509 U.S. 579, 1993.
3. *Daubert v. Merrill Dow Pharmaceuticals.*
4. U.S. Supreme Court ruling, *General Electric Co. v. Joiner*, (96–199) 78 F. 3d 524.
5. U.S. Supreme Court ruling, *Kumho Tire Co. v. Carmichael*, (97–1709) 131 F. 3d 1433.
6. U.S. District Court ruling, *United States v. Starzecpyzel*, 880 F. Suppl. 1027 (S.D.N.Y.), 1995.
7. Kam, M., Wetstein, J., and Conn, R., Proficiency of professional document examiners in writer identification, *J. Forensic Sci.*, 39, 5–14, 1994.
8. Kam, M., Fielding, G., and Conn, R., Writer identification by professional document examiners, *J. Forensic Sci.*, 42, 778–786, 1997.
9. Kam, M., Fielding, G., and Conn, R., Effects of monetary incentives on performance of nonprofessionals in document-examination proficiency tests, *J. Forensic Sci.*, 43, 1000–1005, 1998.
10. Kam, M., Gummadidala, K., Fielding, G., and Conn, R., Signature authentication by forensic document examiners, *J. Forensic Sci.*, 46, 884–888, 2001.

11. Sargur, N., Srihari, S., Sung-Hyuk, C., Hina Arora, M.E., and Sangjik Lee, M.S., Individuality of handwriting, *J. Forensic Sci.*, 47, 856–872, 2002.
12. U.S. Court of Appeals, *United States v. Prime*, 363 F. 3d 1028 (9th Cir. 2004).

Conclusions and Guidelines for Handwriting Comparison

7

CARL R. MCCLARY

Contents

In the early years of document examination, there existed numerous ways of reporting conclusions of handwriting or hand printing examinations. Included in those reports were findings that expressed a definitive opinion, such as whether or not a subject writer was the author of a questioned writing, and reports that were qualified, expressing opinions of lesser certainty. The very nature of handwriting examination dictates that qualified opinions should be rendered, given that several factors affect the outcome of any given case. These factors include the appearance of disguise, lack of clarity in originals or copies of documents, lack of known writing needed for comparison, and non-contemporaneous known writing.

Examiners from both government and private sectors use a variety of conclusion scales and verbiage. Language was used that had the potential to be confusing or misleading to the reader, such as "The writing of items 2 and 3 is consistent with the writing of John Doe." From this type of conclusion it was unclear whether the document examiner was identifying John Doe as the writer or whether he was expressing some level of similarity. Moreover, there was no publication explaining or defining terms such as *consistency* or any other indefinite term being used in the profession at the time.

In February 1990, a paper entitled "Committee Recommendations: The Standardization of Handwriting Opinion Terminology" (published a year later as a letter in the *Journal of Forensic Sciences*) was presented at the 42nd annual meeting of the American Academy of Forensic Sciences. Written by Thomas McAlexander, Jan Beck, and Ronald Dick, it described a nine-level opinion terminology that the authors felt best reflected the needs of the community. Within this terminology were six qualified opinions being utilized at the time, in addition to the mainstay *identification*, *elimination*, and *no conclusion*. The qualified conclusions were termed *strong probability* (wrote), *probably* (wrote), *indications* (wrote), *indications* (did not write), *probably* (did not write), and *strong probability* (did not write). The positive and negative sides of this scale were balanced and reflected various levels of certainty based on potential limiting factors noted above.

The qualified terms using probabilities are described as being equated to a confidence scale, and users were reminded by the authors that "probability in handwriting opinions is not a statistical measurement but a measurement of the examiner's confidence, based on scientific principles and experienced judgment, that the opinion rendered is correct. This is true because probability relates to qualitative as well as quantitative processes."[1]

In 1995, the ASTM E30.02 Questioned Document Subcommittee drafted and passed as standard terminology E1658, *Standard Terminology for Expressing Conclusions of Forensic Document Examiners*,* based on that *Journal of Forensic Science* letter. Included in the original standard is a suggestion that it also may be used in examinations other than handwriting and that there is no requirement of a document examiner to use all of the levels of opinion. It is recognized that some laboratories use five or seven levels, typically omitting the "indications" or "probably" terms. Some laboratory systems, such as those in the U.K., use a conclusion scale based on Bayesian theory.

References

1. McAlexander, T.V., Beck, J., and Dick, R., The standardization of handwriting opinion terminology, *J. Forensic Sci.*, 36, 313, 1991.

Section IV

Handwriting Examination and Comparison

Identification of Handwriting

8

DAN C. PURDY

Contents

Writing is a conscious act. Still, through repeated use, the actual formation of each letter and word becomes almost automatic, so that the experienced writer concentrates most of his or her conscious thought on the subject matter rather than on the writing process itself. Thus, writing comes to be made up of innumerable subconscious, habitual patterns, which are as much a part of the individual as any of his or her personal habits or mannerisms.

However, writing is more than a set of subconscious habits. It is a living, gradually changing product of the writer and is far from a mechanical reproduction prepared by the complex human mechanism of muscles and nerves that are called into play to produce it. It is influenced by a mental picture of copybook form, modified by individual taste and the writer's ability to imitate that which is in his or her mind. The physical and mental conditions that prevailed at the time of writing also may influence a person's writing. Whether handwriting is a criterion of personality is debatable, but that it is individual to each and every person is an established fact. Therefore, it can be identified, and the identification is based on all of the elements that combine to create the individuality.

The factors used to identify a person's writing are numerous. Some are more important than others, but conditions that surround each particular case also can influence which elements are given the greatest identifying value. Even highly skilled writers can produce inferior handwriting if the writing conditions are unfavorable. For example, writing prepared on an uneven surface in the back seat of a moving taxi, where poor lighting and distractions might further impair the writer's ability, will likely display erratic and tremulous movements. Although these qualities are expected in documents produced in a hostile environment, their presence in writing produced under more agreeable circumstances will certainly arouse suspicion concerning the authenticity of a document.

When attempting to identify the author of a handwritten document, all characteristics of both the known and questioned writings must be considered. The basic writing habits in both groups must agree if they are the work of the same writer. The questioned handwriting should be free from evidence of tracing or simulation and contain significant unique characteristics that are represented in the known writing without any fundamental differences. The application of sound reasoning in handwriting comparison involves determining what significant similarities and fundamental differences are present. Differences between the questioned and specimen handwriting provide evidence that they were produced by two writers — unless the divergencies can be logically explained or reconciled by facts surrounding the preparation of the questioned document, the quality of the specimens, or other circumstances.

Several repeated, fundamental dissimilarities are sufficient to establish without a doubt that two writings are not the product of a single person. However, under no circumstances can identity be conclusively established by a few unusual characteristics. Instead, a significant combination of features must always be in agreement without any fundamental dissimilarities so that any chance of accidental coincidence can be excluded. Handwriting identification, therefore, rests not only on a combination of identifying attributes — a condition that always must be fulfilled — but also on a coexistent lack of basic divergencies between the questioned and standard writings.

Figure 8.1 Handwriting of a highly skilled writer is marked by freedom, rhythm, and clarity.

8.1 Writing Forms and Qualities

What are the characteristics that serve as the basis of handwriting identification? In the eyes of most laymen, unusual letter formations are highly prominent. While certainly worthy of consideration, these elements do not represent the entire spectrum of features taken into account during a handwriting comparison. Letter formations might well be compared to the physical appearance of a person — definitely a means of recognition, but not a complete description. Regardless of physical appearance, each individual has typical manners of action and speech that help to identify him or her. In describing handwriting, the qualities of movement are even more essential to accurate identification than letterforms.

Writing is far from a lifeless form. Every specimen reveals an animation that is individual to its writer and reflects the pen movements that produced it.[1] At one extreme is smooth, continuous, rhythmic, rapidly executed writing, filled with grace and poise, and artistically shaded with points of emphasis that display freedom of movement characteristic of a highly skilled penman. Movement of the writing instrument is generally governed by the muscles in the forearm, although the wrist, hand, and fingers also may play a part in controlling the pen (Figure 8.1).

In direct contrast is hesitant, interrupted, laborious, slowly executed writing produced by an unskilled or uncertain hand. Its erratic pen strokes, angular connections, and irregular movements are typical of those for whom writing is a difficult physical or mental task. The unsteady appearance of the writing is mainly due to finger movements utilized by the writer to direct the writing instrument (Figure 8.2).

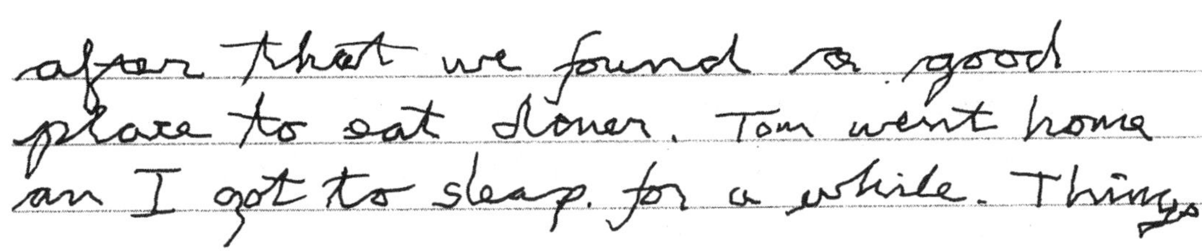

Figure 8.2 Handwriting of an unskilled writer appears slowly and awkwardly executed.

There are other qualities of writing movement. Pen strokes can be written precisely or carelessly, or they might be reworked or retouched in an effort to perfect and improve legibility. Some writers have acquired the habit of adding strokes to certain letters that can carry considerable weight in a handwriting comparison (Figure 8.3). Movement of the pen does not necessarily begin or terminate at the start or finish of a word. With most writers, the pen is in motion before it first makes contact with a document and continues after it is lifted from the paper. These conditions are reflected by tapered initial and terminal pen strokes at the beginning or end of words. In another class of writing, the pen is carefully placed on the document before any writing movement is undertaken and remains in contact with the paper until after each word is completed. Handwriting belonging to this group is characterized by blunt or bulbous initial and terminal strokes. The writing of most people lies between the two extremes, displaying varying degrees and combinations of these qualities.

Animation in handwriting is closely related to the physical processes involved. Depending on the skill and training of the writer, as well as his or her natural inclinations, writing is executed by movement of the fingers, wrist, and arm, either individually or more generally in varying combinations. There is a relationship between the skill of the writer, the speed of writing, the shape of letters, and the manner of execution. The slow, measured writing of the beginner, involving a heavy pen pressure, is generally produced by unskilled finger movements, whereas the rapid, smooth flowing writing of the skilled penman is accomplished by arm movement tempered with unconscious finger impulses at points of shading or abrupt turns. Of course, finger writing is not always crude nor does all arm motion writing result in a highly skilled cursive style. As mentioned above, most writing is not executed by one type of movement. The average writer employs a combination of these movements to achieve his or her particular quality of writing.

Many writers hold and handle the writing instrument in a very personal manner, which may affect the appearance of their writing. Most modern writing instruments, particularly ballpoint and soft-tip pens as well as the pencil, do not reveal clear-cut evidence regarding the position of the pen relative to the paper or the line of writing. However, Vastrick[2] reported that sister lines appearing on zinc oxide-coated paper, no-carbon-required (NCR) paper, and certain business forms can indicate the handedness of a writer and the orientation of the writing instrument relative to the paper. This is in contrast to the nib pen, especially the semiflexible point of the fountain pen, which produces significantly different shading patterns depending on the relationship of its tip to the paper. While unusual pen positions can be recognized when examining writing produced by a nib pen, such determinations are rarely possible when handwriting has been produced by a pencil, ballpoint, or soft-tip pen.

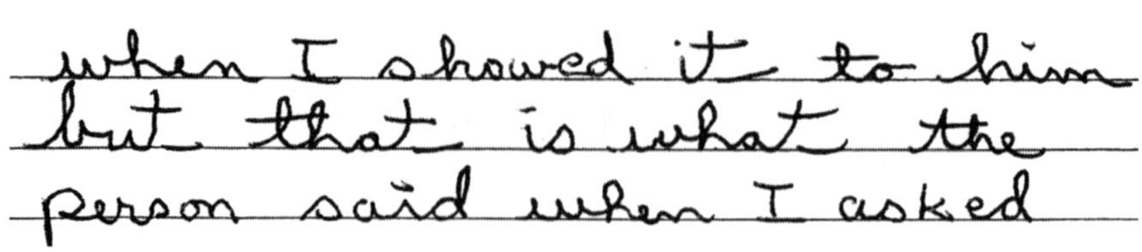

Figure 8.3 The addition of a horizontal stroke to the letter *t* when it is the last letter of a word is an important identifying feature of this individual's handwriting.

Some attributes of letter formations are influenced by the writing system that a writer learned and others are the result of individual taste. Handwriting is not an inherited trait, although personal skill or lack of coordination may be. For example, the theory that dyslexia and other disabilities are inherited and influence the formation of certain letters is supported by research. Remedial action can correct some of these manifestations. Similarities in handwriting produced by family members do sometimes exist when a writer imitates the characteristics of another family member or even a respected acquaintance. This tendency often occurs during adolescence when the writer is developing and experimenting with his or her handwriting. The letters may be angular or symmetrically rounded; they may be tall and slender or short and squatty. The shape of letters can be either artistic or grotesque, and they may be crowded together or spread apart, depending upon the nature of their connecting strokes. Particular letters may be conventional or unusual in style, and some writers use flourishes or ornamentations to further personalize their writing.

Line quality is perhaps the most important characteristic used to describe handwriting. Albert S. Osborn[3], who authored *Questioned Documents* in 1910 and is recognized by most as the father of forensic document examination in North America, introduced the term to embrace all of the factors related to movement that can be derived from a study of writing. It encompasses many features that stem from the dynamic processes used to guide the writing instrument as it moves across the paper. These elements include pen pressure or emphasis, speed of execution, pen lifts, the consistency and uniformity of the writing, rhythm, and writing skill. Handwriting that exhibits poor line quality throughout may have been produced by inexperienced writers or those who suffer from a physical or mental disability. In other cases, the writing may include a combination of rapidly executed pen strokes together with hesitations or tremor that occur haphazardly throughout the document. The random appearance of these opposing patterns is what distinguishes genuine writing from that which is fraudulent in nature. Spurious writing that has been painstakingly constructed to simulate the handwriting of the victim usually contains flaws or irregular writing movements in areas that present the most difficulty for the forger to imitate. These include graceful and artistic pen strokes, intricate patterns, or connected curves that are normally executed rapidly. Forgers are often so intent on copying each letter that they are unable to imitate the spontaneous quality of a genuine writing. The result of such failed attempts is immediately obvious when the questioned writing is examined under magnification (Figure 8.4).

The names given to parts of letters often originate from their resemblance to familiar objects, such as trough, hook, shoulder, or buckle. For example, the cursive letter *w* consists of two joined bowls or troughs, and an *a* can have a cap formation on its top. Other terms used to describe parts of letters appear in Figure 8.5.

Figure 8.4 In a contested holographic will, careful retouching, pen lifts, and hesitations are indicative of a failed attempt at simulating the testator's writing.

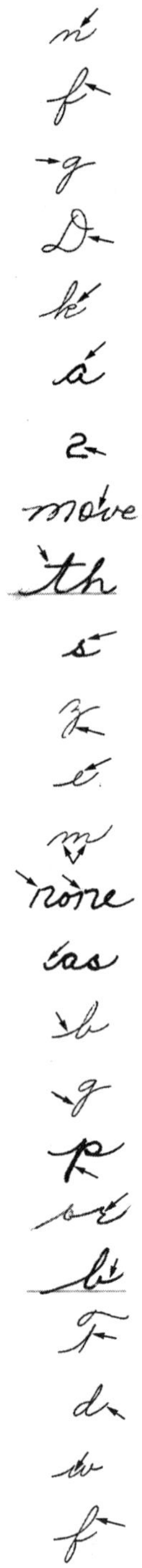

Arch	Convex curved strokes found in certain letters.
Ascender	That part of a letter that rises above the baseline.
Body	The remaining parts of a letter after the initial/terminal strokes and upper/lower extensions are removed.
Bow	A curved stroke generally aligned in a vertical direction.
Buckle or Knot	That part of a letter that ties itself to the staff.
Cap	That stroke in a letter that crosses over the top of the staff stroke.
Compound Curve	A curved stroke whose radius changes direction.
Connecting Stroke	A stroke joining two letters together.
Crossbar	A stroke that intersects the staff or main portion of a letter.
Cusp	The crescent or crown of a letter.
Descender	That part of a letter which drops below the baseline.
Eyelet	A circular movement whose center can be either open or closed.
Foot or Feet	The parts of a letter that touch the baseline.
Hiatus	A gap within a letter when the writing instrument leaves the paper.
Hook	A curved or angular projection usually occurring in an initial or terminal stroke.
Initial Stroke	The first writing movement of a letter.
Lower Loop	A loop situated primarily below the baseline.
Retrace	A stroke within a letter that follows the same course as a preceding stroke.
Shoulder	The flat or sloping part of a letter.
Spur	A small angled appendage at the start or end of a stroke.
Staff Stroke or Stem	A stroke forming the backbone of a letter.
Terminal Stroke	The stroke which forms the final portion of a letter.
Trough	A hollow or concavity between two raised portions of a letter.
Upper Loop	A loop situated primarily above the baseline.

Figure 8.5 Handwriting experts often use special terms to describe parts of certain letters.

8.2 Handwriting Features

No study of handwriting is complete without consideration of all of its factors, for together they describe the most fundamental properties of writing by an individual. They comprise both obvious characteristics of the writing, such as letter formation, and less conspicuous elements, such as pen pressure and the alignment of parts of a letter relative to the letter as a whole. Features considered when conducting a handwriting comparison include, but are not limited to, the following components.

8.2.1 Abbreviations

Writers occasionally use abbreviations for words that appear frequently in their written work. These include day abbreviations (e.g., Mon., Tues., etc.), month abbreviations (e.g., Jan., Feb., etc.), and title abbreviations (Rev., Dr., Cst., etc.), to name just a few. Apart from individual features related to the formation and execution of letters in these abbreviations, some writers adopt conventions that are grammatically incorrect, for example, using incorrect contractions to denote the possessive, as in "Science should know it's limits" or "You're computer is too slow." Abbreviations throughout a written passage can serve as useful identifying features — especially if they are incorrect, such as using "blgd." to denote building instead of the correct abbreviation "bldg."

8.2.2 Alignment

The baseline in a handwritten document is the ruled or imaginary line on which the writing rests. Some individuals have acquired the habit of placing all written words above or below this baseline. In most cases, however, only certain letters or letter combinations are misaligned (Figure 8.6). As McClary[4] reports, the position of the paper relative to the writer's arm can be the cause of alignment problems.

8.2.3 Arrangement

When comparing questioned and specimen extended writing, many factors other than pure handwriting characteristics can and should be considered. The arrangement of writing on a sheet of paper may be as individual as the writing itself. Margins, spacing, address placement, crowding, insertions, and alignment can all be personal habits of a writer; spelling, punctuation, phraseology, and grammar can indicate the writer's mother tongue or aid in identifying the author of a questioned document (Figure 8.7). Every one of these factors can be personal and important when identifying the writer of a contested document.

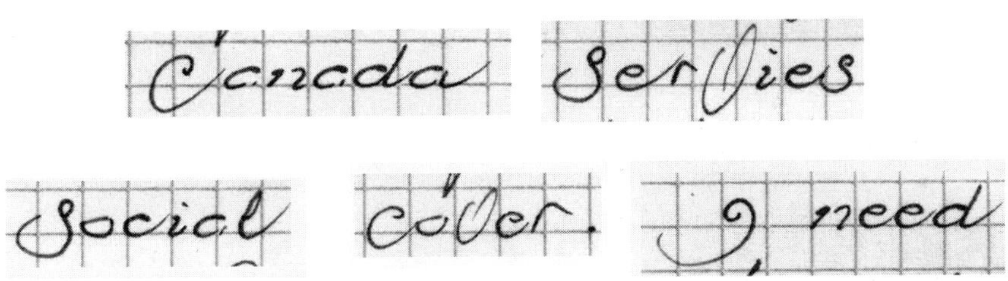

Figure 8.6 The letters *C*, *c*, *l*, and *v* written by this individual all rest below the ruled baseline.

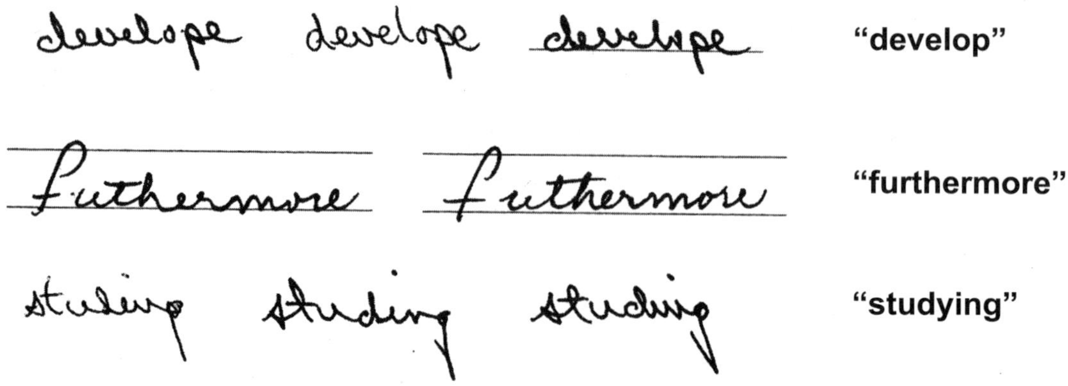

Figure 8.7 Misspelled words can be of assistance in identifying the writer of a contested document.

8.2.4 Connections

Strokes connecting adjacent letters or adjoining words in a handwritten document can be quite unusual. Individuals who produce writing that is compact often use short connecting strokes to join words and letters together, while writers who employ long connectors expand their writing beyond what can best be described as normal limits.

8.2.5 Initial Strokes

Copybook forms prescribe the proper length, shape, and starting position of initial strokes for cursive letters, with most letters beginning on the baseline. Many individuals have modified this element of their writing to suit their personal tastes. Some writers will start the initial strokes of their letters well below the baseline, while others will commence them above the recommended starting position (Figure 8.8). There also are writers who use long and short initial strokes, depending on the particular character and its position in a word.

8.2.6 Pen Lifts

Pen lifts occur when the writing instrument is lifted from the paper. The presence or absence of interruptions in pen strokes is often an important consideration when determining the genuine or fraudulent nature of a handwritten document. Natural handwriting

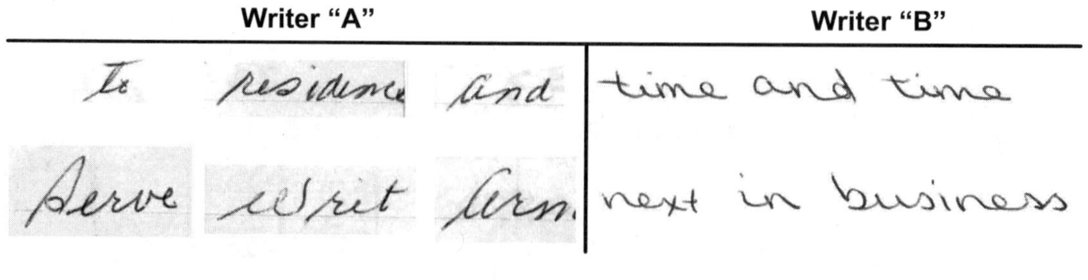

Figure 8.8 The character of initial strokes can be an important identifying feature. Initial strokes of most letters written by writer A commence below the baseline, while letters written by writer B lack any initial upstroke.

produced largely by hand and finger movements is more apt to contain pen lifts than writing that is largely executed by the wrist and forearm. In the first instance, movement of the pen is limited by the reach or extension of the writer's fingers. Once the fingers are fully extended, the pen must be lifted from the paper so the hand can be moved before continuing with the next series of pen strokes.

Pen lifts that always occur at the same place within certain letters can be quite significant. A special type of pen lift called a hiatus occurs when the writing instrument is lifted from the paper at the conclusion of a letter, leaving a short gap in a word before the next letter is started. Such breaks or interruptions are individual habits of a writer if they occur throughout a body of writing.

8.2.7 Pen Pressure

Pen pressure is the average weight or pressure unconsciously applied to the writing instrument during the act of writing. The operative words in this definition are *average weight*, which implies it is the average force with which the pen contacts the paper that determines whether a body of writing was written with light, medium, or heavy pen pressure.

Writing produced with a nib pen will clearly show the effect of applying different amounts of pressure to the writing instrument as it passes along the paper. The more pressure applied by the fingers and hand, the greater the tips of the nib will separate.[5] Occasionally a furrow will be visible in the surface of the paper that is proportional to the amount of pressure applied to the pen. The depth of the furrows can easily be seen if the pen stroke is examined microscopically while illuminating the document with oblique lighting.

When dealing with documents written with ballpoint pens, greater pressure applied during the writing process will cause the tip of the pen to push further into the paper, resulting in a slight thickening of the line and a depression of paper fibers under the written stroke.

Regardless of the type of writing instrument used to prepare a document, a careful inspection of the questioned handwriting can reveal important details that speak volumes concerning the writing process. Heavy, consistent pen pressure frequently associated with slowly executed, drawn movements will quickly become apparent when dealing with forged signatures, as will other deficiencies, including tremor, retouching, pen lifts, and irregular pen strokes.

8.2.8 Punctuation

Punctuation marks, such as periods, commas, exclamation points, colons, and semicolons, are signals that help readers understand the meaning of documents and read them more effectively and efficiently. Rules governing the proper use of punctuation appear in writing style guides.

Punctuation marks can appear profusely throughout a questioned document or they may not appear at all. Writers typically follow fixed patterns when writing personal and business letters or addressing envelopes. The FDE should pay close attention to where they are used as well as their shapes and positions relative to the baseline.

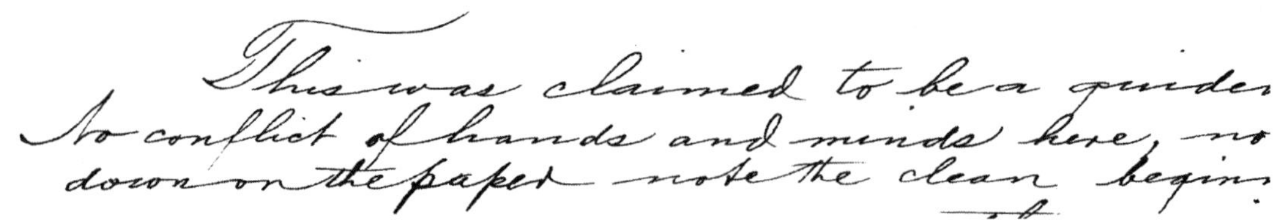

Figure 8.9 Writing by a skilled penman will be pleasing to the eye and exhibit a recurring series of impulse movements known as rhythm.

8.2.9 Rhythm

Rhythm is defined as regularity in the curvature of the writing, taking into account spacing characteristics and the graphic patterns established as the pen repeatedly returns to the baseline. The skilled writer will usually produce handwriting that is pleasing to the eye and that contains a recurring series of stress and impulse movements throughout (Figure 8.9). Writing that is irregularly spaced or that contains erratic movements exhibits poor rhythm due to abnormal contraction and release movements produced by the hand. Consequently, poor rhythm is one feature that distinguishes freely executed handwriting from the crude product of a tracing or freehand simulation.

8.2.10 Shading

Shading, as opposed to the unconscious application of pressure to the writing instrument, is the conscious or voluntary act of applying pressure to the pen while completing certain strokes. However, the effect of applying pressure to the writing instrument is the same in each instance. When a nib pen is used, pen strokes executed with heavy pressure will be wider than strokes written with less pressure. In the hand of a skilled penman, a nib pen will produce shading that gives the writing an artistic quality that is pleasing to the eye (Figure 8.10).

One might think the non-flexible quality of the ballpoint pen will produce pen strokes entirely devoid of shading or evidence of pen emphasis; however, this is not entirely the

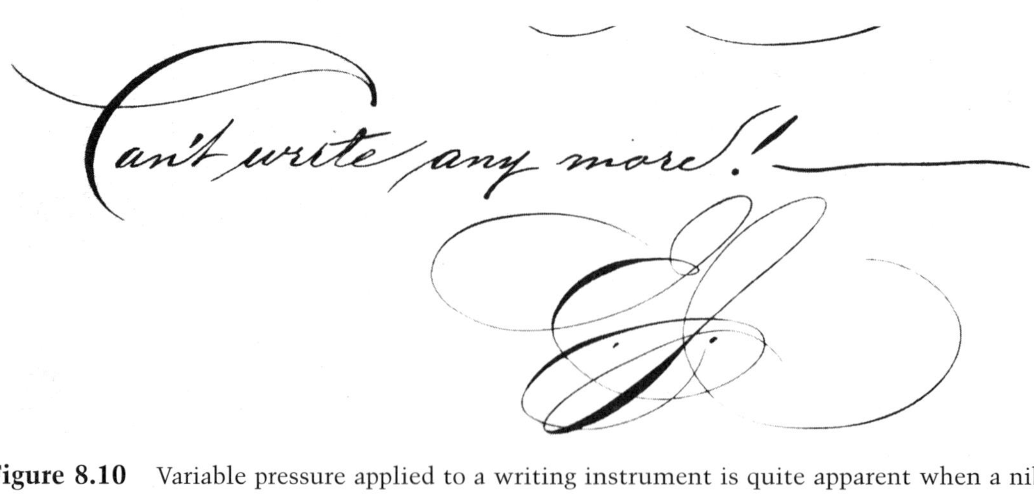

Figure 8.10 Variable pressure applied to a writing instrument is quite apparent when a nib pen is used. Thickening of pen strokes produced by the flexible nib is also referred to as shading.

case.[6] Even though ballpoint pens produce lines that have more uniform widths, some writers nevertheless execute writing that clearly reveals differences in writing pressure.

An assessment of shading in a handwritten document not only encompasses the amount of pressure applied to the pen, but also the region where shading begins and ends. Like all learned behavior, shading is an acquired writing habit developed through years of repetition and practice.

8.2.11 Size

Unless confined to a restricted space or written at different times, the size of writing by most individuals is rather consistent. The size of a writing can be influenced by the circumstances surrounding the writing act. For example, Alford[7] reports that approximately 65% of writers reduce or increase the size of their writing as a means of disguise. This result was also found in a study conducted by Hardy et al.[8] Although changes in the size of natural writing by the same individual occur from time to time, the relative size of capital and lowercase letters is usually constant.

8.2.12 Slope

One of the most obvious features in an individual's handwriting is its slope or slant. Writers who intentionally alter the slope of their writing cause a dramatic change to the appearance of letters. This is why modifying the slope of handwriting is a common method of disguise. While gross changes in writing slant are relatively easy to achieve, slight changes in the slope are exceedingly difficult to maintain and are rarely consistent throughout an extended passage of writing.

8.2.13 Terminal Strokes

Like initial strokes, terminal strokes also are prescribed by copybook forms. Individuals who acquire a personal style of writing invariably deviate from letterforms dictated by the copybook and will add curled strokes, extensions, or embellishments to the ends of certain letters. These writing features are quite consistent and, therefore, warrant consideration when comparing two groups of writing.

8.2.14 Tremor

Tremor can be due to involuntary shaky movements of the hand as the forger slowly copies a victim's letterforms. In these instances, tremor is likely to occur in letters or letter combinations that require the greatest skill to produce. Consequently, broad curves and intricate writing movements should always be inspected microscopically for irregular thickening in pen strokes and deterioration in line quality, as they represent the greatest challenge to the forger.

Tremor also can result when writers are attempting to alter or disguise their normal handwriting habits.[9] Disguised writing does not have the spontaneous appearance of naturally executed words and letters. The practice of concealing one's natural writing habits while simultaneously writing in a completely different manner requires tremendous concentration and effort if the deception is to avoid detection.

Genuine writing formed under unusual circumstances can contain tremor as well. It appears in varying degrees if a writer is suffering from a loss of muscular control due to a debilitating illness or old age. Those who suffer from diseases like Parkinsonism may find temporary relief by taking prescribed medication.

One of the most challenging problems confronting the FDE is determining the cause of tremor in a questioned document. Some important factors to consider include where the tremor occurs, the extent to which the pen deviates from its normal path, and the writer's condition when the document was allegedly written.

8.3 Writing Variation

Identifying the person who wrote a questioned document depends on the similarity between writing habits manifest in the writing and those found in specimens written by a particular person. While weighing the evidence, consideration must be given to writing variation. Since variation is an integral part of natural handwriting, no two samples of writing prepared by one person are identical in every detail. The extent of variation differs among writers and, consequently, natural variation forms an important element in the identification of handwriting.[10–12] In some cases, variation is slight and occurs only in minor details of the writing; in other instances, the formation of letters and words will vary widely about a master pattern (Figure 8.11). Variation is due principally to the lack of machinelike precision in the human body, but it is also accentuated by external factors, such as writing position, the type of writing instrument used, and the amount of care given to the act of writing.

Writing variation also can be influenced by the physical and mental condition of the writer. This is why fatigue, intoxication, drug use, illness, and nervousness are often accompanied by a deterioration in the quality of writing commensurable in degree with the intensity of the cause. Declining health, advanced age, and the quantity of writing prepared during a period may introduce greater variation in documents written at widely separated dates.

Variation does not preclude the identification of handwriting. In fact, variation around the basic qualities of the handwriting establishes an additional factor that serves

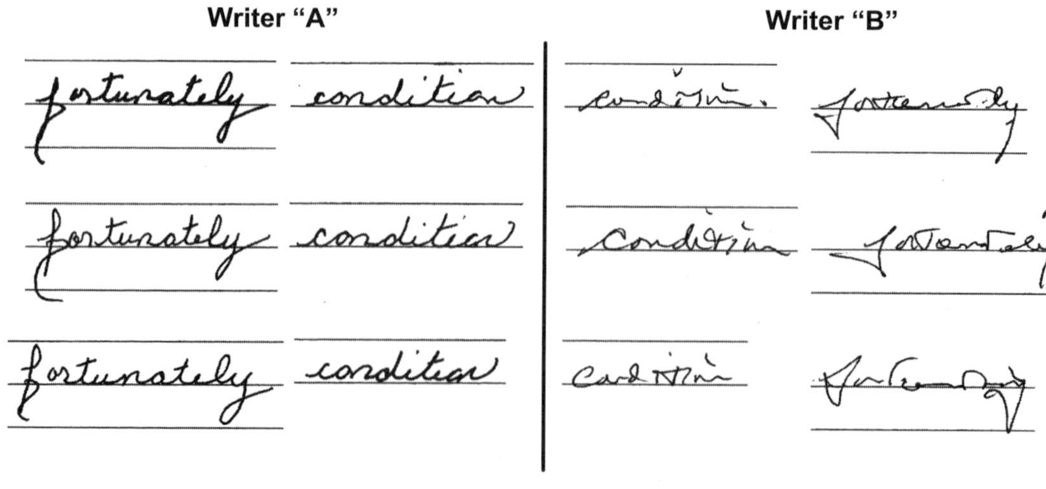

Figure 8.11 Natural handwriting produced by writer A is more consistent and has a narrower range of variation than writing produced by writer B.

to personalize and individualize writing.[13] Thus, handwriting is most likely to be identified when the questioned writing and specimen material are contemporaneous and produced under similar circumstances. Establishing the source of writing becomes a process of determining its fundamental qualities and habits, together with an accurate range of variation over which the writing fluctuates. Not only is it necessary to demonstrate that the questioned writing has the same qualities and habits as the known writing, but also that deviations from basic patterns within the unknown writing can be predicted from variations that occur in the standards. All this presumes that the standards are truly representative of writing prepared under the conditions that prevailed when the questioned material was executed.

8.4 Class and Individual Characteristics

Traditionally, features that are considered when conducting a handwriting comparison fall into two general and somewhat overlapping groups of characteristics. Class characteristics, as the name implies, are writing traits common to a group or specific class of writers. The term often relates to the particular copybook form or style of letters taught in the classroom or to writing features that develop from occupational requirements, educational demands, or cultural influences. Individual characteristics, on the other hand, occur when a letter departs from its usual copybook form. These elements of writing constitute the basis of all handwriting identifications and include such features as the size of letters, the ratio of parts of a letter to the letter as a whole, unusual pen lifts, and the misalignment of letters relative to the baseline, to name just a few.

Prior to 1980, handwriting was usually taught in public schools from published systems of letter designs called copybooks. Students learned to write by imitating letterforms constructed by Palmer, Zaner-Bloser, D'Nealian, MacLean, or other creators of copybook styles (Figure 8.12). Fifty years ago, the system of handwriting taught to a person could often be accurately predicted through a study of the class characteristics present in a body of writing. However, recently this has become a much more difficult and oftentimes

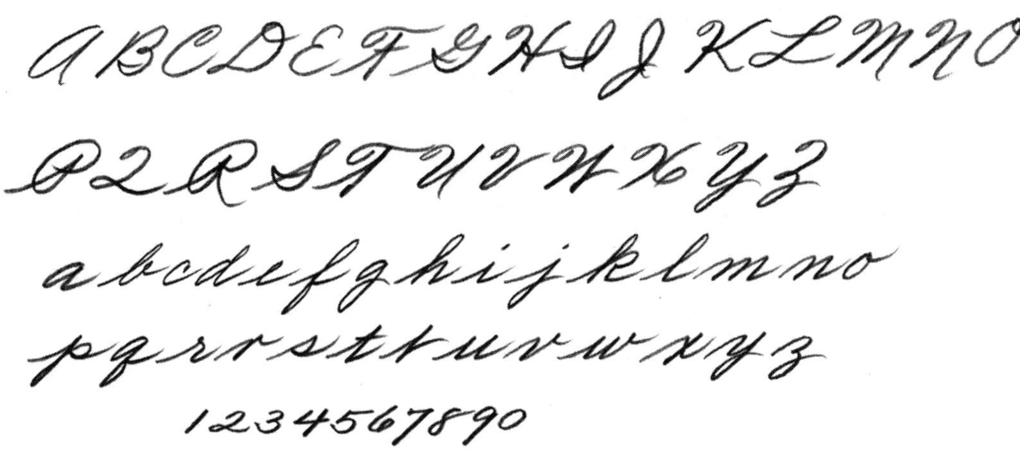

Figure 8.12 Copybook forms, such as the example above, are aids used to teach students to write letters clearly and legibly.

impossible task due to the considerable latitude teachers have been given in deciding what writing system or method to use in their classrooms. While a few schools still strictly enforce the use of a particular copybook, more often students are instructed to imitate letterforms written on the blackboard by the teacher in his or her normal handwriting. The main goal of modern instruction methods is teaching students to write quickly and legibly. The general lack of prescribed copybooks exposes students to different writing systems and teaching methods during the time they are learning penmanship. Consequently, only rarely is it possible to identify the system or method of writing that was taught to an individual by examining his or her handwriting.

In the past, accountants, bankers, and draftsmen were given strict instructions on how to write letters and numerals. Each character was constructed in a particular fashion using the same number of strokes, written in a particular direction and order. Less attention is now given to such matters than was the case during the last century. Since the advent of the computer, professionals tend to rely more upon software programs to label drawings, make numerical entries, or produce daily reports.

Since the writing of many people contains the same class characteristics, these features are given little weight when identifying who wrote a questioned document. It must be emphasized that individuals do not retain many characteristics associated with the style of letters they were taught in school. Instead, handwriting usually consists of personalized letterforms accompanied by certain idiosyncrasies and preferences that give an individual's handwriting a distinctive and appearance. Some of these manifestations are quite unusual and can carry considerable weight when conducting a handwriting comparison.

Rather than label writing features either "class" or "individual," FDEs usually assign significance to each feature based on how rarely it occurs in writing by the general population and how consistently the feature appears in repetitions of the same letter or word written by a particular individual. Of course, special care must be exercised when examining foreign handwriting. The most common error made by inexperienced or unqualified FDEs is assigning a great deal of weight to an unusual class characteristic simply because it originates from a writing system the FDE is not familiar with.

The cornerstone of handwriting identification is the presence of a significant combination of unusual habitual features in both the questioned and specimen handwriting. However, if two documents were *not* prepared by the same writer, this can often be established by the presence of significant or irreconcilable differences.

8.5 Foreign Handwriting

A dramatic increase in immigration during the past 100 years has increased the ethnic diversity of our communities and reflects the tendency of the world's population to be more mobile. This trend increases the likelihood the FDE will encounter documents written in a foreign language or alphabet.[14]

The examination of documents written in a foreign script (e.g., Arabic, Chinese, Cyrillic, Urdu, etc.) must be approached with a great deal of caution. Some FDEs believe it is more appropriate to refer such examinations to colleagues familiar with the foreign system of writing or who have dealt with similar matters previously. If the FDE feels up to the task, the analysis should only be carried out after studying writing systems and teaching methods used in the foreign country. Access to handwriting produced by individuals educated in the same geographic area will ensure the appropriate weight is placed on features that may be unfamiliar to the FDE but which are common to a large group of foreign writers. The FDE also might find it of benefit to work with a translator or consult others knowledgeable about the particular foreign writing system or teaching methods used.

When conducting comparisons of foreign writing, attention must be focused on acquiring proper specimen materials. While it is occasionally possible to compare similarly constructed letters from different alphabets (e.g., Latin and Cyrillic), this often results in the expression of qualified opinions. Preferably, exemplars should be written in the same script as the questioned document.

Fewer problems are encountered when examining signatures written by foreign writers, especially if the questioned signature is stylized and written in a spontaneous manner. Stylized or abbreviated signatures have little, if any, relationship to the system or method of instruction taught to the writer. Instead, their identification is based more on an appreciation of the intricate details, habitual writing movements, and indistinct letter designs than on the formation of legible letters.

8.6 Identification in a Practical Situation

Provided a disputed document contains an adequate amount of developed handwriting, only one person will produce writing with that particular combination of characteristics. While this premise expresses the distinctiveness of handwriting, it is also important to consider some other fundamental postulates.

First, adequate samples of known handwriting executed under the same conditions that existed when the questioned writing was produced should be available for study and comparison. While they do not have to represent all of the writer's handwriting habits, the exemplars should contain individual handwriting attributes that correspond to those found in the questioned writing, and they should accurately depict the slight modifications of letters attributed to normal writing variation.

It is also important to consider what extent of agreement is necessary before it can be established that the writer of the exemplars wrote the questioned document. Of course, the known and questioned writing must contain the same identifying elements. One has only to examine repetitions of the same words or letters in the known material to realize that the disputed writing does not need to be a perfect duplicate to have been prepared by the same writer. Rather, the specimens must incorporate those personal habits and qualities found in the questioned writing that have been discussed in the preceding paragraphs of this chapter. Factors that should be considered when obtaining handwriting specimens are covered in Chapter 12.

With due consideration of the writer's established range of variation, the habits and qualities of the known and questioned handwriting must be in agreement. In other words, the questioned writing must be within the reach or scope of this person. If the known and unknown writing contain a sufficient number of distinctive writing characteristics so that the likelihood of accidental coincidence can be eliminated, *and* provided no basic or fundamental differences exist between the two sets of writing, then it is safe to conclude that the questioned and specimen handwriting were written by the same person. Handwriting charts illustrating individual characters, letter combinations, and words that appear in both groups of writing are often used to demonstrate the basis of an expert's opinion (Figure 8.13).

FDEs are occasionally asked how many points of identification are necessary to establish that two writings were written by the same person. In order to accomplish this task, the document examiner must consider how frequent each feature is encountered in writing by the general population. This expertise is largely acquired through years of full-time study combined with the examination of handwriting produced by literally thousands of writers.

Since the early 1990s, courts in the U.S. have expressed a desire for handwriting identification to be less subjective and more precise. Several studies have been conducted to measure the frequency of certain handwriting characteristics encountered in writing by the general population.[15] The results of these studies have shown that some writing features are far more unusual than others. Most handwriting experts would agree these significant characteristics are key to every handwriting identification.

Establishing what agreement is needed to identify the writer of a questioned document has not been substantiated and probably never will be due to the very nature of handwriting. Handwriting identification encompasses not only an assessment of letterforms, which are relatively easy to evaluate, but also less measurable characteristics such as skill, style, freedom, movement, emphasis, and spacing. The latter qualities, which pertain to all the writing on a document, are difficult and often impossible to quantify.

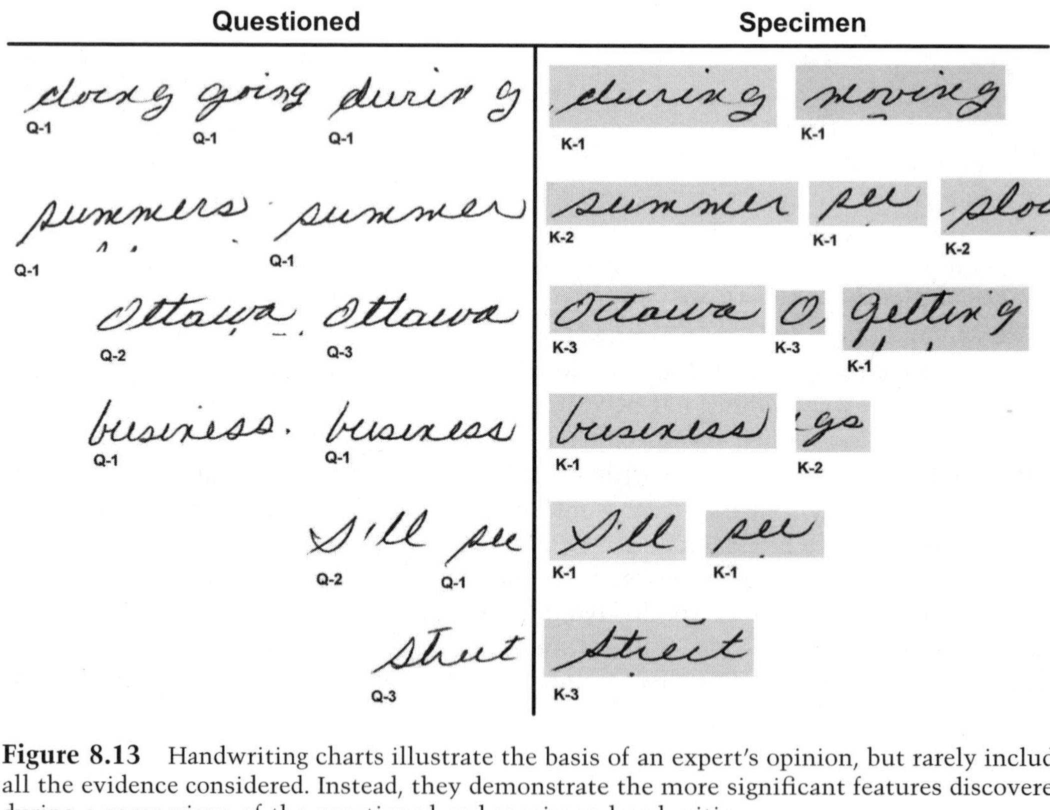

Figure 8.13 Handwriting charts illustrate the basis of an expert's opinion, but rarely include all the evidence considered. Instead, they demonstrate the more significant features discovered during a comparison of the questioned and specimen handwriting.

8.7 Non-Identity of Writing

Writings executed by two different people may be entirely dissimilar, or they may be very much alike but not identical. Although many writers have certain habits in common, every writer has developed personal peculiarities that mark his or her writing. These individualities, many of which are found in the inconspicuous details, distinguish authorship by two people who write very much alike.

Repeated small differences are sufficient to establish clearly that writings are the work of two individuals even though they may contain a considerable number of general similarities. This is often a major concern to the layman, but of less importance to the trained FDE. In situations where similarities clearly outnumber the differences, the existence of a few fundamental, repeated differences can be overwhelming and controlling. Everyone seems to appreciate that two writings are not by the same individual when there is a vast number of differences, but just a few fundamental dissimilarities may not seem to warrant the same finding. Nevertheless, they do. If two writings are by a single person, then no fundamental differences should exist. Conversely, if there are any basic dissimilarities that cannot be accounted for by a logical, commonsense explanation, then the two writings must have been prepared by different writers.

8.8 Typical Problems Involving Writing Identification

The identification of handwriting assumes importance under a diversity of circumstances. Despite the widespread use of the computer, complete handwritten documents are still frequently encountered in laboratory casework. From time to time, questions arise as to who wrote an entire document (Figure 8.14) or who wrote a few key words inserted in an

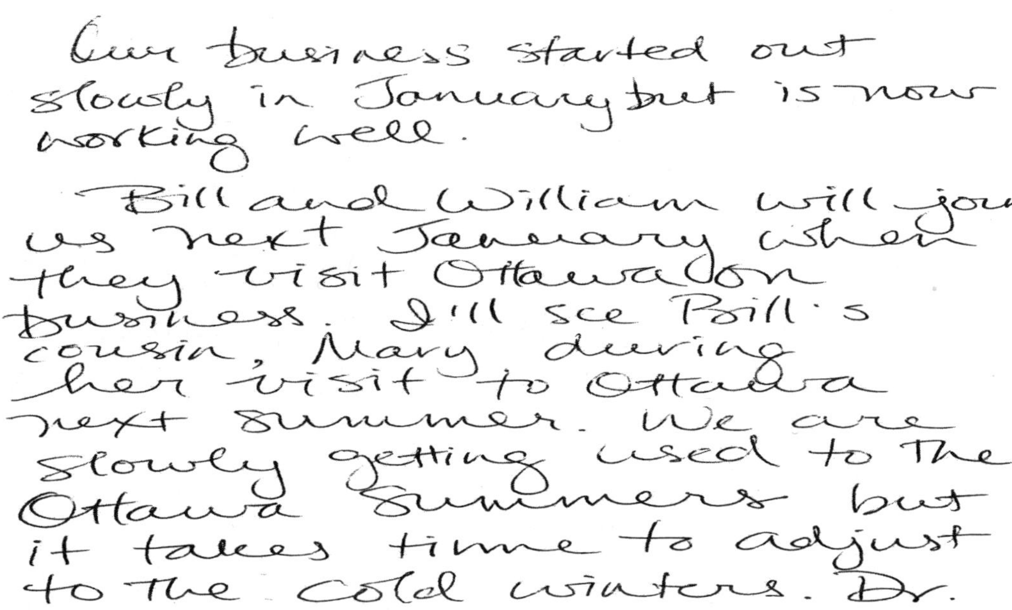
Our business started out slowly in January but is now working well.
Bill and William will join us next January when they visit Ottawa on business. I'll see Bill's cousin, Mary, during her visit to Ottawa next summer. We are slowly getting used to the Ottawa summers but it takes time to adjust to the cold winters. Dr.

Figure 8.14 In some cases, all the writing on a document can be suspect.

otherwise uncontested document (Figure 8.15). It is surprising, moreover, that individuals often deny extensive writing they have prepared. Under such circumstances, it may be essential to establish the facts through a handwriting comparison. There are other instances when the person who wrote a document is no longer available to identify it. It then becomes necessary to establish proof by comparing the questioned document to specimen material from personal and business records. If, on the other hand, someone is wrongfully accused of writing critical material, this misidentification can be corrected by proper study and comparison with his or her writing.

Important documents, such as wills, receipts, agreements, contracts, notes, and checks, are in many cases entirely handwritten. These documents can play a role in either civil or criminal litigation, at times a controlling part, so proof of their identity by expert testimony can be fundamental to uncovering the facts. In criminal prosecutions, the purpose of an examination can be to identify the writer of a ransom letter, bogus check, or anonymous letter. Accurate testimony concerning the nature of these and other written documents can certainly uncover their history or origin. The interests of justice are usually best served by having suspect documents inspected by a qualified FDE.

8.9 Holographic Wills

One of the most important identification problems involving extensive handwriting arises in will disputes. In many jurisdictions a holographic will is a valid instrument, and as such, does not require any witnesses. If a dispute arises concerning the authenticity of a will,

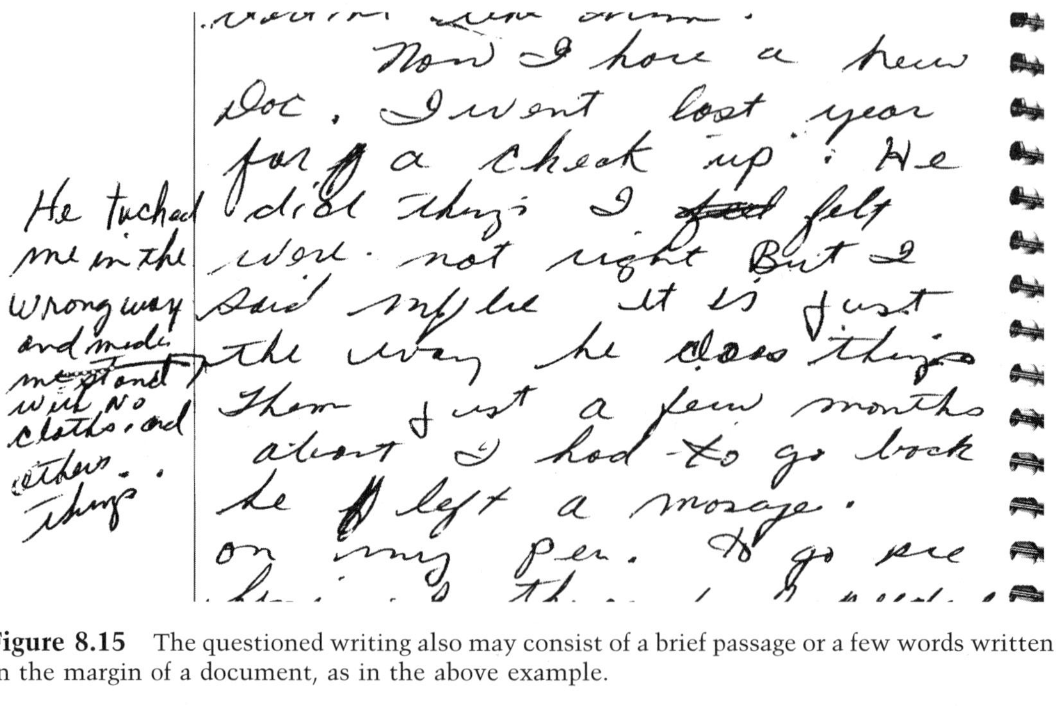
Now I have a new
Doc. I went last year
for a check up. He
did things I felt
were not right But I
said myself it is just
the way he does things
Then just a few months
about I had to go back
he left a mesage.
on my pen.

He tuched
me in the
wrong way
and made
me stand
with No
cloths, and
others
things

Figure 8.15 The questioned writing also may consist of a brief passage or a few words written in the margin of a document, as in the above example.

proof is often obtained by retaining an FDE who is able to study the entire document, including its signature, and determine whether the decedent did, in fact, write the entire instrument.

If a holographic will is drawn well before the time of death, when the testator is still active and in sound health, identification problems are usually not extremely complex. Of course, good specimens of known writing must be obtained for comparison purposes. Writing skill can decline dramatically with failing health, and the FDE should always try to obtain information about the health of the writer, the conditions that prevailed when the will was written, and whether prescribed medications might have influenced the writing act.

When dealing with multi-page wills, attention should be focused on the line quality of writing on separate pages to ensure there is no evidence of substituted pages. Changes in the line quality of the writing, the use of different writing materials, crowded text, and other incongruities may be evident if an entire page or paragraph has been added to an otherwise genuine document.

Occasionally, there may be cases where the decedent has not prepared a will until reaching an advanced age. The detrimental effects of sickness and poor mental health often further complicate matters. Under these circumstances, the will may be comprised of decidedly decrepit handwriting. Limited comparable standards are often part of such problems and further hinder the comparison process. Therefore, opinions justified by the evidence can vary in degree of certainty from a categorical identification to probable or occasionally inconclusive results.[16] Even qualified opinions, when taken along with other evidence, can be helpful in establishing the authenticity or fraudulent nature of a contested will.

8.10 Inhibiting Factors

The scientific examination and comparison of handwriting generally leads to definite findings. Nevertheless, restricting conditions can prevail that prevent the FDE from reaching a positive conclusion. In these cases, the evidence might only justify the expression of a qualified opinion.

While an insufficient amount of known writing is the most common deficiency, standards written under very different circumstances from those that prevailed when the questioned document was written can also limit the findings. Small bits of writing other than signatures, especially a few inserted words either in a near-copybook hand or very carelessly written, can also cause the FDE a great deal of difficulty (Figure 8.16).

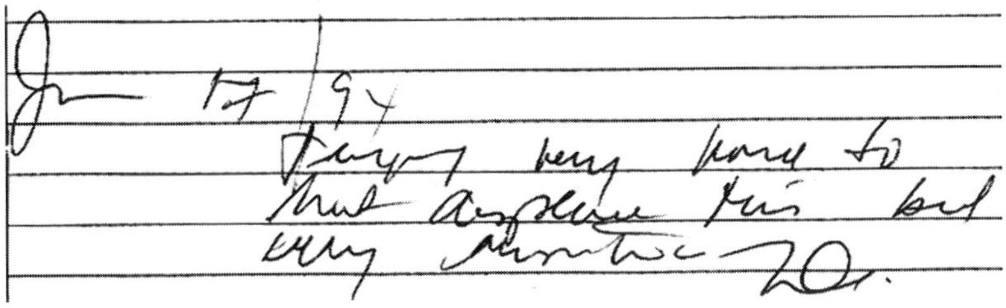

Figure 8.16 The examination of short passages of writing can pose problems if they are illegible and written in a rapid, careless fashion, as in this questioned notation taken from a patient's medical file.

Specimen handwriting produced long before or after the purported date of the questioned document may not adequately represent the writer's identifying features and preclude the expression of a definite yes or no answer. Qualified opinions are also apt to result when a document has been carelessly written with extreme haste, when a writing has been prepared under adverse conditions, or when the writer is suffering from a severe illness. These exceptional cases, however, represent a small fraction of comparisons that result in clear-cut conclusions.

8.11 Disguise

Disguise is the deliberate attempt to remove or modify all or some of one's normal writing habits. In this chapter, discussions will be limited to cursive or semicursive writing rather than disconnected hand printing. Disconnected and seemingly simplistic printed letterforms are occasionally employed as a means of disguising writing habits;[17] however, these attempts are usually unsuccessful. A full discussion of disguised hand printing is covered in Chapter 11.

A favorite practice of the anonymous letter writer is to employ disguise as a means of avoiding detection (Figure 8.17). Perfect disguise, of course, would ensure complete protection, but flawless disguise is itself extremely difficult in an extended text. Although it is relatively simple to change one's writing habits in a short passage, the task of maintaining an effective disguise grows more difficult with each additional word. When as much as a page of writing is disguised, it is generally true that the writer's normal habits are only partially veiled. Under these conditions the disguise may not be sufficient to preclude an identification.

It is understandably difficult to employ a consistent method of disguise when documents are written over an extended period. Even if the writer retains a copy of each disguised document, attempts to reproduce the same unfamiliar style of writing usually fail and are scarcely ever the same.

Just why is disguise so difficult and what causes so many attempts to fail? If forgers are going to successfully conceal their handwriting habits, two difficult steps must be carried out. First, they must have a thorough knowledge of all the identifying details of their writing, along with the significance of each. This condition alone is seldom fulfilled. Assuming it has been accomplished, for it is certainly within the reach of a clever individual willing to use every means at hand to avoid detection, then a second and decidedly more difficult step lies ahead. This involves the complex task of eliminating the identifying habits

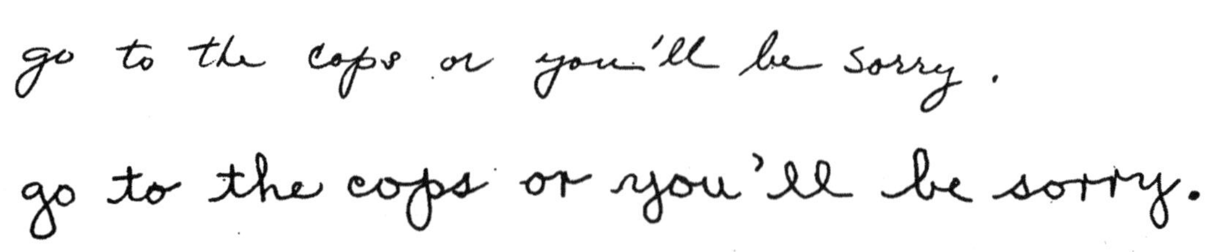

Figure 8.17 The top line is written in a writer's normal style of handwriting. The bottom line of disguised writing by the same individual contains different letterforms of the *r* and *s*, a more vertical slope, and a slower, more rounded style of writing.

of the forger's writing while simultaneously adopting an entirely new set of writing features that describe the victim's handwriting.

Writing habits, especially those of the experienced writer, are very strong and firmly implanted. When attempts are made to discard them, the usual result is an elimination of the prominent characteristics while numerous unconscious, but individual, writing traits are left unaltered. What is then believed to be an effectively disguised writing is actually filled with small but significant personal habits that may be sufficient to identify the writer.

Disguised writing usually contains a distinctive element that distinguishes it from normal writing. It often contains evidence of a conflict — a struggle between the natural tendency to maintain persistent habits and the conscious effort to suppress them. As a result, disguised writing tends to be less skillfully executed than normal writing by an individual. Irregularities and inconsistencies often appear throughout a passage of extended disguised writing, including hesitations, variations in slant, grotesque letterforms, patched or overwritten letters, and slowly drawn strokes. Prominent features, such as the shape of capital letters and the slant of the writing, generally experience the most marked changes, while less prominent though equally important identifying factors often retain their usual character. The identification of disguised writing is only possible if unnatural portions are recognized and properly evaluated, leaving the undisguised segments to form the basis of a handwriting comparison.

Disguise may be accomplished by writing with the hand opposite to that which is habitually used. This can be a very effective disguise as long as standards of wrong-handed writing are not available. Opposite-hand writing can sometimes be inferred from its relatively low degree of writing skill. Once a suspect is located, steps should be taken to obtain writings executed with both hands whenever possible.

A small group of people can write with the same ease and skill using either hand. These ambidextrous writers have practiced and developed their writing to such a degree that writings produced by their left and right hands do not contain features associated with disguise. In spite of a developed skill to write with both hands, writing done with the right hand differs in many ways from writing done with the left. The truly ambidextrous writer is generally known to have this unusual ability and must have written extensively with both hands in order to acquire and maintain the skill. The problem then becomes one of procuring proper standards written with the left and right hands when dealing with suspect writers who demonstrate an ability to use both hands with equal facility.

8.12 Collective Handwriting Comparisons

Preceding sections of this chapter deal with the comparison of questioned handwriting with several documents bearing specimen handwriting. Since most conventional handwriting comparisons fall under this category, it is the usual method followed when identifying the person responsible for writing or signing a contested document. Occasionally, the FDE may be asked whether several documents were written by the same person or different writers. In such cases, a conclusion must be based solely on an intercomparison of the handwriting with little information concerning the writer's range of variation.

The process of intercomparing writing on several items within a group of documents is actually a relatively common occurrence. The FDE must perform this essential step before relying on the reported source of specimens submitted for examination. Regardless of how much care was executed while collecting sufficient specimens, most FDEs will attest that documents occasionally surface that were produced by someone other than the suspect. These so-called standards must be excluded before comparing the questioned and specimen handwriting. Failure to do so would compromise the veracity of any conclusion based on such contaminated specimens.

Detectives investigating bank holdups, check frauds, or a series of anonymous letters may lack clues regarding the individual or individuals who committed these crimes — especially during the early stages of their inquiries. At times, the same person or gang may have committed several offenses in the same or neighboring jurisdictions. In such cases, handwriting evidence associated with seemingly unrelated crimes can show that several documents actually originate from a common source. These findings can save considerable time and provide significant clues that lead to the identification and conviction of those responsible.

Determining if several questioned documents were written by the same person can be an onerous task even when adequate amounts of specimen writing from a likely suspect are available. In these cases, the FDE must judge whether observed differences between the questioned and specimen material are due to variation, disguise, or different authorship. The problem becomes more complex and difficult when the comparison involves only questioned writing. The latitude afforded by variation within the specimen material does not occur in these cases. Writing features are usually assessed in their existing context, and the FDE must decide if words or letters in two suspect documents are similar or different and to what degree. In anonymous letter cases, a particular word or letter combination might appear several times within the same document. This is a tremendous advantage when intercomparing handwriting, as it can provide a means of determining the writer's range of variation.

When faced with an intercomparison of several handwritten documents, it is often difficult to know where to begin such an examination. The solution is found by spreading out the documents so all of the writing is visible. Writings that are markedly similar and that might be the product of one writer will soon become apparent. Once the questioned documents have been sorted by their conspicuous features in this fashion, each group can then be examined to determine if the two most similar documents can be proven to have been written by the same person. Factors considered should include not only letterforms, but also the general arrangement of writing.

Once two or more documents are proven to be written by the same writer, common characteristics will emerge that can be used to associate other questioned documents with the group. Any writings that cannot be identified with the group must be excluded. If there are not sufficient similarities to prove two or more documents were written by the same hand, perhaps only a qualified opinion can be supported by the evidence.

Since handwriting intercomparisons are based on an interactive process, the exclusion of any document at a later time will require the examination to be repeated so the possible effect of the change can be taken into account.

In summary, it is very important to proceed slowly and cautiously when conducting collective handwriting examinations — especially those that involve only questioned writings. All too often, the FDE is presented with a handful of contested documents and spurred on with encouraging remarks that other witnesses can prove the writing was produced by the same person and all that is needed is the FDE's expert opinion to confirm what is already known. The FDE must resist exhortations by case contributors who are quick to alter the course of an investigation should convincing handwriting evidence be presented to the contrary.

8.13 Handwriting Classification Systems

The check passer, fraud artist, bank robber, and other criminals often follow the same modus operandi while carrying out their crimes. A person who repeatedly passes stolen checks might use the same false name, the bank robber might use the same wording when drafting a holdup note, or a blackmailer might always prepare ransom notes in the same manner. These habits can identify suspects who return to their "profession" following a period of incarceration or after residing in another region of the country.

The creation of handwriting databases has proven to be a successful means of dealing with increased incidents involving check fraud, bank holdups, and anonymous letters.[18,19] Such databases are regularly searched to link evidence from a series of unsolved crimes or to identify the individual responsible for writing a particular questioned document. Associating handwritten documents from otherwise unrelated investigations is a unique benefit of these central repositories. The collaboration and sharing of information by investigators from different jurisdictions can result in clearing a significant number of unsolved crimes.

Historically, collections of fraudulent documents were held by regional police agencies in independent systems that were each maintained and searched manually. In 1936, the Federal Bureau of Investigation created the National Fraudulent Check File as a central repository for bogus checks cashed in all parts of the U.S. and its territories. This concept of a central repository also led to the development of an Anonymous Letter File, a Bank Robbery Note File, and a National Stolen Art File to cope with serial offenses involving documents.

During the mid-1980s, Dr. Peter Nicholson[20] developed a system for classifying block capital hand printing while working at the Metropolitan Police Laboratory in London. This system relied upon operators classifying each letter in a hand printed document by the number of pen strokes used in its construction, as well as the pen paths followed when constructing certain characters.

Another system developed by the Arkansas State Crime Laboratory is based on the eight handwritten letters *a, d, f, g, i, k, r,* and *t.* Handwritten documents from cases submitted to the laboratory are classified and compared to other samples filed in a database. Each hit is compared to the questioned document by a handwriting expert to determine if a match exists. A research paper by Leung et al.[21] describes a system based on samples obtained from Chinese writers, and another project conducted by Saxena and Singh[22] deals with the classification of Devnagari script. These and other studies illustrate an international interest in the development of classification systems for handwriting.

Even assisted by a manual classification system, determining whether a handwritten document matches others in a database can be a challenging and time-consuming task. The complexity and effort required to both maintain and search a manual system increase exponentially as the number of writers and documents increases. Many of these difficulties have been addressed with the advent of the computer and advances in information technology.

In 1977, a group within Germany's Bundeskriminalamt commenced a research project called Forensic Information System for Handwriting (FISH).[23] A prototype was completed in 1989, and following several modifications and improvements, FISH proved capable of classifying handwriting by either using text-independent features or relying on a series of interactive, computer-aided measurements performed by an operator. The text-independent component is based on pattern recognition, where a handwritten document is evaluated and compared to a standard before storing each measurement as a pattern difference. The interactive approach looks at writing in a more traditional sense by considering such qualities as slope, height, width, upper and lower extensions, distance between baselines, oval height, and the shape of loops. The FISH system is presently used by German authorities as well as certain federal agencies in the U.S. to consolidate and associate unsolved cases, to link handwritten documents, and to assist investigations involving missing and exploited children.

During the 1990s, the Netherlands Institute for Forensic Examinations and Research (NIFO) in collaboration with the Institute of Applied Physics developed an automated handwriting analysis system called SCRIPT.[24] Like FISH, all operations are carried out on digitally scanned images of handwritten documents. In addition to searching handwriting collections for samples that match a particular questioned document, SCRIPT also has proven beneficial for training FDEs by providing quantitative information about the frequency of certain writing features that are encountered in the data set, as well as studying the effects of natural variation.

Recently, the Center of Excellence for Document Analysis and Recognition (CEDAR) has expressed interest in classifying handwriting using computer algorithms to extract information derived from features similar to those considered by FDEs.[25] Data from this research have led to measuring the relative significance of various handwriting features as well as providing scientific support for the admissibility of handwriting evidence in the courts. A program developed at CEDAR utilizes 11 macro-features and 512 microfeatures to characterize any quantity of handwriting. The results of research by CEDAR illustrate how information technology can assist the FDE by narrowing large database searches to smaller, more manageable sets of writing.

Locating similar words, letters, and letter combinations when large numbers of questioned and specimen documents are involved can be a daunting task. The software program Write-On© was developed by Pikaso Software, Inc., to help handwriting experts examine such cases.[26] After scanned images of all questioned and specimen documents have been associated with transcribed contents of their texts, all variations of a particular word or letter combination can be quickly located and presented in juxtaposition for viewing by the FDE (Figure 8.18). In this way, Write-On© provides a convenient means of establishing the extent to which two groups of written documents are similar or different, and assessing the range of natural variation within the questioned or specimen material.

The results of character string searches (Figure 8.19) and the frequency with which certain features occur in the questioned and specimen writing (Figure 8.20) can be determined from reports based on the data set for a particular case. Simply put, Write-On© quickly searches large quantities of documents thoroughly and generates reports that can present the basis of a handwriting comparison in a convincing and comprehensible manner.

Write-On© also can be of assistance when examining documents bearing non-Latin alphabet writing. In such cases, difficulties are often encountered when searching several

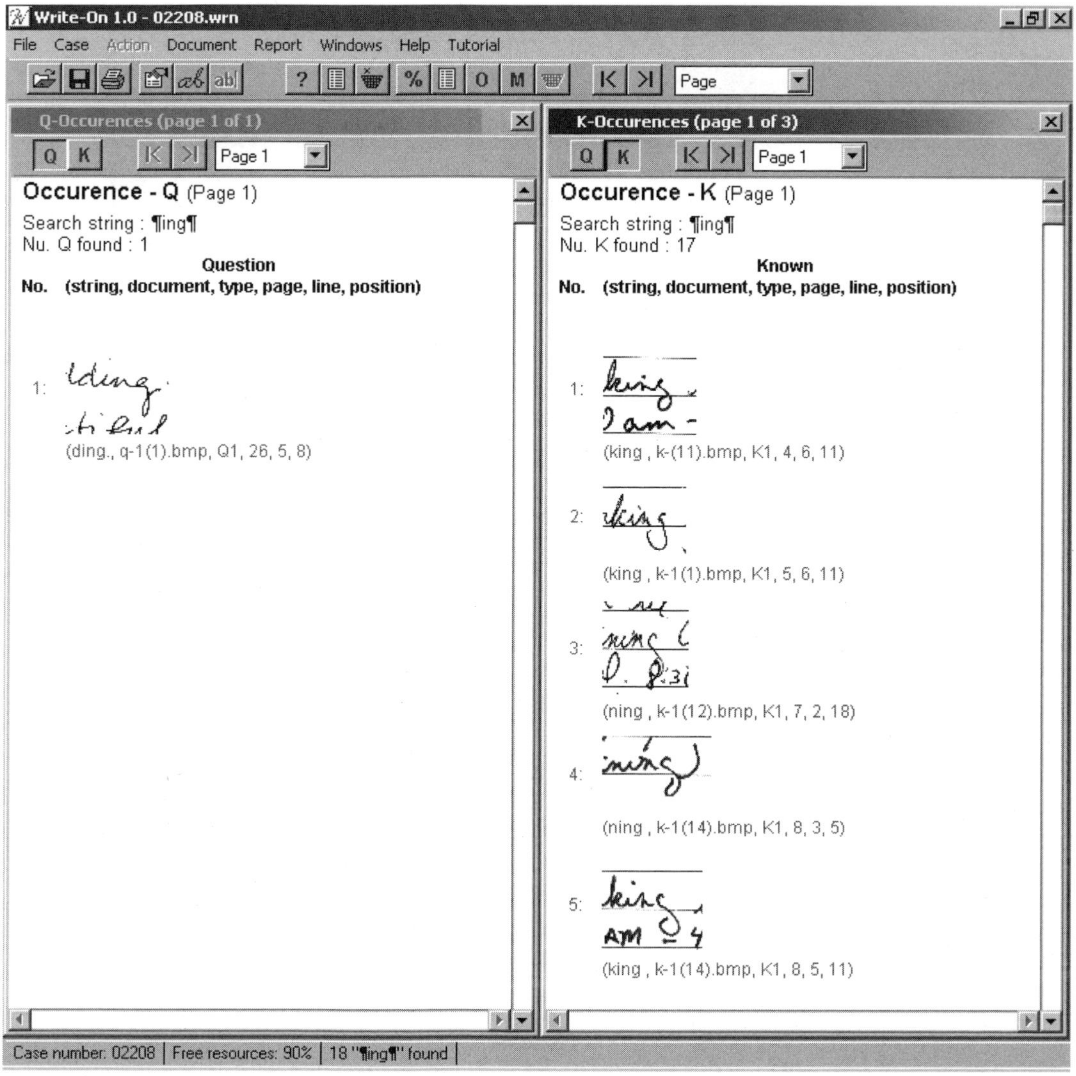

Figure 8.18 A search of several documents for the letter combination *ing* using Write-On© shows 1 example from the questioned documents (left) and 5 of 17 examples from the specimen material (right).

documents for repetitions of the same character in an unfamiliar script. Once translated, words from all documents are associated with their respective images. Write-On© then can be used to quickly locate and bring into view words or characters in the questioned and specimen material that correspond to their equivalent English translations.

The aforementioned classification systems definitely help the FDE cope with large quantities of handwriting. However, it is important to recognize that none of these modern developments are intended to supplant the role or skills of the handwriting expert. The discriminating eye and sharp mind of the FDE are still required to see and correctly interpret physical evidence in handwriting cases. Only through logical reasoning and the application of scientific principles by a qualified expert can the authorship of a contested handwritten document be accurately established.

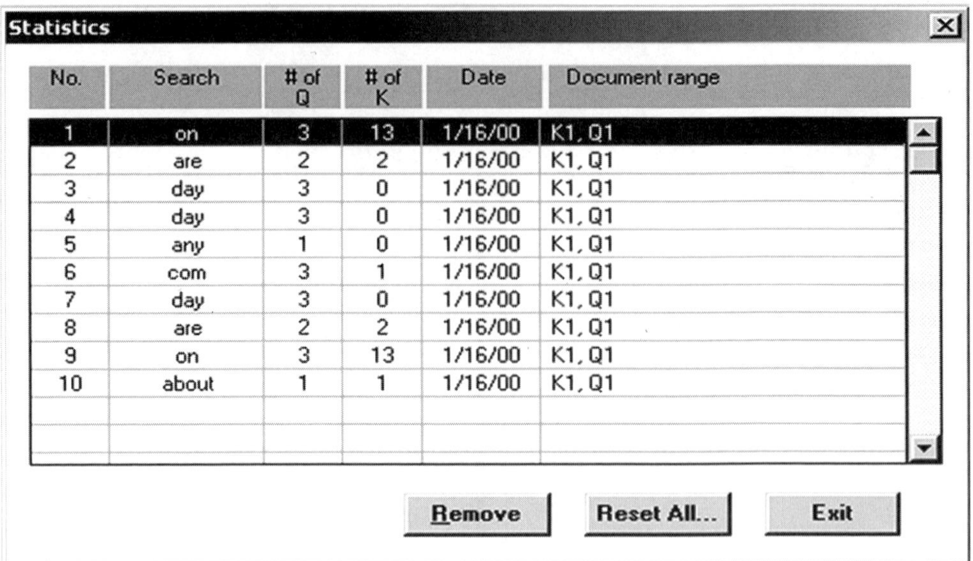

No.	Search	# of Q	# of K	Date	Document range
1	on	3	13	1/16/00	K1, Q1
2	are	2	2	1/16/00	K1, Q1
3	day	3	0	1/16/00	K1, Q1
4	day	3	0	1/16/00	K1, Q1
5	any	1	0	1/16/00	K1, Q1
6	com	3	1	1/16/00	K1, Q1
7	day	3	0	1/16/00	K1, Q1
8	are	2	2	1/16/00	K1, Q1
9	on	3	13	1/16/00	K1, Q1
10	about	1	1	1/16/00	K1, Q1

Figure 8.19 Write-On© statistic reports show the results of word and character string searches.

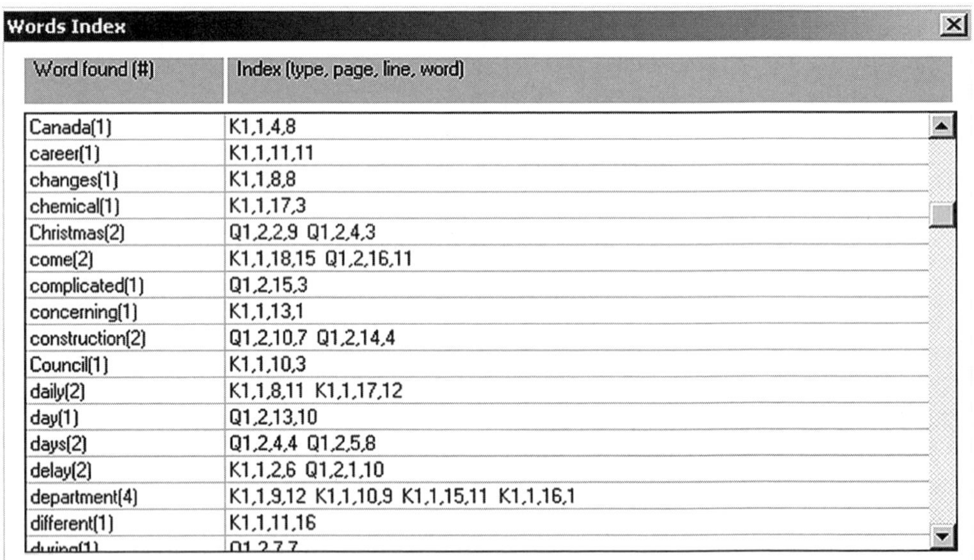

Word found (#)	Index (type, page, line, word)
Canada(1)	K1,1,4,8
career(1)	K1,1,11,11
changes(1)	K1,1,8,8
chemical(1)	K1,1,17,3
Christmas(2)	Q1,2,2,9 Q1,2,4,3
come(2)	K1,1,18,15 Q1,2,16,11
complicated(1)	Q1,2,15,3
concerning(1)	K1,1,13,1
construction(2)	Q1,2,10,7 Q1,2,14,4
Council(1)	K1,1,10,3
daily(2)	K1,1,8,11 K1,1,17,12
day(1)	Q1,2,13,10
days(2)	Q1,2,4,4 Q1,2,5,8
delay(2)	K1,1,2,6 Q1,2,1,10
department(4)	K1,1,9,12 K1,1,10,9 K1,1,15,11 K1,1,16,1
different(1)	K1,1,11,16
during(1)	Q1,2,7,7

Figure 8.20 Write-On© word index reports show the location and number of times words appear in the questioned or specimen material.

8.14 Conclusion

Handwriting identification involves the discovery and study of characteristics found in both the questioned and specimen material. In each case, features that are common must be distinguished from those that are abnormal or represent the unusual. The FDE must also recognize elements of the writing that are disguised or that have been deliberately and consciously changed. The extent of writing variation within a questioned document as well as the range of variation within the specimens of a particular writer must be determined.

If the unknown writing is not disguised, it should contain writing habits and qualities that the writer unconsciously employs. Similarities between the questioned and natural writing need not be identical in the sense that the two sets of writing can be matched or superimposed bit by bit; nor, on the other hand, can differences between the disputed and standard writing far exceed the variation customarily found in the known writing. There must be sufficient agreement of individual writing characteristics in both the known and questioned handwriting to establish beyond all doubt that both sets of writing must have been written by the same person.

Ensuring that conclusions are fully supported by the evidence is an important part of the FDE's work. It requires a complete study of both the questioned and known writing. Ample and accurate exemplars are essential requirements as are a full understanding of a person's writing ability and a careful analysis of what is observed. In every problem, the search for differences must accompany the search for similarities, and before one reaches an opinion of non-identity, any apparent differences must be evaluated in the light of known and possible writing conditions to ensure that they are not simply variables due to external factors. Unusual writing conditions, age, illness, and disguise are just some of the complicating factors that must be considered. In other words, the identification of handwriting is far from a simple or cursory operation.

References

1. Smits-Engelsman, B., Van Galen, G., and Meulenbroeki, R., Handwriting and its temporal evolution: a process oriented perspective, *J. Forensic Doc. Examination*, 9, 27–44, 1996.
2. Vastrick, T.W., Illusions of tracing, *J. Forensic Sci.*, 27, 186–191, 1982.
3. Osborn, A.S., *Questioned Documents*, 2nd ed., Boyd Printing Co., Albany, NY, 1929, pp. 273–290.
4. McClary, C., Write Down to the Wire or a Study of Baseline Alignment in Signatures and Handwritten Sentences, paper presented at the American Academy of Forensic Science meeting, Seattle, WA, 1995.
5. Tytell, P., Pen pressure as an identifying characteristic of signatures: verification from the computer, *J. Am. Soc. Questioned Doc. Examiners*, 1, 21–31, 1998.
6. Hilton, O., Characteristics of the ballpoint pen and its influence on handwriting identification, *J. Criminal Law Criminol. Police Sci.*, 47, 606–614, 1957.
7. Alford, E.F., Disguised handwriting: a statistical survey of how handwriting is most frequently disguised, *J. Forensic Sci.*, 15, 476–488, 1970.
8. Hardy, H.J. et al., Dynamical Features of Disguised Handwriting, paper presented at Gesellschaft fuer Forensische Schriftuntersuchung (GFS) meeting, Luzern, Switzerland, 1997.
9. Vastrick, T., The Skill Level of Pseudo-Tremor, paper presented at the American Society of Questioned Document Examiners meeting, Ottawa, Canada, 2000.
10. Osborn, A.S., *Questioned Documents*, 2nd ed., Boyd Printing Co., Albany, NY, 1929, pp. 205–236.
11. Harrison, W.R., *Suspect Documents*, Frederick A. Praeger, New York, 1958, pp. 297–305.
12. Morris, R., *Forensic Handwriting Identification: Fundamental Concepts and Principles*, Academic Press, New York, 2000, pp. 132–135.

13. Muehlberger, R.J., Variation: A Measure of Genuineness, paper presented at the ASQDE meeting, 1982.

14. Stangohr, G.R., Comments on the determination of nationality from handwriting, *J. Forensic Sci.*, 16 (3), 343–358, 1971.

15. Aloha, N.M., Classification and frequency of occurrence of specific number styles, *Can. Soc. Forensic Sci. J.*, 33, 13–22, 2000.

16. McAlexander, T.V., Explaining qualified handwriting opinions to the jury, *Int. J. Forensic Doc. Examination*, 5, 20–21, 1999.

17. Keckler, J.A., Felonious disguise: a study of the most commonly used modes of disguise adopted by convicted felons, *Int. J. Forensic Doc. Examiners*, 3, 154–158, 1997.

18. Livingston, O., Milwaukee fraudulent check filing, *Identification News*, August 1957.

19. Livingston, O., Milwaukee fraudulent check filing, *Identification News*, November 1957, pp 2–3.

20. Nicholson, P.J., A system for the classification of block capital letters, *Int. J. Forensic Doc. Examiners*, 5, 138–145, 1999.

21. Leung, S.C., Cheung, W.L., and Fung, H.T., A comparative approach to the examination of Chinese handwriting. Part 5. Qualitative parameters, *J. Forensic Sci. Soc.*, 33, 9–19, 1993.

22. Saxena, H.M. and Singh, M., Classification of the writing elements in Devnagari script, *J. Forensic Sci. Soc.*, 32, 143–150, 1992.

23. Philipp, M., Expected Future Developments in the Forensic Information System Handwriting (FISH), paper presented at the 4th European Handwriting Conference for Police and Government Handwriting Experts, London, 1994.

24. Kroon, L., Computer-Aided Analyses of Handwriting, the NIFO-TNO Approach, paper presented at the 4th European Handwriting Conference for Police and Government Handwriting Experts, London, 1994.

25. Srihari, S. et al., Individuality of handwriting, *J. Forensic Sci. Soc.*, 47 (4), 1–17, 2002.

26. Mohammed, L., Write-On©: a new tool for handwriting comparison, *J. Am. Soc. Questioned Doc. Examiners*, 2, 104–109, 1999.

Identification of Signatures 9

HOWARD C. RILE, JR.

Contents

When a person signs his or her name, he or she is carrying out one of the most common writing acts. With time and usage, a signature can become highly individualized, often in a style quite distinct from that of the writer's other handwriting. The complexity and consistency of an individual's signature does not reflect his or her education or intelligence. How each person writes depends upon the combined effect of a number of factors. Most marginally educated individuals produce a deliberate, awkward signature, lacking in skill and freedom, but nonetheless individual to the writer. With more proficient writers, the entire makeup of the signature assumes greater skill, movement becomes less primitive and elementary, and many letter designs approximate, to a greater or lesser degree, the copybook writing that the individual was taught. Just how a person writes his or her name depends upon numerous factors, such as muscular control, coordination, health, age, the frequency at which he or she is called upon to write, writing instrument, etc. Over time, all of these factors are blended into an individual's signature; a writer is almost never aware of any of these factors. Instead, through repetition, the writer executes an automatic formation of strokes, which usually but not always tends to assume a consistent pattern. Each signature varies to some degree from a master pattern and from other specimens of his or her signature. For the vast majority of individuals, signing one's name is a habitual act. The act of reproducing this piece of writing called a signature requires a minimum of concentration. Individuals can usually multi-task when signing their signature. The degree of variation among genuine signatures for each individual can range from slight to extreme. This is referred to as a range of variation and is an expected feature.

The simplest dictionary definition of a signature is a person's name as written by that person. This simple definition does not include requirements for legibility, consistency, complexity, writing instrument, writing position, etc. While all of these elements are significant in determining authenticity, they are not required elements of a signature by this definition. Also not included in the basic definition are requirements for intent, capacity, and personality idiosyncrasies. By personality idiosyncrasies is meant the totality of qualities and traits that are peculiar to an individual, such as behavior, interpersonal skills, moral values, etc.

A legalistic definition of a signature is found in *Black's Law Dictionary*:

> The act of putting one's name at the end of an instrument to attest its validity; the name thus written. A signature may be written by hand, printed, stamped, typewritten, engraved, photographed or cut from one instrument and attached to another, and a signature lithographed on an instrument by a party is sufficient for the purpose of signing it; it being immaterial with what kind of instrument a signature is made.[1]

This definition allows for the signature to be reproduced other than by the person's hand, using a name other than his or her given name, and for it to be a copy and not an original. This definition also includes the intent of the writer, the location of the signature on the document, and would allow someone else to create the person's signature.

For the purposes of this discussion, only the simple definition will be used. The issues of capacity, intent, and personality idiosyncrasies will not be considered. The intent of the writer and his ability to understand (i.e., capacity) are determined by the trier of fact — a judge or jury — based on testimony from witnesses other than the document examiner. The determination of personality idiosyncrasies (graphology), if even possible, is irrelevant to the question of authenticity. In any case, personality assessment of signatures or handwriting is not a skill or even a prerequisite for a competent FDE.

The use of the term *forgery* or *forger* in this chapter and other chapters in this book needs to be clearly understood. Strictly speaking, forgery is a legal term requiring an element of intent to defraud. The ultimate decision whether a particular disputed signature

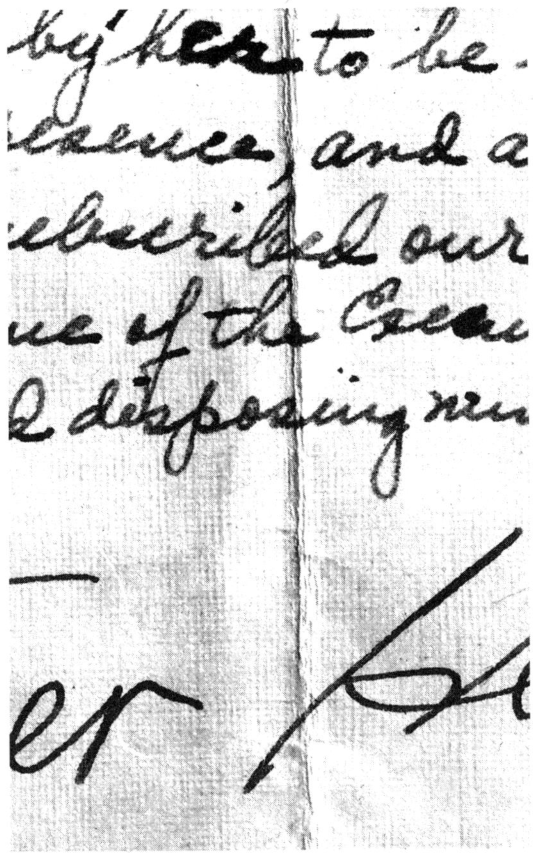

Figure 9.1 The document in question was a purported handwritten will. The text of the will was written by a nephew, who was the beneficiary. The will was witnessed by two friends of the nephew. The signature was found to be genuine. The will had been folded into quarters. The pen used to write the will was a fountain pen. The ink was a very free-flowing liquid ink. The examination revealed a difference in the ink line over the fold, through the signature and the written text. The document had been signed in blank by the decedent before being folded. The nephew wrote the text of the will after the paper was folded. Despite the fact that the signature was genuine, the document was ruled to be a forgery.

or document is, in fact, a forgery is a determination made by the trier of fact. It is not within the purview of the document examiner to determine intent. Not all non-authentic signatures are signed with the intent to defraud, e.g., spouses authorized to endorse checks for deposit. On the other hand, authentic signatures are many times found on documents determined to be forged, e.g., where critical entries were altered or inserted after the document was signed or where other pages of the document were replaced (Figure 9.1).

For the purpose of this discussion, the words *forgery* and *forger* will be used as terms of art to facilitate the discussion. The use of these terms implies that the signature or document has been established as not genuine and that actual fraud was determined. It is inappropriate for a document examiner to use the term *forgery* in either a report or testimony. Usage of *forgery* and *forger*, particularly while testifying, gives the appearance of lack of understanding of the role of the FDE.

9.1 Genuine Signatures

The identification of signatures constitutes a specialized branch of handwriting examination. Fundamentally, the identification principles set forth for general handwriting are the same, but certain factors require greater emphasis. Standards of known signatures may contain elements not common to other classes of the person's writing. An individual's signature many times is atypical of his or her general writing. An individual's signature is modified to suit a person's own needs and abilities. The extent of these modifications can, in some instances, be very significant over a relatively short period or, conversely, remain relatively consistent over a long period. The complexity and consistency of an individual's signature can only be evaluated by obtaining a truly representative sampling of all variations of his or her signature. The identifying attributes that are given closest study in signature problems may not always receive the same special consideration in the identification of general writing.

A signature is a combination of a rather limited number of letters or symbols. Because of its frequent use, it becomes with almost all individuals automatic habitual writing. With those who do not write as fluently, the measured, conscious writing act produces a more primitive writing movement, which is of significant identifying importance. Here again, writing movement is a basic consideration, as discussed in the following sections. Thus, while the design of letters may be the eye-catching feature, the factors related to the execution of the signature are the cornerstones of accurate identification.

While it is assumed that most individuals have the skill or ability to acquire and maintain a consistent signature, this is not always true. If an individual has a very simple signature that lacks consistency, forming a definite opinion of authenticity may not be possible. Individuals who, for whatever reason, do not develop and maintain a minimally complicated and consistent signature do not have a piece of writing that serves the purpose of a signature. The lack of a consistent and at least minimally complex signature allows an individual to deny any signature that he or she wishes. On the other hand, the individual's signature may be simulated relatively easily by another individual.

The skill of execution and the movement of the pen can be ascertained from a careful study of the signature itself. Numerous factors, which in combination fully describe the individual writing movement, must be analyzed:

- Continuous writing or movement interrupted at intervals by either stops or actual pen lifts
- The rhythmic or jerky pattern of writing
- The shading and pen emphasis on particular strokes, as well as the overall writing pressure
- The position of the pen
- The speed of execution
- The smoothly rounded, sharply curving or elliptical, or angular connecting strokes between letters and in turns of letters
- The starting of the initial writing movement before or after the pen makes contact with the paper and the corresponding condition at the termination of the word or at interruption within words
- Habitual re-touching of specific letters

These qualities are dependent on how the writing instrument is moved, whether by action of the fingers, hand, or arm, or a combination of these, and also the developed writing skill or coordination. All of these factors are collectively referred to as line quality, which is a term characterizing the visible record in the written stroke of the basic movements and manner of holding the writing instrument. The significance of line quality is in the degree of agreement between the questioned and known signatures. In some instances, a significant difference in line quality may be explained by, for example, health problems, but generally the differences in line quality indicate different writers.

An understanding of the complexity of the questioned signature can be obtained if the FDE draws or sketches the questioned and known signatures. Subtle movements and design features will be revealed that might not be apparent with a less detailed examination. Sketching of the signature should reveal the difficulties confronting a forger. Sketching is one method used to evaluate signatures. Another method is to prepare composite charts where the questioned signature is placed in juxtaposition with contemporaneous known signatures. This allows for an intercomparison of specific movements, breaks in the writing line, height ratios, connecting strokes, etc.

The assumption is made that what each writer strives for in signing his or her name is to reproduce a particular model or master pattern. This model contains the personal concepts of design, which when repeated and refined over time is intended to distinguish his or her signature from all others. Thus, the elements that define the general model include the design of the characters and their slant; the relative size of upper- and lowercase letters, and positiong of one character to another; ornamentations and flourishes; simplifications of form; and the arrangement of different parts of the signature with respect to its balance. The analysis and understanding of this combination of personal designs, movements, writing quality, skill, and form makes possible the identification of an individual's signature.

If a complex questioned signature agrees with a truly representative sampling of standards in all its identifying elements, an opinion that they are the work of the same writer is justified (Figure 9.2). If, on the other hand, significant differences exist between the known and unknown signatures, they must be the work of two writers. The differences between them have to be of a fundamental nature and not just the minor variations expected to be found between signatures of one writer.

This natural variation plays an important role in an identification. No two specimens of even a minimally complicated signature should be identical. How much signatures vary from one another depends on the individual writer and the conditions under which each was written. A lack of complexity and consistency in an individual's signature defeats the purpose of a signature, and the possibility of an accidental or intentional similarity between the signatures of two or more individuals is increased.

Because of natural variation, more than one known signature must be used to properly evaluate and form a meaningful opinion concerning a disputed signature (Figure 9.3). The methodology involved requires that the known signatures establish with reasonable precision the extent of the individual's variation. It is then necessary to ascertain whether the identifying qualities of the disputed signature fall within these limits. The identification of a signature consists of not matching closely with a particular known signature, but determining that the disputed signature contains the characteristics of and is written in

Figure 9.2 Three distinct signature problems arose in connection with the Loretta A. Byrne will. Two of the three witnesses died before the testatrix, and it became necessary to prove not only Miss Byrne's signature, but also the signatures of the two deceased witnesses, Kathryn R. Dawley and Florence Stewart. The Byrne will signature is shown above two of the many authentic signatures of hers. All of the qualities of execution and habits of form that make up her usual signature are combined in the will signature. The signatures of the witnesses, Kathryn R. Dawley and Florence M. Stewart, appear together between two of Miss Dawley's and two of Miss Stewart's genuine signatures. Both of these women, like Miss Byrne, write a fluent, clear, distinctive signature, and there is absolutely no reasonable doubt as to the authenticity of these two witnesses' signatures.

the same manner as the standards. For an identification, the disputed signature should fit within the extremes of variation established by the collection of known signatures.

9.1.1 Accidental Variations

In the previous paragraph it was presumed that the variations encountered between the questioned and known signatures are found within the natural variations of the writer. This condition exists in the majority of cases. There is another class of variation that is encountered rather infrequently. It is termed a rare or accidental variation. This variation

Figure 9.3 All of the Galatio signatures were signed to payroll receipts by a single employee and demonstrate the wide variations in his signature. Variation occurs in slant, size, alignment with the baseline, connections, and letterforms. His range of variation is unusually wide. The more variable a person's writing, the greater the number of known specimens required to define his habits accurately.

is a divergence that may occur only, for example, once in 100 or more signatures, or possibly not again. A variation of this nature is most troublesome in signature problems. This generally unexplainable variation may be due to a factor such as an unusual writing position, health or physical change, defective pen, etc. There normally is no provable or logical explanation for this variation, especially since the exact conditions under which virtually all signatures are executed are almost never known. If this rare divergence exists between a questioned signature and known specimens, one may still be justified in concluding that all are the work of one person, but at the same time one may be hard-pressed to persuade someone who adamantly believes that the signature is not genuine. Nevertheless, accidental or rare variations do not in and of themselves prove that a disputed signature is not genuine if no other concerns exist (Figure 9.4).

9.1.2 Unusual Genuine Signatures

The principles for establishing that a signature is genuine or identifying its author were considered in the preceding section. The document examiner is usually confronted with problems in which the questioned signature is at least initially presumed to be a routine or typical signature of the individual. These routine signatures appear on letters, applications, checks, legal documents, and other formally or semiformally signed papers. However, there are problems that involve a special class of signatures that create rather unusual technical problems. The principles of identification or the proof of genuineness do not vary, but some elements of these unusual signatures take on greater significance or create certain limitations. Examples would include receipt signatures, those written

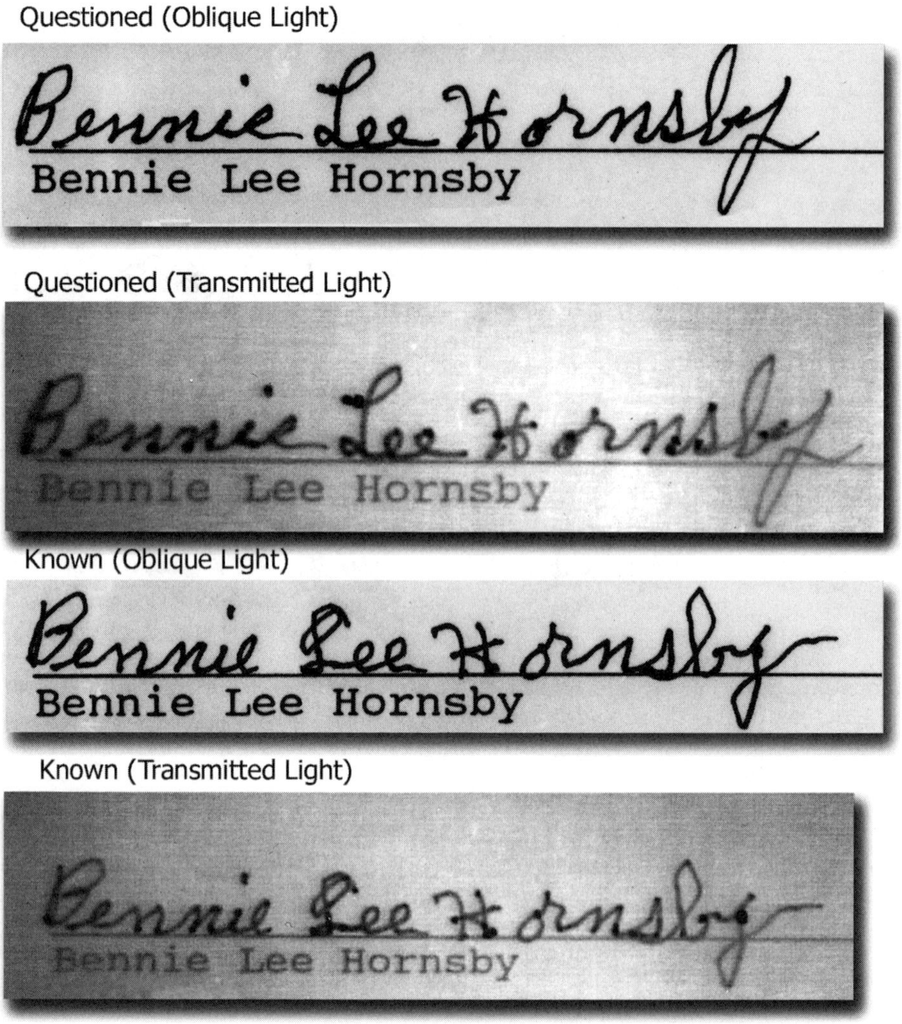

Figure 9.4 Microscopic examination of the signature in question revealed suspicious pooling of ink at points throughout the signature. This pooling of ink was suggestive of a simulation. Verified genuine signatures signed contemporaneously revealed the same phenomenon. Mrs. Hornsby hesitated at specific points as she wrote her signature. This, in combination with the free-flowing type of ink, resulted in the suspicious accumulation of ink. What was viewed initially as a sign of forgery subsequently was determined to be consistent with her normal writing movement and line quality. Additional factors that were significant in determining the authenticity of the document were ultraviolet examination of the paper used, a printing defect in the laser printer, and staple hole patterns. The initial contention was that a page had been substituted and that the signature was a tracing. The lack of the oval at the top of the capital *L* is an example of an accidental variation not observed in any of the submitted exemplars.

under the influence of alcohol or drugs, those written during severe illness, those of elderly persons during declining years of life, signatures signed with the actual intent to deny at a later time, and non-genuine signatures that are identified by the true person as, in fact, genuine.

9.1.3 Receipt Signatures

The term *receipt signature* encompasses a group of signatures that are hastily written under adverse writing conditions to acknowledge, for example, delivery of a parcel, registered mail, or merchandise. It is common to write with the paper resting against a wall, on a handheld clipboard, or on what is referred to as a diad, which is a pressure-sensitive recording device, in any number of awkward writing positions. Many consider this task a nuisance, and its execution is a perfunctory operation. They are frequently dashed off in, at best, a scribble. Details of form are lost because of haste, lack of care, and unfavorable writing position. A series of these scribbled signatures generally does reveal some pattern, but the variation from one to another can be very great. Formal signatures may not be of much value in comparison to this class of writing. Consequently, to resolve the problem accurately, it is necessary to locate known signatures of the same variety. These observations relate to many writers, though not necessarily all, for a percentage of people produce receipt signatures that remain within the writer's normal range of variation. With a less than legible and abbreviated form of signature, a qualified opinion as to its authenticity often has to be made. The qualification is based on a few typical forms in the signature coupled with the extreme freedom of writing. Unlike a retouched simulation, these signatures do not reveal any subtle attempt to improve their defective elements. A significant exception may be an occasionally freely written and not too accurate correction of a badly defective letter.

Starting in the mid- to late 1990s, courier companies adopted technology that did away with paper. What was now used involved a pressure-sensitive device using a stylus rather than a pen. This pressure-sensitive signature is digitally stored in case of a question about delivery or proof of purchase. Some, but not all, diads force the person to write in an unusual position, resulting in a distorted genuine signature (Figure 9.5). Inevitably, questions have arisen concerning the authenticity of these signatures. An evaluation can be made concerning the overall pictorial appearance. Given the inherent limitations, the best that can be expected in terms of an opinion would be one that is qualified.

9.1.4 Signatures Influenced by Alcohol and Drugs

A person decidedly under the influence of alcohol or drugs usually does not write in a normal fashion. While some of these impaired writers can produce an almost normal signature, most people find their writing coordination badly weakened and their signatures erratic and very significantly different (Figure 9.6). Usually the problem relates to alcohol, but it also can occur with some drugs. The defects are very different from the flaws of non-authentic signatures, which will be discussed in a subsequent section. In extreme cases, coordination suffers significantly, with free movement restricted and poor form resulting. The pen seems to have staggered across the paper. With higher blood–alcohol ratios, signatures generally, but not always, tend to be spread out slightly and somewhat larger than normal. If some known writing prepared under the influence of alcohol or drugs is available for comparison specimens, the document examiner's findings can be more accurate. An identification of the signature can be made without these special standards, but the opinion becomes more persuasive and demonstrable when some of the known writings were signed under similar, if not the same, conditions.

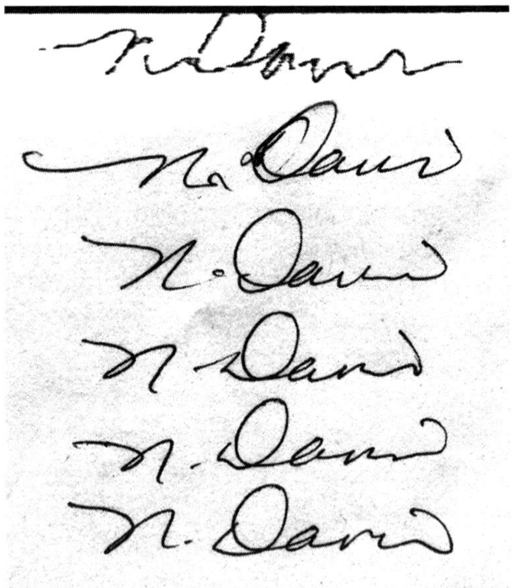

Figure 9.5 The top N. Davis signature is genuine, but was signed on a pressure-sensitive device. The image of the signature was downloaded from the Internet. The significant difference between the top signature and the known signatures is not the result of any deliberate attempt to distort the signature, but rather reflects an extremely awkward writing position.

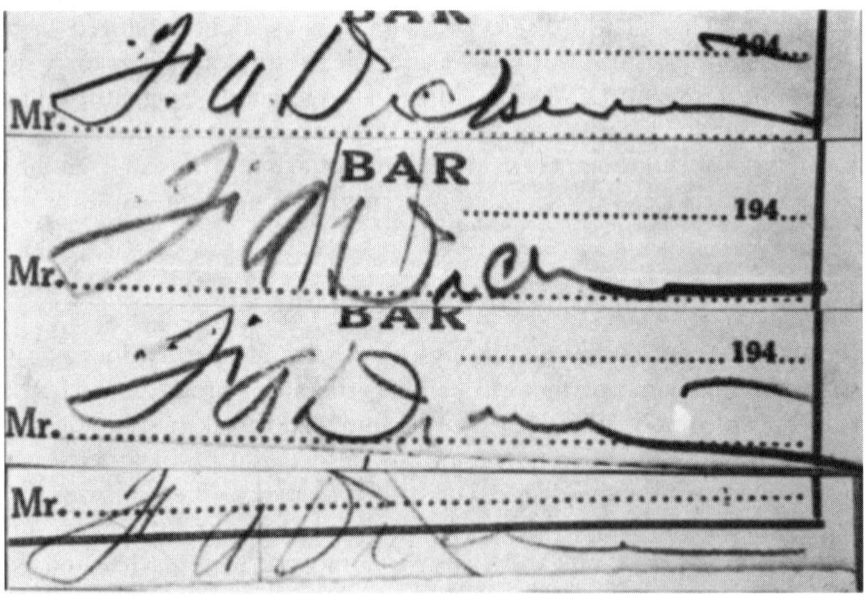

Figure 9.6 The four bar checks were signed by the same person during an evening of drinking. Note the progressive loss of detail from the first (upper) to last (lower).

While it might be assumed that alcohol and drugs would necessarily result in poor-quality signatures, in some instances the result is actually the exact opposite. With respect to alcohol, some individuals' signatures actually improve with the ingestion of a moderate amount of alcohol, specifically in cases of advanced alcoholism when the "sober" signature is tense or tremulous, but becomes more fluent with the relaxing effects of alcohol. The signatures of individuals suffering from Parkinson's disease, for example, can be significantly improved by medication. This has resulted in instances where the disputed genuine signature is decidedly more skillful and has better line quality than a signature written when the writer was not medicated.

9.1.5 Illness and Old Age

Problems involving writing of seriously ill or very weak elderly individuals can involve signatures that are significantly pictorially different. The weakened physical condition and the decline in coordination are important factors that help to create a poorer-quality signature compared to those of an earlier time. The decline in health as a result of either serious illness or old age affects both the design features of the signature and the writing skill and line quality. Irregular strokes, poorly formed letters, and poor alignment and letter spacing are among the features found in this class of signatures (Figure 9.7).

One quality is known as the tremor of old age or weakness of illness, which is found in some, but not all, of these signatures. This tremor or awkwardness is accompanied by a corresponding deterioration in design. Another typical quality is the inability of the writer to get the signature started. There may be one or two false starts made up of weak, erratic strokes that have no connection with the final signature. By the same token, there may be points within the signature or near the end where suddenly the movement is free and relaxed. This odd mixture of awkwardness and apparent ease of execution can initially be perplexing and may be found in some but not all signatures by the person at this time in his or her life.

The collection of exemplars for elderly or infirm individuals is fraught with a special danger. It is not unusual for family members or attendants, with no intent to defraud, to routinely sign for their elderly relative or patient. This may or may not be done with the knowledge of the relative or patient. Many times it is done to get routine bills paid. The family member or attendant initially may attempt to replicate the true person's signature, not to defraud, but merely to ensure that the checks clear the bank. Obviously, this creates a problem for the document examiner. These non-authentic signatures have been presented as verified knowns. When mixed in with authentic exemplars, the unwary FDE could assume that the individual merely has a wide range of variation. This contaminated collection of supposedly credible known exemplars can easily lead to an erroneous opinion. The worst-case scenario occurs when the elderly individual at some earlier point in time stops writing his or her signature altogether, and then the family member or attendant signs all subsequent documents. The standards that are then produced may be these simulations.

Known signatures should be obtained going back several years. Of particular value are exemplars that predate the illness and predate the involvement of an attendant or family member. A significant change in the signatures submitted as known could indicate a serious health issue or the beginning of the non-authentic exemplars. Document examiners familiar with writing of this kind may suspect that a poorly written signature was executed by an ill or aged writer; however, contemporary specimens, if they exist, will be needed to conclusively prove genuineness. The problem involves again searching for signatures written under comparable conditions. Hospital or doctor

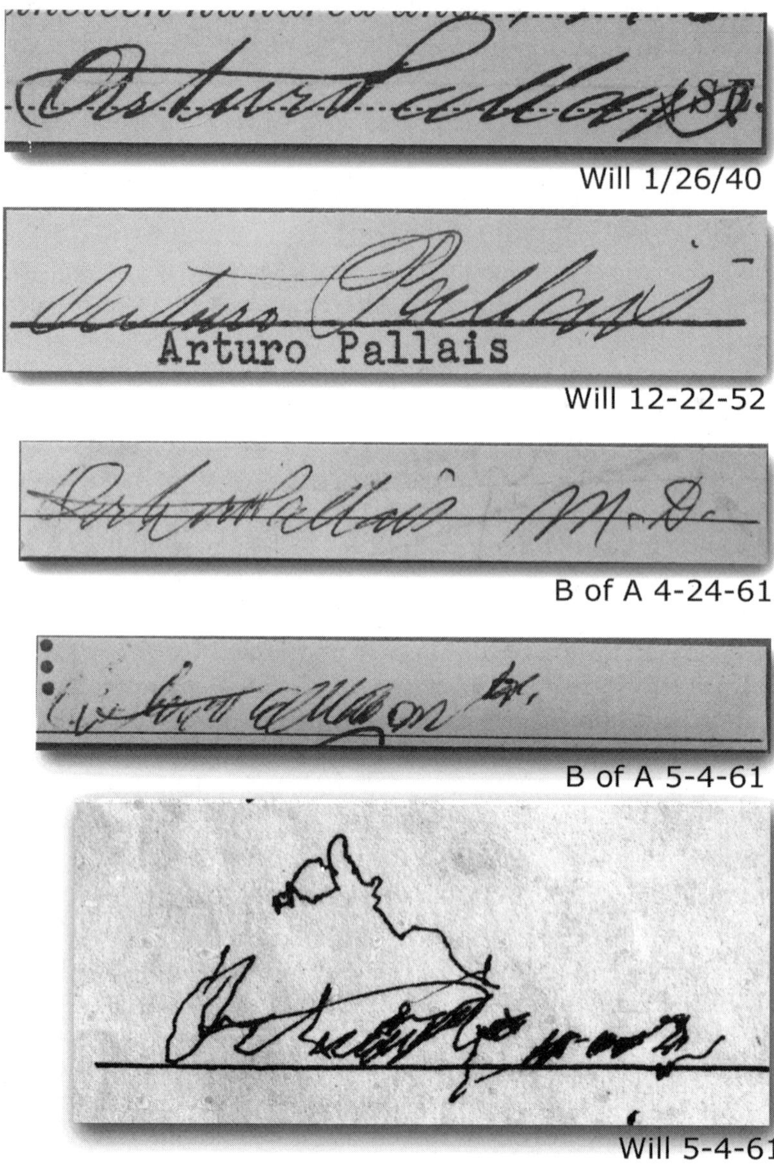

Figure 9.7 The signature in question was on a will dated the day before Dr. Pallais died. The most proximate exemplar signature was on a check signed earlier the same day. Dr. Pallais in good health had a very developed, skillful signature. The May 4, 1961, will signature demonstrates the deterioration in his physical condition. Whether his mental condition also had deteriorated was a question for the court. The court ruled that the will signature was genuine and based on other testimony that Dr. Pallais had capacity at the time.

records can be invaluable because they demonstrate the effects of an illness or treatment upon the person's ability to write and may be witnessed by a person not involved in the lawsuit. Few credible standards will be found for a very sick person who writes only when absolutely necessary. With the elderly, there may be a gradual, extended, though not necessarily straight-line, deterioration of their signature as death approaches. There will be within the progression of deterioration occasional signatures of better quality. Often only a few verifiable known specimens can be located. The final judgment will depend on the document examiner's experience and understanding of these problems in determining the authenticity or lack of authenticity of such a deteriorated signature.

9.2 Forged and Distorted/Disguised Signatures

To a greater or lesser extent, almost all individuals have the ability to distort their handwriting or signature. How successful one is depends on his or her awareness of his or her own writing or signature, the amount of writing or the number of signatures that need to be disguised or distorted, and his or her creativity. Individuals may be motivated to produce a signature that they will deny at a later time. This has been encountered involving government checks, acknowledgment of receipt of property, leases, business contracts, etc. Disguise can be accomplished by changing the slant, increasing or reducing the size of the signature, distorting letter designs, reducing or increasing speed, or any combination of these processes. Disguised signatures that are produced at leisure and with practice usually will be more successful than those that must be produced in the presence of a teller or delivery person (Figure 9.8).

The success of the attempt to disguise is also increased if the true individual is the sole source of standards used for comparison. If the individual has the opportunity to cull or carefully select exemplars submitted to the FDE, one can assume that only those that appear different will be produced.

9.2.1 Non-Genuine Signatures Identified as Genuine

There are occasions when a case will arise involving a disputed signature that truly tests a document examiner. What is being referred to are situations where an individual will, for a variety of reasons, identify as genuine a signature that he or she did not sign. To further complicate the problem, the individual will sometimes submit and identify as genuine simulations that were written by the forger. Types of cases where this confusing scenario occurs include physical examinations for insurance and where others are hired to take tests for educational or professional certificates. In the first instance, individuals want to obtain insurance knowing that they have a disqualifying illness or medical condition, and in the second instance, they are trying to get a better scholastic standing or a professional license or accreditation. This also occurs when individuals wish to establish an alibi or claim pay for work allegedly performed.

As can be expected, these types of problems are difficult to resolve. The questioned signature may be a practiced simulation that is explained as an unusual variation of a genuine signature. As strange and confusing as this scenario sounds, it is based on the experience of a number of document examiners. Short of a confession, the only possible way of resolving this type of problem is to obtain credible contemporaneous normal-course-of-business signatures and apply basic principles in the examination.

Figure 9.8 Despite the significant pictorial difference, the top signature and four known signatures were written by the same individual. Whether Ms. Grignon initially intended to distort or disguise her signature is not known. If all of the submitted exemplars were similar to the four displayed, an opinion of non-authenticity might be given. That opinion would be clearly wrong. Identification, on the other hand, would be difficult to demonstrate. An appropriate opinion based on the material submitted would be either inconclusive or indications of a common author.

9.2.2 Methods of Producing a Forged Signature

To attain his or her goal, a forger must employ one of two techniques to imitate a signature. Equipped with a genuine signature as a model, the forger may choose to imitate it freehand and produce what is referred to as a simulation, or feeling that he or she lacks the requisite skill to accomplish this successfully, the writer may choose to trace the forgery from the model.

Less skilled simulations and tracings are drawings. In drawing, an individual is focused on the process to produce the image and is not relying on a habitual act. With respect to freehand simulations, an individual can practice another individual's signature and, if skillful enough, should be able, over time, to produce more natural appearing simulations. A forger with the requisite skill can eventually produce spontaneous writings that actually include features found in genuine signatures. This phenomenon is observed in what is referred to as bookkeeper embezzlement cases, and also where family members or attendants become proficient in signing for an elderly or infirmed individual.

9.2.3 Traced Forgeries

A tracing is usually produced in one of two ways. The first, which can best be termed the transmitted-light process, is prepared by placing the spurious document directly over the genuine signature and, while directing a strong light through the two sheets from below, tracing the outline on the uppermost sheet. The second technique, which may be designated the carbon outline method, is completed in the following manner. A sheet of carbon paper or similar material is placed between the top sheet, which bears the genuine signature, and the spurious document below. Then the outline of the model signature is traced with a pointed instrument to give a carbon offset on the lower sheet. This outline can subsequently be covered over with a broader ink stroke. Often, when the ink stroke fails to cover the underlying outline completely, an attempt is made to erase the outline. While passable forgeries may be prepared by either of these methods, each has its inherent faults and weaknesses that give it away.

A tracing prepared by the transmitted-light technique has additional potential defects. Even with a strong light and thin paper, some of the subtle details of the model signature are not clear to the imitator as he follows the outline. With thicker paper and weaker light sources, a greater number of fine details are missed. This signature is produced not by a writing process but by drawing. Only the exceptional individual can, with practice, trace accurately with a fast, free pen movement. The faults of a freehand or simulated reproduction also can be found in tracings. The typical traced signature is drawn with a slow, measured stroke, which is usually filled with points of hesitation, uncertain movement, and sudden abrupt turns or jogs. This class of reproduction typically contains a uniformly heavy stroke that lacks natural shading or pen emphasis common to natural writing. It is not unusual for the traced signature to display patching and retouching made in an effort to correct letterforms. Retouching may even be made with the pen moving in the direction opposite to the writer's usual movement. In addition to pen lifts or breaks followed by careful splicing, there also may be indications that the pen stopped in the course of a stroke but remained on the paper before continuing on. The forger must carefully consider each letter before he or she attempts to reproduce it. The result may be pictorially consistent with known signatures, but the line quality is clearly different.

Traced forgeries have been made with outlines other than using carbon. A very light pencil tracing, possibly using the transmitted-light process, can be made and easily corrected if necessary, and subsequently covered with an ink stroke. Tracings have been created with a guideline consisting of a simple writing indented impression that the forger produces by tracing the genuine signature with sufficient writing pressure to produce the indentation on the underlying sheet. Such an outline is difficult to follow and the end result may not be very accurate. In one instance, the outline of a forged signature followed along small pinholes put in the paper. It appeared that these served as a guide for the first half of the signature, but then strangely there was no guide for the second half.

A traced forgery prepared from a lightly drawn outline is a crude imitation at best. Not only are the defects of tracing present, but they are emphasized by the double-tracing process. This act gives two chances for errors and inaccuracies in following the outline. The presence of the lightly drawn outline itself condemns the signature. This outline can be detected where the ink fails to cover it, a condition that is apt to occur at several points. Infrared examination may differentiate the covering ink from the underlying pencil or carbon line, and thus reveal the lightly drawn outline (Figure 9.9). Using microscopes or other specialized equipment, the difference in the ink stroke and lightly drawn outline can be observed. In an effort to remedy this defect, the forger may attempt to erase the lightly drawn outline not covered by ink. The material used for the drawing can be difficult to completely remove, and identifiable traces have been known to remain even after careful erasing. If the material is successfully removed, there may still remain a slight indentation from the pressure of tracing the original outline. Furthermore, the erasing may remove or smear the ink used to overwrite the tracing line. Each step in the preparation of a tracing generally introduces faults in the finished product that serve as the means to show the defective nature of the signature and to establish how it was prepared.

Locating the model signature is of major importance in the investigation of a suspected tracing. When a disputed signature is suspected of being a tracing, any genuine

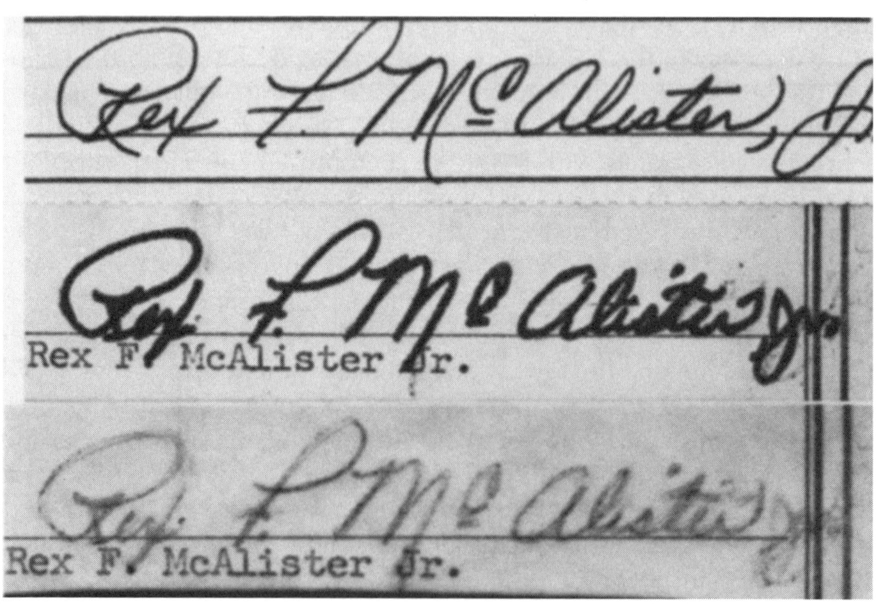

Figure 9.9 The upper signature is authentic; the second is a forgery. The bottom signature is an infrared photograph of the forgery, which filtered out the ink covering the carbon outline.

signature that may have been used as the model should be carefully examined to determine two facts. First, do the alleged model and tracing coincide sufficiently well to establish that the latter was traced from the former? A tracing and the model generally do not coincide exactly throughout. The accuracy of the tracing depends on the skill and care taken by the forger. The strokes of the disputed signature can wander away from those of the genuine and then return to the common track. When this condition exists, the conclusion that the disputed signature was traced from this model is justified. The second condition showing that a genuine signature served as the model can be revealed by a close examination of the genuine signature. Often an indentation, closely paralleling the outline of the signature, results from pressure of the tracing instrument. Furthermore, traces of pencil strokes or pen scratches, when either instrument was used to make the outline of the tracing, may be found around the model signature. The reverse side of the document should also be closely examined for embossing or traces of dye if carbon paper or similar material was used. Of course, a tracing can be proved without locating the model signature, but locating and properly understanding its condition and importance leaves those who stand against the facts with little, if any, defense (Figure 9.10).

In many cases, a forger creates two or more signatures by tracing the outline of a single model. The tracings can be detected in part by the close correlation and constant returning to the same outline. Most of the time, the model is not located, nor is there any outline around either signature, but their near identity indicates clearly how they had been prepared.

Figure 9.10 The signature in question appeared on a will that was typed on the reverse side of a 20-year-old check that had not been negotiated. A microscopic examination of the signature revealed an un-inked line paralleling the outline of the last letters in the last name. A photocopy of the actual model signature was produced as an exemplar. The fact that the two signatures were almost completely superimposable was initially asserted as proof that the traced signature was genuine because the two matched so closely.

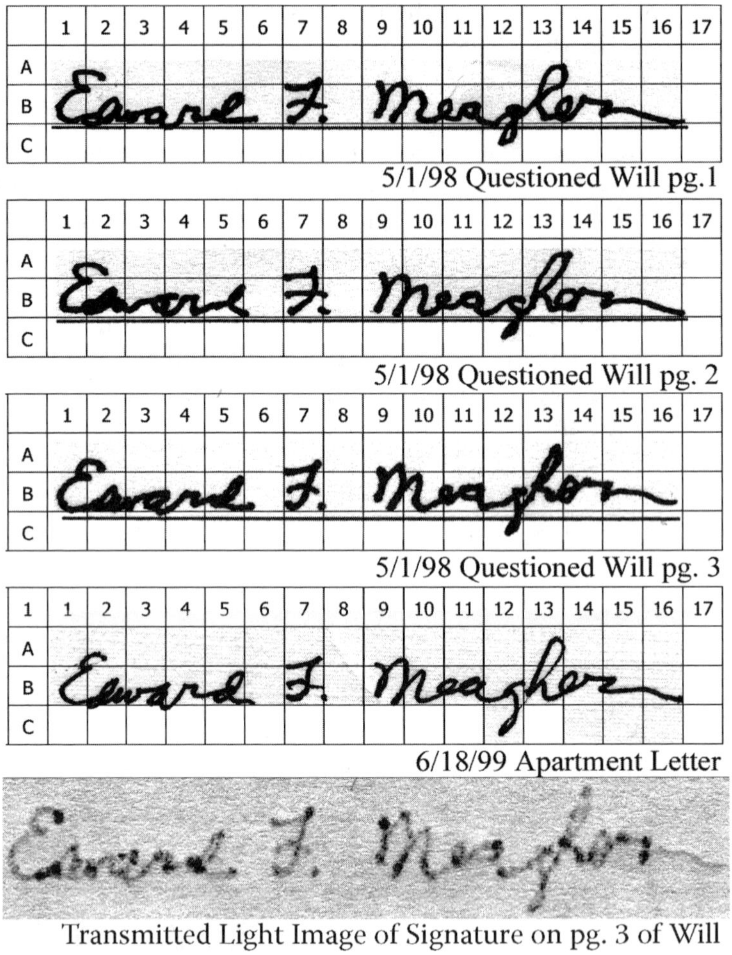

Figure 9.11 Three Edward F. Meagher signatures on the questioned will were written with black liquid ink. The questioned signatures exhibited poor line quality and numerous hesitation points. When examined with transmitted light, a number of the hesitation points were observed as ink dots similar to what was observed in Figure 9.4. The grid demonstrates the remarkable correlation with the apartment letter signature of June 18, 1999. The three will signatures are tracings of the apartment letter signature. In both the questioned Hornsby (Figure 9.4) and Meagher signatures, the same defects in execution were observed, namely, the extra accumulation of ink at the hesitation points. In the Hornsby case, this anomaly was an indicater of authenticity, whereas in the Meagher case, the anomaly is evidence of non-authenticity. Mere observation of this anomaly without correct interpretation would result in an erroneous opinion.

Even if the examination of the original signature reveals convincing evidence of a tracing, the submitted exemplars should be compared to the tracing. Many times the actual model for the tracing is submitted as an exemplar. The forger points out that the questioned signature must be genuine because it agrees so clearly with the exemplars (Figure 9.11).

It is necessary to note that the mere presence of an outline around the signature does not in and of itself establish that the signature is a tracing. This writer has examined an entire holographic will, prepared with a pencil overwritten with ink strokes. The pencil entries had the qualities of genuineness, and the ink overwriting had the classic defects of a tracing. It was ultimately ascertained that a family member had applied the ink overwriting when photocopies did not satisfactorily reproduce the original pencil writing. Needless to say, this explanation was received with great skepticism. Walters and Flynn reported examining a signature with an apparent guideline running adjacent to much of the signature.[2] The questioned ink signature was written with a pen on a photocopy made by a now obsolete process using zinc oxide-coated paper. The questioned signature had the characteristics of genuineness. Experiments and research established that the socket of the ballpoint pen had rubbed the zinc oxide coating on the paper, thus producing the outline. These examples show that an outline itself does not automatically prove that the signature is a tracing. In these instances, there were strange but provable explanations for a drawn outline with a genuine signature. These examples reinforce the concept that mere observation without careful analysis and interpretation can lead to an erroneous opinion.

9.2.4 Simulated Forgeries

A successful simulation of a signature involves a double process. The successful forger must discard all of his or her own writing habits and at the same time assume those unfamiliar characteristics of another writer. Writing is not just one or two simple habits, but a multitude of habits, which vary in complexity and which are routinely woven together into an intricate pattern. The conscious mental task of trying to successfully simulate another's even minimally complicated signature is enormous. The forger must struggle with using a strange writing process in place of his or her own well-established habitual writing movements. (See color Figure 9.12 and color Figure 9.13 following p. 366.)

Practiced freehand simulations are generally the most skillful forgeries. To produce a perfect freehand simulation, the forger must be aware of and reproduce the habits and line quality of the authentic signature. He or she should simultaneously discard all conflicting elements of his or her own writing. The writer must discover from a detailed study of the model signature that is to be imitated all its unique features and must know enough about the corresponding characteristics of his or her own writing to be able to eliminate their influence in the final simulation. This process involves writing the simulation in the same natural way and with the same degree of skill and speed as a genuine signature. Failure generally results because the forger has only a superficial idea of a few characteristics of both his or her own writing and the writing to be simulated, or because his or her skill as a penman fails to measure up to that of the person whose writing is being imitated. Even having circumvented all these pitfalls, the exacting task of execution leaves only a slight probability of success if examined properly by a competent document examiner (Figure 9.14 and Figure 9.15).

The dilemma that initially confronts a forger is that in order to capture the design, one needs to sacrifice speed, and in order to have speed in the writing, he or she will sacrifice the movement, letter design, and other unique features of the true person's signature. A perfect simulation would be one where the reproduction is indistinguishable from genuine signatures, in terms of execution and design. As can readily be appreciated, the less complex and consistent the true person's signatures are, the more easily they can be simulated. While document examiners are not able to determine intent, they can evaluate whether an attempt has been made to pictorially reproduce, i.e., simulate, a genuine signature.

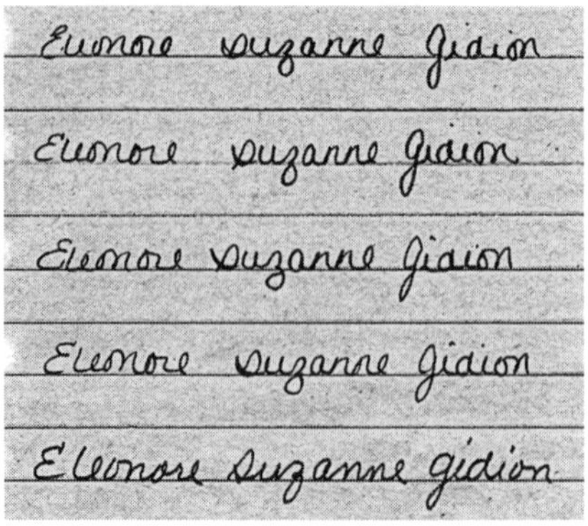

Figure 9.14 One of the Eleonore Suzanne Gideon signatures is a simulation. Her signatures are a relatively simple design written with pencil. Close inspection reveals the discrepancies in letter design and line quality of the fifth signature.

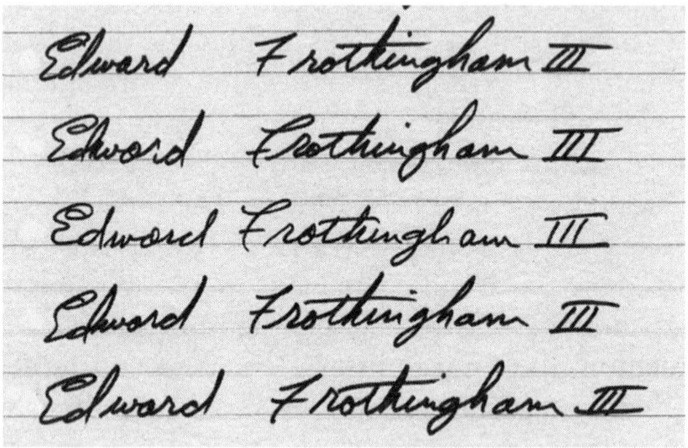

Figure 9.15 One of the five signatures was not written by Edward Frothingham. The pen used was a broad-tip liquid ink pen, which obscures some of the defects that might have been observed if a ballpoint pen had been used. Mr. Frothingham is a skilled writer with a consistent signature. The simulation is spontaneously written and exhibits good line quality in comparison to the genuine signatures. The errors in letter design are only revealed in comparison to numerous examples of his genuine signature. As a general rule, in order to capture the design, it is necessary to sacrifice the speed of the writing, and, conversely, the more quickly the simulation is written, the greater the chance for deviation in letter design. Which is the simulation?

A non-genuine signature can generally be detected because of the inherent defects in how it is written (i.e., line quality). The normal signatures of most writers consist of a relatively consistent speed of writing, skill level, pattern of shading, subtle movement, break points, continuity, and pen position. A simulation generally differs from the genuine signature in some or all of these qualities. Hesitation, unnatural pen lifts, patching, tremor, uncertainty of movement as portrayed by abrupt changes in the direction of the line, and a stilted, drawn quality devoid of free, normal writing movements combine to reveal the defective nature of a poor-quality simulation. Even practiced simulations by a skilled writer will reveal defects in design, if not in execution (Figure 9.16 and Figure 9.17).

By chance, some of the above defects commonly observed in poor-quality tracings or simulations are found in the normal signatures of many unskilled or ill writers. However,

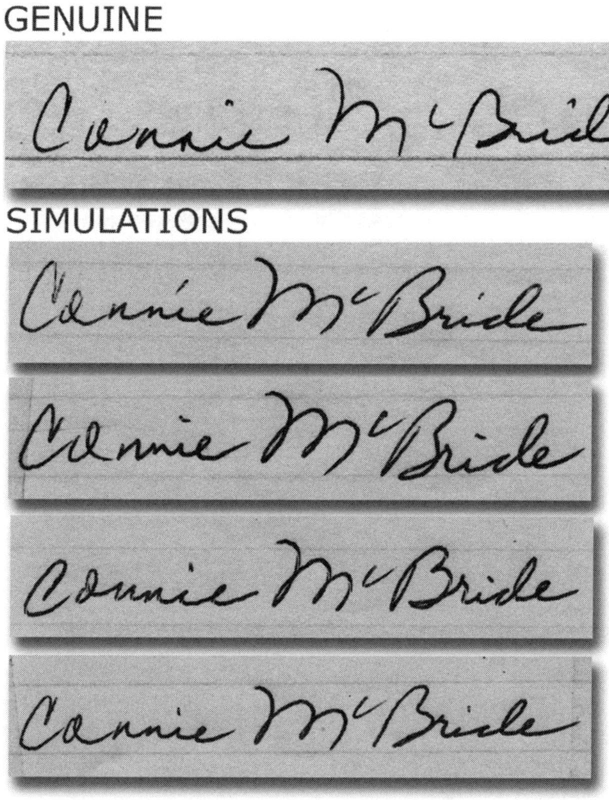

Figure 9.16 A skilled penman spent some time studying the construction of the genuine Connie McBride signature. Displayed is an example of a genuine signature in comparison to four simulations produced by the skilled penman. Subtle defects in design and execution can be observed in the simulations, but it is also apparent that the simulations are improving.

despite what might be believed, replicating an unskilled signature is difficult, particularly if the forger is unaware of its subtleties. The forger may have a greater skill level and may produce a simulation that is superior to the true person's signature. However, the signature of an unskilled or ill writer presents difficulties in trying to replicate.

It is of primary importance, therefore, to ascertain whether the variations between the known and disputed signatures are within the normal writing habits of the individual or, if different, actually point toward non-authenticity. In the usual forgery, the signature tends to be a deliberate drawing in which emphasis is placed on the imitation of characters and design rather than on the movement and writing qualities. In other words, the signature may be consistent in design but defective in execution. With a simulation, therefore, significant differences in the execution as compared with the standard signatures stamp its fraudulent character.

There is a popular belief that an acceptable forgery can be executed by writing upside down, but this is far from the truth. With this process, there is usually a sharp divergence in those qualities of writing, such as stroke direction, tapered strokes, breaks, and ink lay down.

Determining whether a spurious signature was produced as a result of an attempt to trace or simulate may not be possible. The end result of either process will be a signature

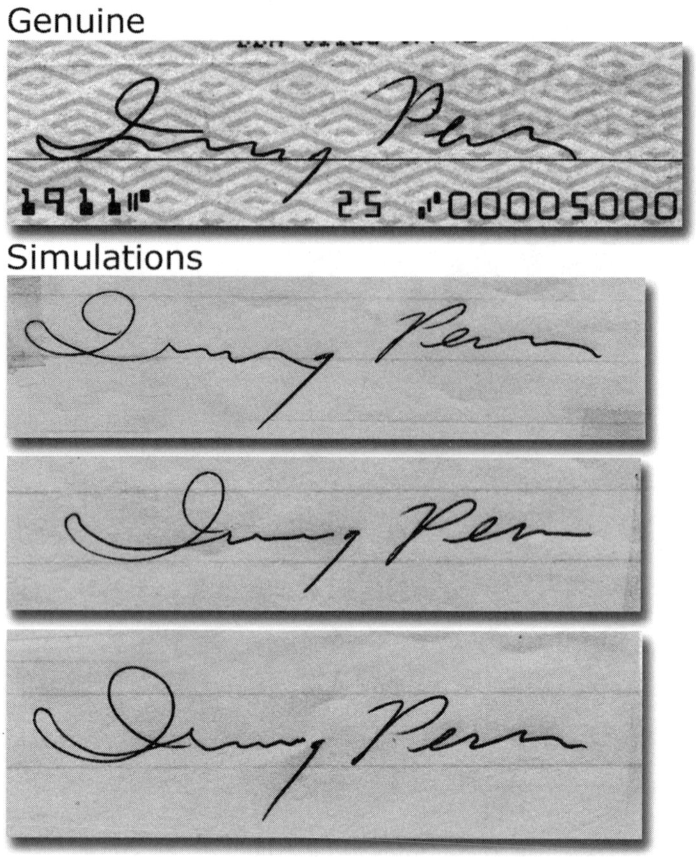

Figure 9.17 Irving Perlitch has a style of signature that is partially decipherable, at least with respect to his first name, but deciphering his last name would not be possible. His signature is moderately complicated. As can be seen, the skilled penman's simulations are improving. Most forgers lack the skill or patience to perfect their simulations. A spontaneously written simulation that captures letter design and movement may not be distinguishable from an unusual variation of a genuine signature, particularly if the genuine signatures are simple in design.

that has defects suggestive of either process. It really is immaterial which method was used if the defects of the disputed signature can be demonstrated.

9.2.5 Spurious Signatures

When the forger has no genuine signature available, he or she may still attempt to commit the fraud with what can be best termed a spurious signature. Without endeavoring to simulate the true person's signature, the writer merely produces the name in his or her normal or a slightly modified handwriting, and contrives some means of passing the document for personal gain before anyone recognizes the fraud. This is the technique of the check thief, the criminal who steals, endorses, and passes government or corporation checks. With signatures of this class, the proof of non-genuineness is rather straightforward once authentic signatures of the named person are obtained. Any similarities between the disputed and known signature are due to a common writing system or pure chance.

Furthermore, the identity of the forger often can be established by a writing comparison, as will be seen later in this chapter.

9.2.6 Spouse's Imitations

One class of questioned signatures encountered is those of a husband written by his wife or vice versa. In most instances, but not all, these signatures cannot be considered forgeries, for often they are written with the knowledge and consent of the other party or with a certain unspoken, tacit understanding. Examples are endorsements to paychecks that are either negotiated at neighborhood stores or deposited in a joint bank account. The other partner merely forgets to sign the check or considers the procedure perfectly proper. Spurious signatures produced by spouses generally bear little, if any, similarity to those of the spouse. There usually is no real attempt to duplicate the spouse's signature except possibly for a few very prominent features.

Problems concerning these signatures often arise, however, as marriages dissolve or when one spouse dies. One spouse may have signed the other's signature on a loan, deed, or stock transfer with or without knowledge or consent of the other partner. After the divorce or during a probate action, controversies arise regarding the signatures, requiring a forensic document examination. Whether an attempt to simulate the spouse's signature has been made can usually be determined and may assist the trier of fact (i.e., judge or jury) in determining intent.

9.2.7 Memorabilia, Artwork, and Historical Documents

These three areas create interesting problems for document examiners. Sports memorabilia, artwork, and historical signatures have great sentimental and financial interest. Evaluating the signatures of historical individuals may not be the best way to evaluate the document. As can be expected, obtaining credible, contemporaneous, reliable examples of the historical person's signature may not be possible. A generally more productive avenue of investigation involves a close study of the materials used to prepare the questioned historical documents. A notorious case from the 1980s, which was referred to as the White Salamander case, involved an individual by the name of Mark Hofman. Hofman diligently researched and obtained appropriate materials to fabricate a wide range of documents supposedly prepared in the 1800s. Some of these bogus documents that he created were particularly damaging to the Mormon religion. Careful examination revealed much of his fraudulent work. For example, while he did duplicate the components of the ink used in the time period, he needed to artificially dry the ink. In doing so, he created a pattern on the ink line that, when examined microscopically, was found to be different than ink lines that had dried normally over the years. Another document case involving a supposed relationship between President John F. Kennedy and the actress Marilyn Monroe was found to be bogus because the typewriter used was not commercially available until 3 years after the date on the documents in question. The supposed Hitler diaries also were found to be bogus because the paper was manufactured using dyes that were not commercially available until after 1945. A good simulation of Babe Ruth's signature (who died in 1938) was written with a roller ball pen that was not available until the 1970s. So, while signatures may have been of some initial interest in the above-cited cases, the actual resolution involved other types of forensic examination.

Evaluating the signature of a sports figure involves the standard methodology; however, there are also special considerations. It apparently was, and possibly still is, a common practice to have autographed items, such as baseballs, routinely signed by other individuals (e.g., bat boys) and then distributed as authentic souvenirs to unsuspecting fans. The person who was designated to sign the sports figure's signature may or may not have attempted to simulate the actual signature of the player. Another consideration is whether the sports figure had developed a specialized signature to be used just for that purpose. Furthermore, unscrupulous individuals who rightly deserve the term *forger* practiced and became very proficient at reproducing credible simulations of the sports figure's signature. All of the above scenarios create significant, if not insurmountable, problems for document examiners attempting to validate sports memorabilia. One survey concluded that 80% or more of sports memorabilia were not, in fact, signed by the sports figure.

Another specialized type of problem involves verification of artwork. The signature or mark used by artists may differ significantly from their normal signatures. An instrument used to produce the signature on the piece of art may also be different than that used for a normal signature. The individuals who forge artwork are many times themselves very talented artisans. As expected, they would have the requisite skill to successfully replicate the signature. To further complicate the problem, the fact that the genuine artist's signature is on the artwork does not necessarily imply that the artist completed the artwork. One famous artist was known to sign canvasses in blank and either sell or give them away.

9.3 Identifying the Forger

After determining that a purported signature is not genuine, the next assignment may be to identify who prepared the forgery. Often, the question cannot be answered, since the processes of simulating and tracing are not writings, but drawings and are also forms of disguise. Identification of the forger is the exception rather than the rule. Seldom do enough of the forger's own writing habits remain to serve as the basis for an identification. Comparison of a forger's own signature to his or her simulation of another person's signature is not generally productive. A credible and comparable sampling of the known handwriting of the suspected forger must be obtained for any chance of an identification. In rare instances, however, the forger may be identified by a number of his or her subtle and possibly highly individual habits that occur in the simulated signature. It must be understood that positive proof of the forger's identity cannot be based upon one or two similarities between the simulation and the suspect's writing. An identification of the writer of a forged signature may be made if the simulation only duplicates the salient features of the true person's signature, but the remainder of the simulation reflects the actual writing characteristics of the forger. The forger's unconscious writing habits can, on occasion, be found in the forged signature.

With spurious or non-imitated signatures, especially those in which the writer did not attempt to change or disguise his natural writing, it is much easier to establish who prepared the questioned signature. In this circumstance it may be difficult to collect adequate known writing upon which to base an opinion. Once these specimens are available, the problem reverts to establishing the link between the writing of the forger and the disputed signature. Many times similarities can be pointed out letter by letter or, even better, by a combination of characters. A major consideration, of course, is that the forger writes with the required skill to have produced the questioned signature. There are many cases in which a suspect has been requested to write the forged name and an identification made from these request specimens. The possibility of disguise in request exemplars must always be considered.

9.4 Assisted or Guided Signatures

In some instances, deathbed signatures, as well as those prepared during serious illness, are written with the assistance of another person. This aid may consist of a helper merely steadying the writer's hand or arm. Sometimes, however, the helper becomes the second writer, actually guiding and more or less controlling the movement of the hand of the primary writer through the writing process. The former type of signature is best described as an assisted signature, while the latter is often referred to as a guided signature. The line of demarcation between the two types is difficult to determine. It is not always clear how much participation can be attributed to the second writer or even if the primary writer contributed at all. Even when the assistance is only minimal, it can introduce constraint and awkwardness in the writing. If the assisting party actually guides the primary writer's hand, he is not just participating, but may be actually controlling the writing. Characteristics foreign to the primary writer's usual habits are introduced. Also, this jointly produced signature displays an irregular appearance with pronounced evidence of conflict and uncoordinated movement brought about by the two writers trying to work in unison (Figure 9.18). Guided and assisted signatures can legally validate a document. However, because of the importance of the document and the abnormal appearance of the signature, serious disputes often arise concerning not only the signature but the process to produce it. The irony is that from a legal perspective, a mark or X, if properly witnessed, is accepted by the courts. There is really no reason to produce a guided-hand signature. Nevertheless, these do occur. The real question most of the time is whether the primary writer had sufficient mental capacity at the time the signature was executed; that is a question the document examiner cannot answer.

The technical problems encountered in these signatures can become extremely difficult. Opinions cannot always be as precise as with other types of signature problems. Assistance is necessary only when the writer has great difficulty in writing alone. In some instances, he or she first attempts to sign unassisted and cannot complete the name (Figure 9.19). When the document demonstrates evidence of such a preliminary attempt, this fragment often aids the FDE in analyzing the entire disputed signature. There is always the question of how much assistance was given. Was it simply steadying an infirm hand or arm? Was there actual help in creating the writing? The testimony of those present, particularly of the assisting person, and the physical evidence within the document must be considered and weighed in combination by the trier of fact. Unfortunately, there is seldom even one other signature available that was prepared in a similar manner with the same person assisting. Even if there is such a specimen, the question remains: Was the degree of assistance the same as on the document now in question?

Many times an allegedly guided or assisted signature resembles a poorly written simulation. Most allegedly guided signatures have, after proper examination, been found to be non-genuine. The process to produce a simulation is significantly different than the

Figure 9.18 The Catherine J. Glover signature on the will dated October 3, 2000, is approximately 150% the size of the most contemporaneous exemplar signatures of Ms. Glover. Her attendant was in the habit of writing Ms. Glover's name. The questioned signature was written in a very awkward but at times spontaneous manner. The *Cat* combination in *Catherine* and the *Gl* combination in *Glover* are more consistent with the attendant's writing than with Ms. Glover's. If the questioned signature was a guided-hand signature, there is little evidence of participation by Ms. Glover.

process that produces a guided-hand signature. A careful and thoughtful evaluation should differentiate them. Almost without exception, the possibility of an assisted or guided signature is not mentioned until after an initial examination has raised serious questions about its authenticity.

The conflicting characteristics of a guided signature result from a combination of factors. They include the strength and writing ability of the primary writer, the extent to

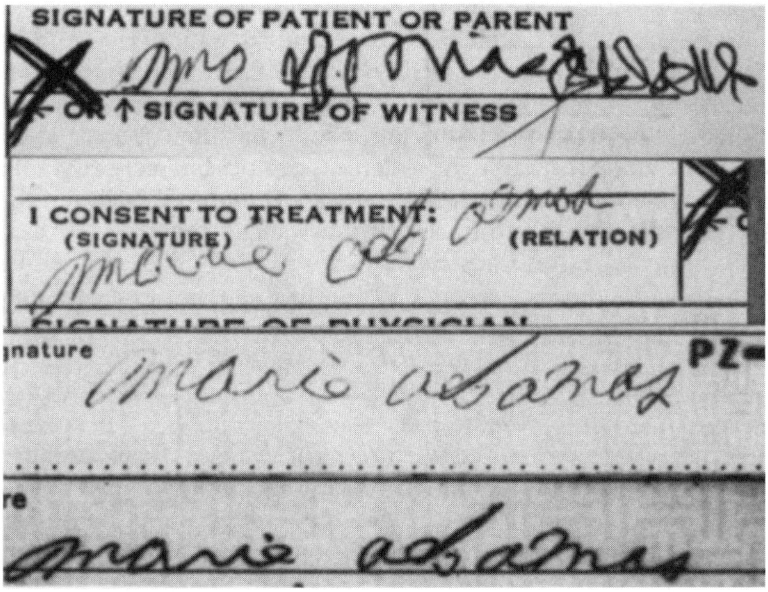

Figure 9.19 Marie Adams started to write the top signature, but stopped after two letters. The remainder of the signature was assisted by an unknown hospital attendant. From other evidence in the case, it seems likely that the signature was completed with a reluctant Marie Adams.

which both the assisting party and the primary writer participate in the joint writing act, and the various physical conditions surrounding the actual execution. These include the positioning of the helper's hand on the forearm or hand of the primary writer, the writer's and guider's positions, and the type of support for the document during execution.

There may be an occasional guided signature that actually turns out to be the handwriting of the assisting party but written in an awkward manner. Such a signature might well arise when a close friend or member of the family is assisting the writer who is, for all intents and purposes, unable to write. Under these circumstances, if he or she trusts the assisting party, he or she may merely hold the pen and relax completely, so that the guider must move the writer's hand with the pen in the execution of the signature. In these cases, the document examiner can only give an opinion that the signatures appear to have been written by the guider. The court ultimately decides the issue of genuineness on the basis of whether the assisting person's and others' testimony appears truthful. The reader can appreciate that with this class of signature problems, each case is unique. The general rules set forth in this section will serve as a guide in determining the basis of the expert's opinion.

9.5 Identification of Initials and Illiterates' Marks

Initials and illiterates' marks can be considered special classes of signatures. The former are used often as shortened signatures in many classes of business activity; the latter are the only signatures available from the illiterate. Since disputes arise about both, further consideration of their identification is in order.

9.5.1 Initials

For many individuals, initials are simply the first letters of the person's given and surname. Many writers, especially those who are called upon to initial documents on a regular basis, develop a personal manner of writing their initials, sometimes significantly different from their signatures. Genuine specimens may be denied, and occasionally the initials of another person are imitated or forged. How the initials are written as reflected in the line quality is generally the controlling factor. Authentic exemplar initials are sometimes very difficult to locate, but are essential when the writer has developed an unusual manner of initialing. If written separately, it may be possible to supplement a limited number of known initials with signatures to solve the problem. Basically, the small amount of writing makes the problem more difficult than signature investigations.

9.5.2 Cross Marks

Under the law, "his mark," properly witnessed, has been accepted as an uneducated or disabled person's signature for generations. While the mark is generally a simple X, each may have some degree of individuality. In fact, two illiterates are unlikely to make identical marks, but individuals who use this class of signature tend to develop a fixed pattern. The usual mark consists of but two or three strokes, and because of its simplicity, it is easily imitated. Often, though, the forged mark is made without reference to a genuine one. Forged marks may contain such common faults as improper sequence of strokes, greater skill of execution than the imitated writer is capable of, defective writing movement, too heavy or too light pen pressure, and a pen position that is inconsistent with the writer's habits. Though this kind of signature is simple, there are many elements to be considered in its possible identification.

On occasion, a deathbed document is authenticated by a mark made by the dying person who under normal conditions would have signed his or her name. Usually, answers to questions about the authenticity of these marks must be ascertained from the attending witnesses. At times, the witnesses are not available or their reliability and integrity are in question. Under these circumstances the document examiner must endeavor to evaluate the questioned mark. Whether he or she can assist in resolving the dispute is problematic.

9.6 Evaluating Non-Original Signatures

The discussion up to this point has been based on the assumption that an original, that is, human-written, signature is being evaluated. Due to changing technology, evolving business practices, and changes in the law, original documents are often not available for evaluation. This is particularly critical when it involves a questioned signature. Depending upon the copy quality of the non-original, the FDE may encounter limitations in evaluating the overall line quality of the disputed signature as well as structural elements.

There are several different methods utilized to transfer signatures from legitimate to fraudulent documents. Because of this relative ease of manipulation, a copy cannot be validated as a true and correct copy of a missing original. On the other hand, if evidence of manipulation is found or if there are fundamental design differences between the questioned and known signatures, then an opinion of non-authenticity is justified. Absence of evidence of manipulation is not proof that a copy is an accurate reproduction of the missing original.

At the time of this writing, the main process used to create bogus copies involves the indirect electrostatic process, in other words, photocopiers and laser printers. Disputes also arise over reproductions of signatures produced by color photocopying, inkjet printing, and rubber stamps. Anyone attempting to evaluate signatures must be able to differentiate between signatures and writing produced by various writing instruments and reproduction processes. Most of the time, telling the difference is relatively straightforward, such as differentiating between photocopier toner and ballpoint pen ink. Problems may occur, however, in differentiating inkjet printing from liquid ink pens and a rubber stamp impression from a broad-tip liquid ink pen.

A copied reproduction of a signature may be consistent in design with exemplars; whether or not the signature was actually written on the document cannot be determined without examining the original. If a careful examination is conducted with appropriate standards, an opinion discrediting the authenticity of the copied reproduction may well be justified. Significant defects in design and construction cannot always be explained as the result of the copy process(es) (Figure 9.20). If there is agreement in overall design and construction, and no apparent defects in execution are observed, it can be stated that the questioned and known signatures are consistent. A qualified opinion may be justified. The document may well be what it purports to be, or the document may be a manipulation. One may not be able to say with certainty either way. If the reproduction, however, is of extremely poor quality, such as a multi-generation copy prepared by one or more different processes, it becomes less likely that even a qualified opinion can be given for or against authenticity.

A careful examination of a disputed non-original signature will sometimes reveal anomalies such as unusual horizontal fragments attached to descending portions of the signature. Occasionally, fragments of a line can be found paralleling portions of the signature, and on occasion, lower descenders of the questioned are found to be shorter than on exemplar signatures. These anomalies can be telltale evidence of a manipulation. The forger may select a signature that he or she wishes to transfer, but the selected specimen intersects either underscored lines or printed text. In this case, the forger generally attempts to remove or clean up these areas of intersection, many times not successfully (Figure 9.21). The fragments of the underscored line or text from the model document are sometimes found on the manipulated document. Depending on the copy process utilized, shadow lines sometimes are observed that result from a slight difference in height between the attached cut-and-paste signature and the underlying page when the bogus copy was made. A more sophisticated method of manipulating signatures involves a scanner and computer software (Figure 9.22).

9.7 Summary

We have seen that a non-genuine signature may be prepared in several ways. To be successful to the extent that it prevents recognition under expert scrutiny, the simulation or tracing must duplicate all the attributes of the authentic signature to a degree that differences are few and can be attributed to chance or to the normal range of variation of the genuine signatures. Not only the design of letters but the qualities of writing movement must be

Figure 9.20 The original quitclaim deed is unavailable and is presumably lost. The image of the questioned signature was produced from a third-generation duplicate of the microfilm that was prepared at the time the deed was recorded. The copy of the questioned signature was at least four generations removed from the original document. The significant differences in letter design, proportional heights, spacing, and overall construction cannot be explained as defects caused by the copy processes.

accurately reproduced. The problem involves imitating simultaneously a number of characteristics rather than each one individually. Depending on the complexity of the signature and the skill of the forger, this task can be and usually is extremely difficult. Almost without exception, all non-authentic signatures contain flaws in both execution and design. Recognition of these flaws requires close examination of both the questioned and known signatures in order to accurately recognize and demonstrate the flaws. The ability to recognize and correctly evaluate disputed signatures requires not only the theoretical knowledge but also the practical experience gained by good-quality training and extensive experience.

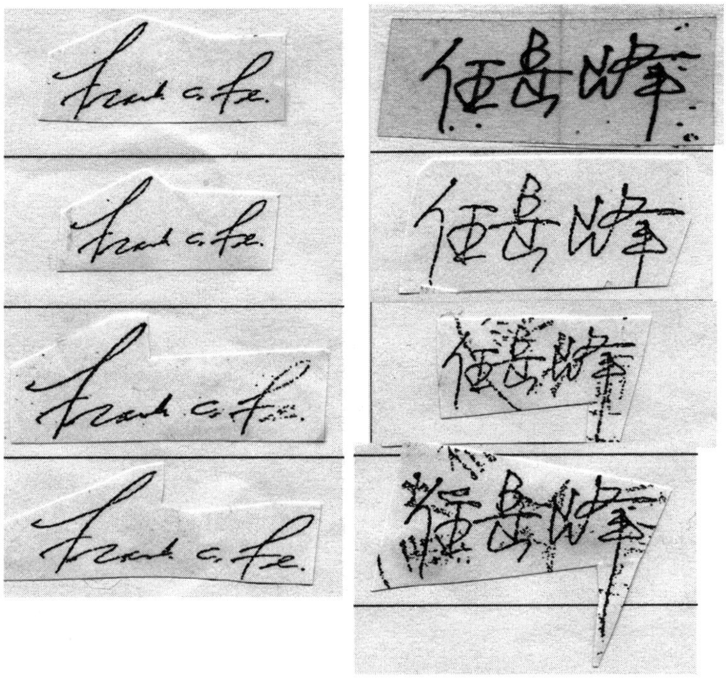

Figure 9.21 Displayed are two individual signatures that were used to create, in total, seven bogus documents. Altogether, five individuals' signatures were involved. The original model document was located. Minor differences were observed in the questioned signatures. Also, changes in size were observed. Excluding these minor differences, the signatures are identical.

With reliable standards, determination of non-genuineness normally should not be difficult. Careful examination should reveal differences even between two writers whose signatures are not highly individualized or who have had their formative training in the same writing system. Short names and an attempt to imitate a prominent element create similar designs, especially in the eyes of laypersons. The subtle differences, not the similarities, should separate the authentic from the non-genuine.

The identification of a signature or proof of its genuineness assumes prominence in many legal controversies. It is not a rare event for an individual to deny having signed a contract, deed, receipt, or virtually any class of document. He or she may raise a cry of forgery or may remark, "Well it looks like my signature, but I never signed that document." If genuine, this fact must be established. Disputes over signatures arise when the writer has died or is otherwise not available to authenticate them. Possibly, there are no witnesses or the witnesses themselves are dead. The authenticity of all of the signatures on the will must be demonstrated. These conditions are encountered with contested wills or in claims against estates, in which almost any kind of legal document may figure. Because of a combination of circumstances, it becomes necessary to establish the genuineness of a signature through technical study and at times to demonstrate the fact in court (Figure 9.23).

Signatures to wills, deeds, notes, contracts, and checks are more frequently forged than signatures to other documents, but there is probably no class of document in which there

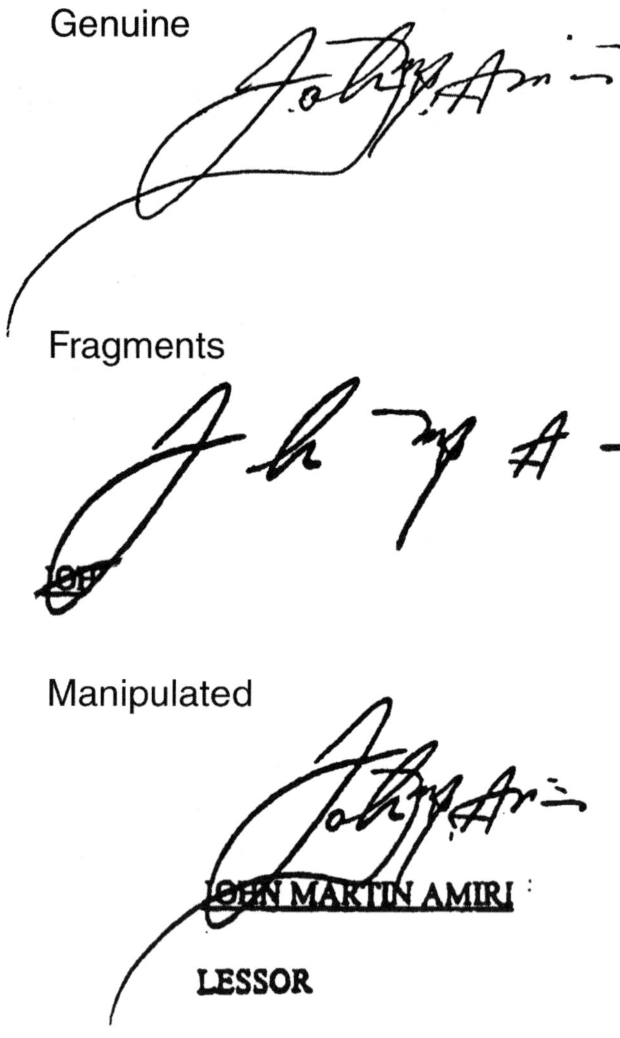

Figure 9.22 A more sophisticated manipulation was created with the use of a scanner and computer software. A genuine signature from a letter dated November 20, 2002, was used to create the purported signature on a fax dated January 10, 2004. Five elements of the genuine signature were isolated and rearranged. Additional modifications were made on the letter *o*, the left side of the capital *A*, the *m* in the last name, and the sweeping stroke going through the first name. Using this technique, a forger must be careful not to create a bogus signature that could be discredited because of significant design differences.

has not at one time or another been a forgery. Scientific examination is generally the chief means of detecting these frauds.

Whether these questions can be accurately answered generally depends on how many authentic signatures are available. With a formally written signature in dispute, that is, one executed with normal writing care, a sufficient number of genuine signatures almost always permits a clear-cut answer. If the disputed signature is written with little or no care, such as a hurried pencil-written receipt signature, there is less likelihood that the opinion will be a definite one. Here the greatest difficulty is that genuine signatures written under similar conditions may be limited in number and hard to locate. The disputed signature

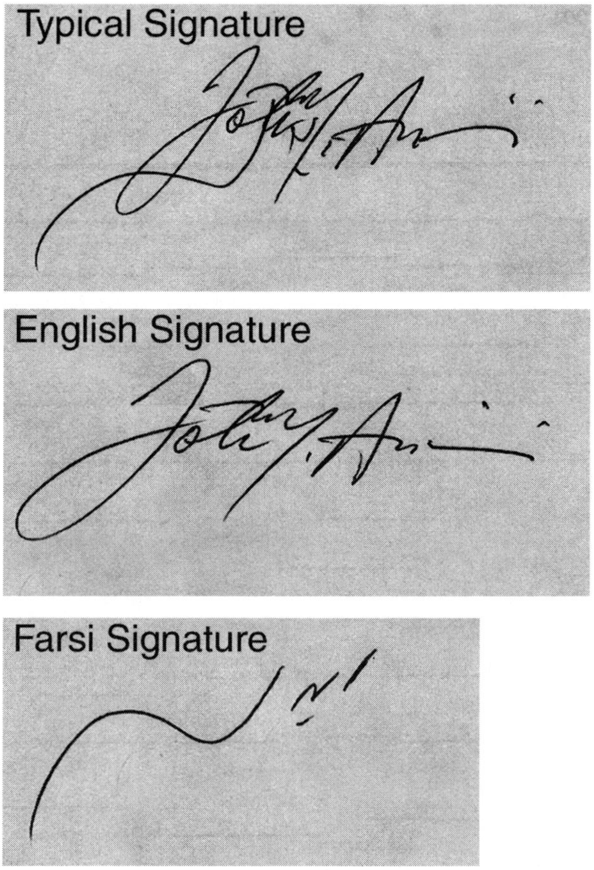

Figure 9.23 John M. Amiri has a unique signature in that it is composed of both an Anglicized and Farsi signature. The Anglicized signature is written first, left to right, and the Farsi signature is written second, right to left. The Farsi signature is inserted in the middle of his Anglicized signature. In order to successfully simulate this signature, the forger must be aware of not only the elements of the signature, but also the direction of the strokes. With careful examination, the stroke direction can be determined from ballpoint written signatures.

may be so badly scrawled that it contains few identifying characteristics. Similar limitations may be encountered with signatures of the seriously ill or those of the infirm or aged. In such a case, only a qualified opinion can be given. Fortunately, the vast number of disputed signatures are a part of important papers, and the issue of their genuineness can be definitively determined through technical study.

This discussion is intended to sensitize the reader to the complexities that are involved in the correct evaluation of signature problems. The methodology in conducting the examination requires the gathering of appropriate materials, careful observations of details, and the correct interpretation. While the ultimate opinion is subjective, the process follows a scientific methodology. The accuracy of a subjective determination is improved with appropriate training and experience. Conscientious document examiners continuously monitor their methodology. Studies have shown that consultation among experienced FDEs significantly reduces errors.

References

1. Black, H.C., *Black's Law Dictionary*, 5th ed., West Publishing Co., St. Paul, MN, 1979, p. 1239.
2. Walters, A. and Flynn, W., The illusion of traced forgery on zinc oxide-coated photocopy paper, *Journal of Police Science and Administration*, 2, 376–380, 1974.

Identification of Hand Printing and Numerals 10

JAN SEAMAN KELLY

Contents

10.1 Hand Printing

Hand printing, also known as manuscript writing, was brought to the U.S. by several England educators shortly after 1920 and adopted for use in a number of private and specialized schools in the New York and Boston areas.[1] It has been taught as a writing style in the majority of U.S. urban schools' primary grades since the mid-1940s. Prior to the introduction of manuscript writing, hand lettering was taught to those in professions such as drafting and engineering. Charles W. Reinhardt devised and published a system of rapid single-stroke lettering for engineers based on simplified forms.[1] Modern manuscript writing evolved from Reinhardt's formal hand printing system and is taught to U.S. children entering kindergarten or first grade.[2]

The primary source of class characteristics of any writing style originated from the copybook form initially taught to the writer. However, in the contemporary school systems, class characteristics in hand printing are not limited to copybook format. Generally, the teaching of handwriting is not as regimented as it was when whole school systems designated a particular handwriting copybook.[3] Little time is given in school for children to work on their penmanship skills. Deviation from copybook style and limited supervised practice produce the individualized combination of habits necessary for an identification or elimination of a writer.

10.1.1 Research Focused on Hand Printing Characteristics

From the 1940s to 2004, a great deal of work has been presented and published regarding class and individual characteristics of hand printing. This research has provided data on class characteristics by recording the frequency of occurrence of individual letter formations. Orville B. Livingston authored one of the first articles discussing the frequency of certain characteristics in pen printing. He examined 200 cards and recorded a frequency tabulation of certain letter formations; for example, 92% of the 200 writers wrote an uppercase *W* with a small center.[4]

The Metropolitan Police Forensic Science Laboratory in London established one system of classifying block capital hand printing. This system assesses frequency of occurrence by focusing on letter types and not the variation of a particular letter. In this classification system, letter types were classified based on the number and order of pen strokes in the individual characters of 140 writers. The block capital *E*, for example, was grouped according to which of the three cross strokes was written first.[5]

The Metropolitan database contained 1000 writers by 1990.[6] In her article, Dr. Audrey Giles reported that the database had been used in casework and proved beneficial in establishing frequency of occurrence of a single letterform or a combination of letterforms. Since the database only records the structure of the letterform, the FDE must still conduct an examination to determine the shape of the letter, height and width proportions, and how the character relates to the surrounding letterforms. These additional characteristics determine the individuality of the printing, allowing for an identification or elimination of a writer.

Geographical location is another source of class characteristics. Much has been researched regarding class characteristics in the hand printing of U.S. and non-U.S. writers. One of the first studies conducted by the U.S. Immigration and Naturalization Service (INS) used 2500 documents to determine frequency of occurrence of individual letterforms of writers from Latin American countries.[7] This 1993 study established that seemingly individual characteristics were class and not unique features, and that geographical location impacts the frequency of occurrence of a few of the class characteristics.[7] For example, the two-circle formation for 8 was common for all of the countries and should be considered a class characteristic, while the Greek *E* was very common in the samples of writers from Columbia, but uncommon in the samples from Dominican Republic writers.[7]

An additional 20,000 hand printed documents that included writers from numerous countries were utilized in two subsequent INS studies. Ziegler and Trizna[8] focused on African hand printing that incorporated eight new letters and all but seven of the letterforms used in the 1993 research. They confirmed that letterforms thought to be uncommon were not, and selected characteristics appear with differing frequencies in the various countries. Trizna and Wooten later looked at characteristics in the hand printing of individuals from Bangladesh, Cyprus, Egypt, India, Israel, Jordan, Kuwait, Pakistan, Saudi Arabia, Syria, and the United Arab Emirates.[9] The 1996 work revealed that letterforms identified from previous regional studies also appeared in the specimens from the Middle East and the subcontinent (Figure 10.1).[9]

10.1.2 Considerations in the Examination of Hand Printing

Hand printing contains the individuality necessary to allow for an identification or elimination of a writer. The absence of elements such as connecting strokes does not prohibit

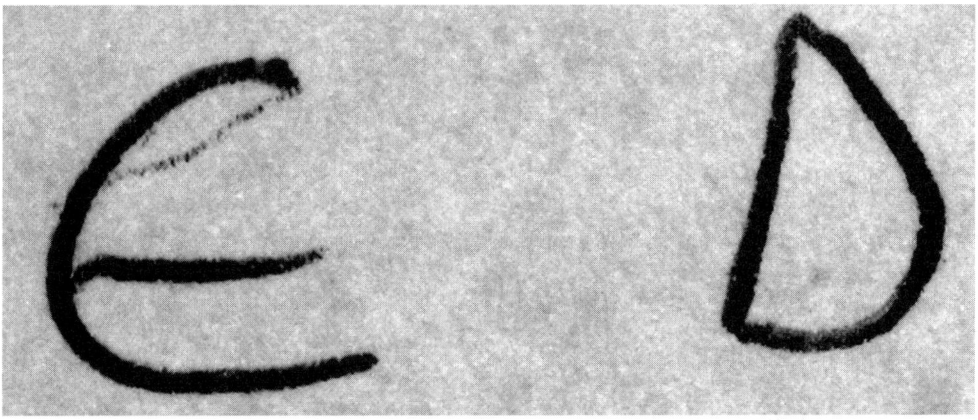

Figure 10.1 According to Trizna and Wooten[9], certain letterforms should be considered class and not individual. The *E* made like a *C* with a middle crossbar and the *D* formed as a triangle with a sharp point at the top of the staff are two such letters that were repeated by writers in more than one study.

a thorough analysis of other important characteristics. The early works of Osborne, Hilton, and Conway propose using the same examination methodology as for handwriting. Hilton wrote:

> The conclusion that a single writer prepared two sets of lettering is thus assured when all the habits of the disputed matter are found in the known specimens, while at the same time the existing variations between the representative known forms and the disputed lettering can be accounted for by the normal variation common to that individual's lettering. Futhermore, a sufficient number of individual habits must be common to the two sets of specimens, questioned and standard, so that the likelihood of both coming from the same origin is so great that it becomes a virtual certainty. By the same token, fundamental differences, especially consistently repeated ones, establish that the two sets of lettering are by different writers. Instead of handlettering being exempt from technical identification, as many believe, it is as highly individual as a person's cursive writing or signature and, therefore, is subject to identification.[10]

An identification or elimination is based on a combination of habits within the individual's writing or hand printing. Because of this basic premise, it is obvious that the databases listing class characteristics based on construction of a feature provide limited assistance in the examination. Determining a writer's natural variation of each feature within the combination of his or her writing habits provides individualization of the hand printing. When one considers the infinite number of variations of one feature in combination with the infinite number of variations of the other features in the same set of writing, the determination of authorship is truly based on the individualizing characteristics within the set of writing.

The requirements for a hand printing examination include known exemplars that are *comparable* to the disputed text, *adequate* in amount, and *timely* or contemporaneous

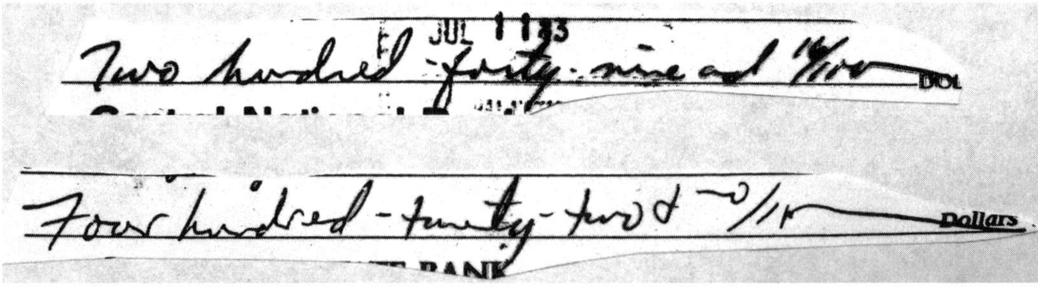

Figure 10.2 Both writings were written by one writer. The top entry was written when the writer was 28 years old. The bottom entry was written 15 years later.

(CAT). The questioned text may be all uppercase, all lowercase, or a combination. It may be a hybrid of hand printing and cursive writing forms. The request exemplars must reflect the same type of letterforms. Common sense dictates that uppercase hand printing is not comparable to lowercase, nor are cursive letters comparable to hand printed forms.

Adequate refers to the number of known exemplars needed to allow the FDE to determine the writer's range of variation. It is not possible to set a definite number of known exemplars. Factors such as length of the questioned text, the writer's cooperation in providing comparable exemplars, complexity of letter formations, and natural writing variation affect the number of known exemplars deemed adequate for comparison to the disputed material. For example, 20 to 30 repetitions of the letterforms in a 10-word robbery note may be adequate for examination in one case, while 2 or 3 pages of dictated text of a disputed 5-page letter may be adequate for that particular examination.

Timeliness refers to the time frame of the known exemplars to the questioned document (Figure 10.2). Contemporary exemplars should reflect the writer's skill and range of variation in relation to the time frame of the disputed text. This can be a significant factor in cases where there is deterioration in a writer's writing due to a decline in health, chemical or alcohol abuse, or writing immaturity.

Once the FDE has CAT known exemplars, a determination as to naturalness of the questioned and known texts must be made. The examination is less complicated when both texts reflect hand printing that is naturally and freely executed. The FDE examines the questioned and known documents separately in order to determine the combination of characteristics or habits present in each set. Letter construction, internal proportion (size and shape) of each letterform, ratio (height relationships), line quality, format or arrangement of text, and punctuation are a few of the characteristics to be considered.

Internal consistency within a set of writings is another characteristic that further aids in the determination of range of variation. Internal consistency is defined as an element of handwriting when the same or similar strokes or sequences of strokes occur in different letters.[11] In his study on this topic, G.A. Dawson examined samples by 47 writers who were requested to hand print the London letter.* Ratio was the first characteristic studied. He found consistency in the relative heights and sizes of the letters having an upper or lower extender, and those having only a middle zone (Figure 10.3). He also found that the ratio transferred from one letter to another; for example, the

*The London letter is an 80 word paragraph containing all the numbers, and letters are in upper- and lowercase.

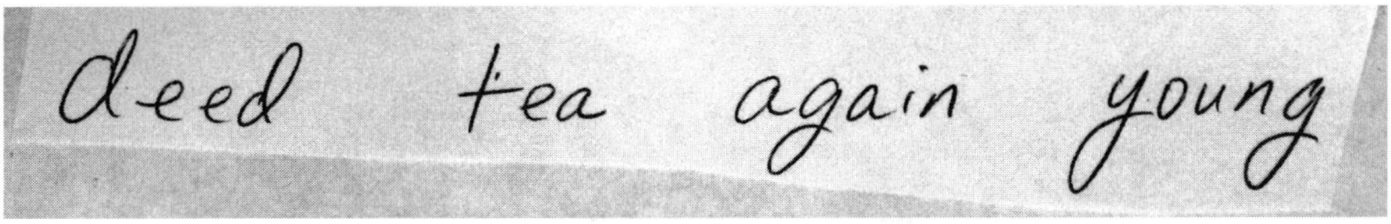

Figure 10.3 Internal consistency in the formation of the bowl of the *a*, *d*, and *g* is present. Internal consistency in formation between similar letter formations is one characteristic an FDE considers in the determination of authorship.

length of the upper extender for *d* was the same in the other letters that have upper extenders.[11] The presence or lack of internal consistency within the set of writings is another characteristic that further individualizes a writer's hand printing.

Sometimes a determination of authorship is not possible. It cannot be stressed enough that the hand printed text must be adequate in length and contain a combination of individual habits to allow an FDE to identify or eliminate a writer. Due to its brevity, authorship of the disputed text "Hi" cannot be determined. Even though these two letters may contain individualizing characteristics, the fact that there are only two letterforms prevents an FDE from identifying or eliminating a writer to the exclusion of all others.

10.2 Numerals

10.2.1 Considerations in the Examination of Numbers

The identification of numerals is closely related to that of hand printing in that it is also disconnected. Three fundamental factors are involved in the identification of numerals: form, writing quality or movement, and variation. A unique combination of these factors provides the basis for an identification or elimination. Fundamental differences in either form or movement are the means for distinguishing numerals prepared by a different writer.

Identification of numerals rests not only on their basic design, but also on subordinate factors of design. These factors include initial and ending strokes, ornamentation, any simplification of design, overall slant, and the interrelationship of component parts of the compound digits (2, 3, 4, 5, 6, 7, 8, and 9). Size, space, presence of connected strokes between numerals, baseline alignment, and positioning of multiple numbers contain the combination of habits that are identifiable to one writer. This section of the chapter will briefly discuss research that provides the data on class characteristics and variation of numeral formations, information that assists the FDE in establishing the significance of the features observed in the numerical forms.

As children, we are taught how to write numbers through formal training using a copybook style. Divergence from the copybook system introduces individualities into the number form. However, not all divergencies are individual in nature. Some are more common than others.[12] Just as with hand printing, class characteristics are sourced not only to copybook styles, but also to geographical and cultural influences.

Figure 10.4 Seaman Kelly observed five different styles of the numeral 8. In this illustration, one writer used two forms in a telephone number.

Figure 10.5 The following percentages of writers were found to use a tent-shaped 4 in their naturally written numbers: 6.6% (Horan and Horan), 5.6% (Horton), and 9% (Seaman Kelly).

10.2.2 Research on Class and Individual Characteristics of Numerals

Schuetzner compiled the class characteristics of numbers within each copybook system by U.S. and non-U.S. writers.[3] Her work catalogs each variation of a number according to the similarity with another system and the differences from the Palmer system. This compilation provides the starting point of numeral construction that writers are initially taught.

Seaman Kelly's 1999 study[13] researched the habits observed in naturally written numbers. The frequency of number formation, percentage of occurrence of connected numbers, and internal consistency between numbers of similar formation were determined from 200 writers. The participants ranged between 16 and 81 years of age and resided throughout the U.S. Each numeral was repeated 138 times in differing locations within a combination of numbers.

The variety of formations in each numeral establishes the identifying element. Determination of construction or formation of the number is critical in the examination process. It provides the basis to establish how much the numeral deviates from the copybook style. The Seaman Kelly and the Giles[14] studies examined a writer's variation in his or her execution of a particular numeral. For example, six different formations of the numeral 5 and five of the numeral 8 were observed in both studies (Figure 10.4), indicating that the construction of a number occurs in a predictable manner.

Horan and Horan,[15] Horton,[16] and Seaman Kelly[13] established a percentage range in identifying the frequency of occurrence in individual number formations. In Horan and Horan's study, 6.6% of their 625 participants formed a tent-shaped 4 and 68% used an S formation for 8. Of Horton's 580 writers, 5.6% used a tent-shaped 4 and 76% used an S formation for 8. Seaman Kelly's percentages for 200 writers included 9% using a tent-shaped 4 (Figure 10.5) and 66% using the S formation for 8.

Seaman Kelly's study included the frequency of occurrence of internal consistency. Twenty-five percent of the 200 writers used the same ending formation for the numerals

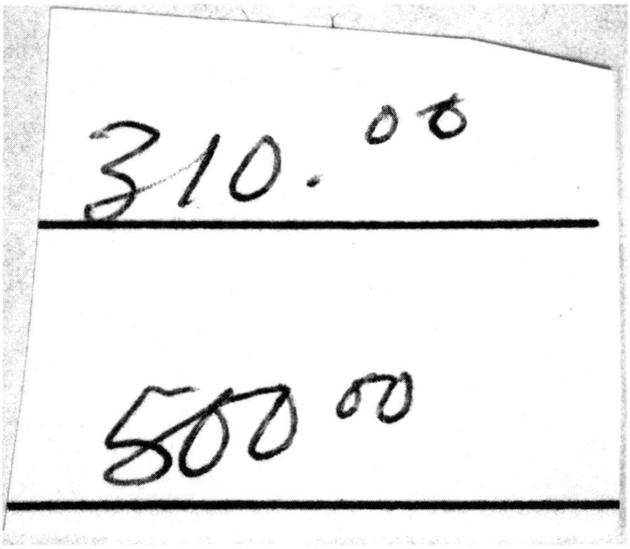

Figure 10.6 Twenty-five percent of the participants in the Seaman Kelly study used the same ending formation for 3 and 5. This illustration shows one writer who loops the terminal stroke through the bottom bowl of the 3 and 5. The terminal stroke continues to taper until it connects to the following number.

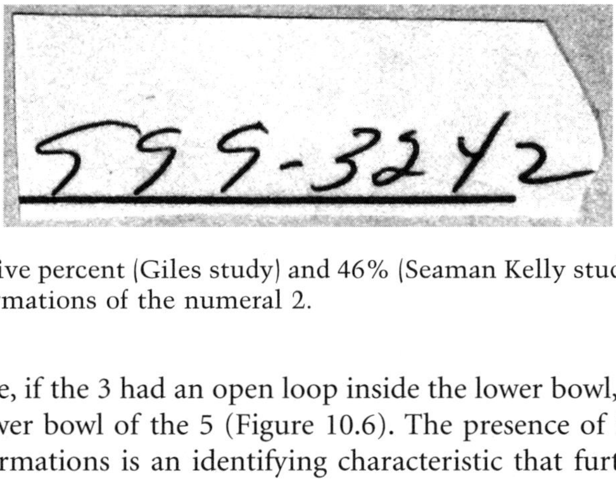

Figure 10.7 Fifty-five percent (Giles study) and 46% (Seaman Kelly study) of the participants used two or more formations of the numeral 2.

3 and 5. For example, if the 3 had an open loop inside the lower bowl, this movement was replicated in the lower bowl of the 5 (Figure 10.6). The presence of internal consistency between number formations is an identifying characteristic that further aids an FDE in determining authorship.

Conway observed that it is not difficult for many writers to develop several classes of numerals and occasionally to use them interchangeably without being aware of their own variations.[17] The use of two or more numeral formations may be a part of a writer's variation. In her research of 100 writers, Giles[14] found that over 48% used two or more forms of 6, 55% used two or more forms of 2, and 43% used two or more forms of 9. Seaman Kelly's study recorded 46% used two forms of 2 (Figure 10.7), 38% of 9, and 37% of 7.[13] These data strongly suggest that an FDE exercise caution in eliminating a writer in cases where both the questioned and known exemplars of numerals are equal in skill level. The importance of numerous known exemplars, both request and course of business, cannot be overemphasized. If the request exemplar reflects a writer's use of only one formation of a numeral, course of business exemplars

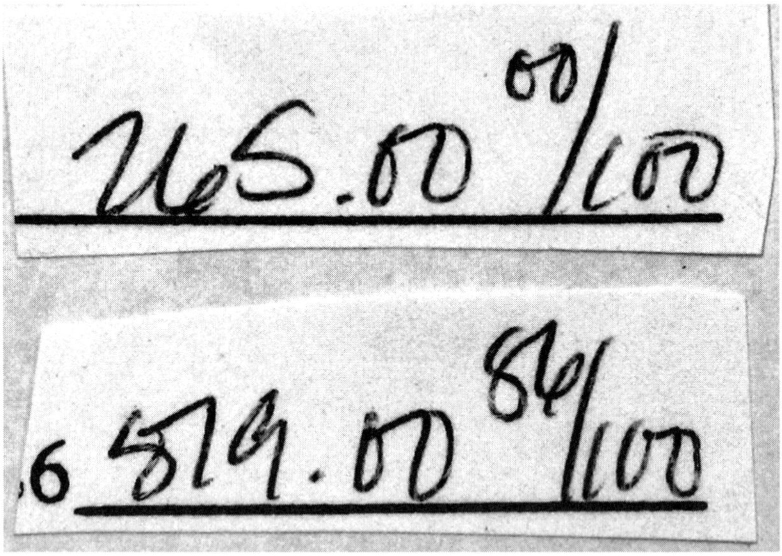

Figure 10.8 Twenty-two percent (Seaman Kelly study) connected three or more sets of numbers involving 2, 5, and 8.

may provide enlightenment as to the writer's habit when it comes to his or her variation of number forms.

Tying one number to another with a connecting stroke is an individual characteristic and can be significant because the writer is not taught to connect the numbers. Seaman Kelly's study assessed the frequency of occurrence of connected numbers. Of the 200 writers, 35% connected at least one combination of numbers; the first number being a 5, 8, or 2. Sixteen percent (16%) connected two sets of numbers with the 5 and 8 or the 8 and 5 combination occurring most frequently. Twenty-two percent (22%) connected three or more sets of numbers, with 2, 5, or 8 (Figure 10.8) being the first number.[13]

10.3 Conclusion

Hand printing and numerals are as identifiable as handwriting. The same examination methodology is used for all three. The techniques used in examining hand printing have progressed over the last half century as new data were brought forth. Research studies recording the frequency of occurrence (class characteristics) of certain letterforms, proficiency testing, and Dr. Moshe Kam's validation tests of FDEs outperforming laypersons in the identification/elimination of hand printing provide objective documentation that the forensic document examination profession uses scientific principles to reach conclusions of authorship.

References

1. Godown, L., Systematic Hand Printing, paper presented at the American Society of Questioned Document Examiners Annual Meeting, Memphis, 1952.
2. Ramsey, S.L., A study of the evolution of handwriting from grades three to six, *J. Am. Soc. Questioned Doc. Examiners*, 1, 32–39, 1998.
3. Schuetzner, E.M., Class characteristics of hand printing, *J. Am. Soc. Questioned Doc. Examiners*, 2, 5–33, 1999.
4. Livingston, O.B., Frequency of certain characteristics in handwriting, pen-printing of two hundred, *J. Forensic Sci.*, 8, 250–259, 1963.
5. Nicholson, P.J., A System for the Classification of Block Capital Handwriting, paper presented at the International Association of Forensic Science, Vancouver, 1986.
6. Giles, A., Increasing the Level of Objectivity in Handwriting Examinations, paper presented at the American Society of Questioned Document Examiners Annual Meeting, San Jose, CA, 1990.
7. Berthold, N.N. and Wooten, E.X., Class characteristics of Latin American hand printing, *J. Forensic Doc. Examiners*, 4, 135–151, 1998.
8. Ziegler, L.F. and Trizna, L.A., African Hand Printing, paper presented at the American Society of Questioned Document Examiners Annual Meeting, Long Beach, CA, 1994.
9. Trizna, L.A. and Wooten, E.X., Hand Printing of the Middle East and the Subcontinent, paper presented at the American Society of Questioned Document Examiners Annual Meeting, 1996.
10. Hilton, O., *Scientific Examination of Questioned Documents*, Elsevier Science Publishing Co., New York 1982.
11. Dawson, G.A., Internal Consistency in Handprinting, paper presented at the Annual Meeting of the American Society of Questioned Document Examiners, Savannah, GA, 1986.
12. Hilton, O., Individual or class characteristics in foreign numbers. Some observations of numbers from Austria.
13. Seaman Kelly, J., Habits observed in naturally written numbers, *J. Am. Soc. Questioned Doc. Examiners*, 5, 58–66, 2002.
14. Giles, A., Figuring it out, *J. Am. Soc. Questioned Doc.*, 2, 62–64, 1999.
15. Horan, J. and Horan, G., A Study of Numbers, paper presented at the 10th International Association of Sciences Meeting, Oxford, 1984.
16. Horton, R.A., A study of the occurrence of certain handwriting characteristics in a random population, *Int. J. Forensic Doc. Examiners*, 2, 95–102, 1986.
17. Conway, J.V.P., *Evidential Documents*, Charles C. Thomas Publisher, Springfield, IL, 1959.

Disguise in Hand Printing and Numerals

11

JAN SEAMAN KELLY

Contents

A forensic document examination normally consists of the comparison of naturally written questioned and request text to determine if they are of common authorship. Occasionally, an FDE receives a case in which he or she observes distortion in either the disputed text or the request exemplar. Some causes of distortion in the writing can be a temporary or permanent health problem, missed medication, writing position, chemical or alcohol abuse, or disguise. Examinations involving documents containing distorted text are discussed in the chapters on handwriting and signatures. This section will discuss intentional distortion, i.e., disguise of hand printing and numerals.

The difference between disguise and distortion is intent. In disguised writing, the writer is making a conscious effort to change his or her letter formations in order to avoid being identified. However, in distortion, the writer is not intentionally altering his or her writing. Distorted writings share most of the same characteristics of disguise. Fluency is one feature that may be present in distorted text, but absent in disguised writing. Inconsistency and lack of fluency present red flags to the FDE that the writing may be disguised as opposed to distorted. For example, tremor and pen stops tend to occur at inappropriate locations on a character when they are a result of disguise and not distortion.

A conclusion as to authorship of hand printing and numerals is based on the writer's combination of habits or characteristics within the set of writing. A conclusion most likely cannot be reached if the disguised hand printing is either a drawing or illegible to the point of being a scribble. However, the fact that the writing is disguised does not necessarily negate the possibility of determining authorship.

The quality of the disguise will determine the writer's success in preventing a forensic document examination of the disputed text. There are people who choose to use hand printing as their means of disguise because they believe it cannot be identified. Others, however, produce disguised hand printing by attempting to change from a usual to an unusual manner of writing.[1]

Past research has found that disguise more often occurs on the dictated exemplars rather than on the questioned document. It is fortunate for society that the majority of criminals focus their attention on obtaining their ill-gotten goods (merchandise or money) and not the act of writing. Forged checks, for example, usually contain natural writing on the face detail.

11.1 Disguised Hand Printing

11.1.1 Methods of Disguise

Numerous papers (published and unpublished) and chapters in books have been written regarding the characteristics writers commonly use when attempting to disguise their handwriting or hand printing. Inconsistency in letterforms, tremor (poor line quality), use of alternate letterforms normally not used by the writer, angularity of letter formations, numerous individual strokes (stick printing), the overwriting of letters, use of the unaccustomed hand, use of squared forms, use of grotesque forms, and intentional misspellings are characteristics commonly associated with intentional disguise (Figure 11.1). Utilizing one or more of these techniques alters the pictorial appearance of the hand printing.

Davidson and Keckler categorized the use of certain characteristics in their study of 50 writers.[2] Change in slope was utilized by 23 of the 50 participants. It was also the most difficult characteristic for the writer to maintain throughout the exemplar. Approximately half of the writers in this study disguised their hand printing by changing the size of the letterforms and spacing, lengthening or shortening the upper and lower extensions, and changing terminal strokes of the upper- and lowercase letters. Of the 28 writers who utilized this latter characteristic as a form of disguise, change to the terminal stroke occurred more in the capital letters *D*, *H*, *R*, *U*, and *Y*. The most commonly changed lowercase letters were *d*, *e*, *g*, *h*, *l*, *m*, *n*, *p*, *q*, *s*, *u*, and *w*. Lengthening the terminal strokes and adding curves or embellishing the letters were the most common methods for changing the terminal strokes. Not surprisingly, 46 of the 50 writers changed their capital letters by using a block style of printing, shortening or lengthening the beginning and ending strokes, or embellishing the letters. Only 35 of the participants changed their lowercase letters. Of the 35, 10 changed from either all uppercase to lowercase letters or vice versa.

Formal education influences the characteristics chosen by the writer. Hull's study[3] of 160 participants revealed those with fewer than 12 years of education were more likely to use hand printing as their method of disguise. Twenty-eight percent of this group was least likely to change their numeral formations. Change in number forms was used as a method of disguise by 81% of the participants with a higher education degree. Alternate capital letters were used by 96.9% of the 160 participants, while only one writer used change in height ratio.

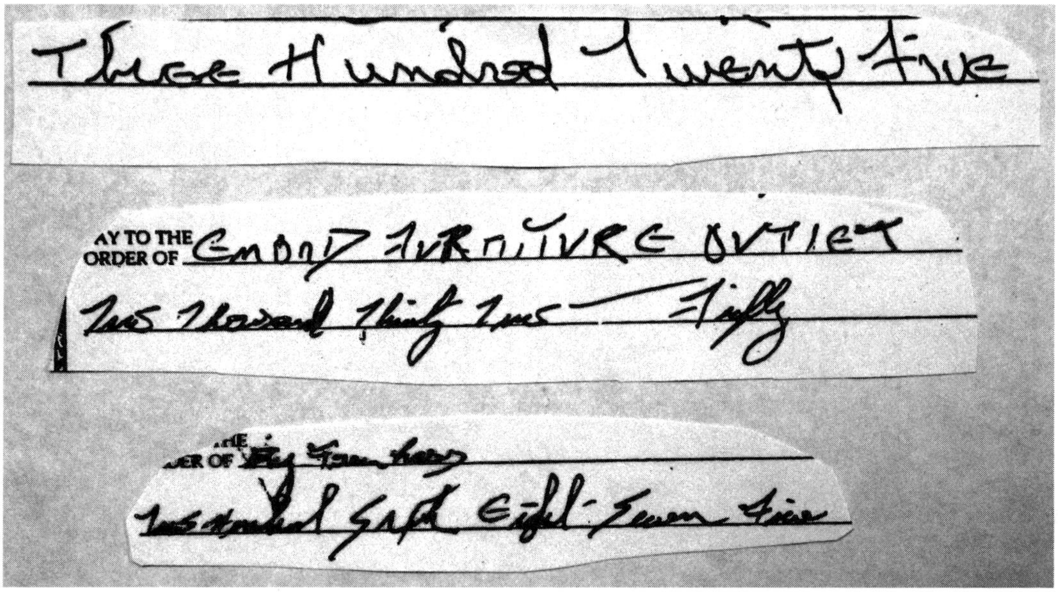

Figure 11.1 Portions of request exemplar provided by a forgery suspect. Stick printing, grotesque letter formations, tremor, and alternate letter formations are a few of the characteristics that indicate the writer provided disguised exemplars. In cases such as this, having course of business writing will assist the FDE in determining if the request exemplars are disguised.

Hand printing using the unaccustomed hand is another method of disguise. This method usually affects the skill level of the writing, thereby creating a drastic difference in pictorial appearance of the letters. Dawson's study[4] confirmed past research regarding the use of the unaccustomed hand as a method of disguise. His study was comprised of 20 participants who were instructed to write the London letter once with the right hand and once with the left hand. Hand printing produced by the unaccustomed hand (sometimes referred to as offhand writing) is characterized by loss of fluency, and the letters are generally larger than their right-handed counterparts. Other characteristics observed in this study include the following: rounded portions of letters such as *a*, *d*, and *g* were jagged and subject to abrupt directional changes in pen movement; tick marks were more prevalent; vertical or horizontal strokes used in the staffs and crossbars of letters like *b*, *d*, *f*, *h*, *l*, and *t* were curved, straight, or wavy; and most crossbars were sloped haphazardly (Figure 11.2).

Even though the unaccustomed hand affects the pictorial appearance of the writing, the examiner may still be able to conduct a complete examination. The writer using the unaccustomed hand will still produce familiar letterforms because it is extremely difficult to focus attention on producing a readable product and invent new letter formations.[5] The writer's combination of habits is still present within the set of hand printing, making it possible for the FDE to conduct an examination and reach a conclusion.

11.1.2 Disguised Hand Printed Exemplars

The suspect may disguise his or her request exemplar in an attempt to prevent the FDE from identifying him. If it is believed that the suspect writer is providing disguised

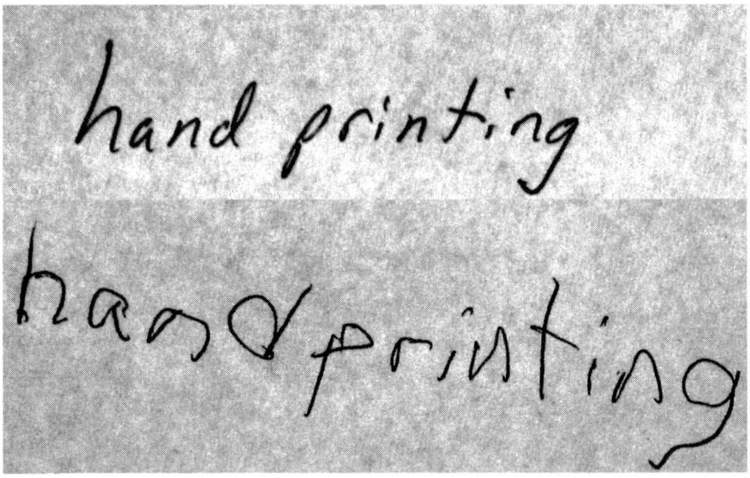

Figure 11.2 Unaccustomed, or offhand, hand printing will contain characteristics of tremor in the upper or lower extenders and the bowls of letters, angular formations, and haphazard crossbars on letters, such as *t* and *f*. The hand printing on the top was executed with the right hand; the hand printing at the bottom was written with the offhand.

exemplars, the FDE or investigator will need to obtain significantly more dictated exemplars. Justification for the increase in the amount of exemplars is due to the difficulty in maintaining disguised hand printing. To execute successful disguise, the writer has to focus complete attention on deviating from his or her natural writing. Awkward pen position, odd positioning of the request exemplar while the writer is completing it, heavy pen pressure creating a deep indentation in the writing line, and tremendous amount of time used by the writer in completing the personal information area of the form are hints that the writer is disguising his or her exemplar. When any of these observations are made, the request exemplar process may require a longer time frame than normal. The detective or document examiner must try to determine if the observations are a result of dealing with a person who is uneducated, a low-skilled writer, or producing disguised writing. Careful attention to the writer completing the exemplar is mandatory. Only by close observation will the detective or investigator be able to determine if the writer is attempting to disguise his or her writing since the author's natural letter formations will appear throughout the exemplar. Writing habits are neurologically fixed, and it is very difficult to override this biological setting to maintain the disguise. If disguised writing appears, the FDE or detective will need to take additional exemplars until the writer returns to natural (undisguised) hand printing or there are adequate samples of naturally hand printed letterforms comparable to the questioned text throughout the disguised request exemplar.

In cases where it is suspected the writer used his or her unaccustomed hand on the questioned document, the request exemplar should also include samples of unaccustomed hand printing. If the writer refuses to voluntarily provide hand printed exemplars with his or her offhand, it may be necessary to obtain a court order. If unaccustomed hand printed exemplars cannot be obtained, the FDE will need to include in his or her demonstration these features: overall size of the letters (larger than normal); shift in slope; and whether vertical and horizontal strokes are curved, straight, or wavy, which will change the pictorial appearance of the letterforms commonly found in hand printings done with the unaccustomed hand.[6] Demonstration of hand printed exemplars executed with the writer's accustomed hand should be convincing evidence that the minute characteristics identifiable to the suspected writer are present in the questioned text.

Course of business exemplars bearing hand printing of the suspected writer may be of assistance in determining authorship. Employment and rental applications are good sources of natural hand printing. If the course of business documents reflect natural hand printing, they provide objective documentation that the request exemplars submitted are either natural or disguised. By having request exemplars that lack consistency in letter formations throughout the exemplar (disguised hand printing) and course of business exemplars (natural hand printing), the FDE can safely report that the request exemplars are disguised.

11.2 Disguised Numerals

The appearance of disguise in numerals is less than the occurrence in hand printing and handwriting. The infrequent use of disguise in numbers may be due to the general misconception that numbers are not identifiable. In addition to this belief, the writer's focus tends to be more on successfully disguising text, thus overlooking the numbers.

Disguise of numerals may not always be recognized. Inconsistency, tremor, awkwardness, and lack of varying pen pressure provide grotesque forms commonly associated with disguise. A limited presence of numbers in the questioned text or a more subtle method of disguise, such as use of an alternate number form, may make it difficult for the FDE to determine if he or she is examining disguised numerals.

11.2.1 Preferred Method of Disguise

Davidson and Keckler[2] assessed the method of disguise used by 50 writers in their 1986 study. In their study, they noted disguised numbers in both handwriting and hand printing. In the participants' handwriting exemplar, 28 writers' method of disguise included changing their numbers. The most commonly changed numbers were 2, 3, 4, and 8. Most of the writers attempted to form the copybook style of the number rather than their natural informal style. In the hand printed exemplars, the writers changed their numbers by writing more slowly and produced more block-type rigid numbers. The most commonly changed numbers in the hand printed text were 4 and 8 (Figure 11.3).

Seaman Kelly[7] researched the preferred methods of disguising numbers by 200 writers. The participants were instructed not to use their off-hand or to change writing instruments. The methods of disguise listed in order of preference are (1) use of alternate number forms (Figure 11.4), (2) increased size of the numbers, (3) formal version of the writer's carelessly written natural number, (4) change of slant, and (5) change of pen direction in the formation of the number (Figure 11.5).

Use of an unfamiliar formation, whether it is handwriting, hand printing, or numerals, produces inconsistency in any extended exemplar. Even though inconsistency was present at the beginning of the disguised exemplars in the Seaman Kelly study, it lessened in severity and there was an increase in fluency toward the end of the exemplar. This result indicates that if the writer practices the alternate form, he or she will eventually be able to execute the new numeral with minimal evidence of disguise.

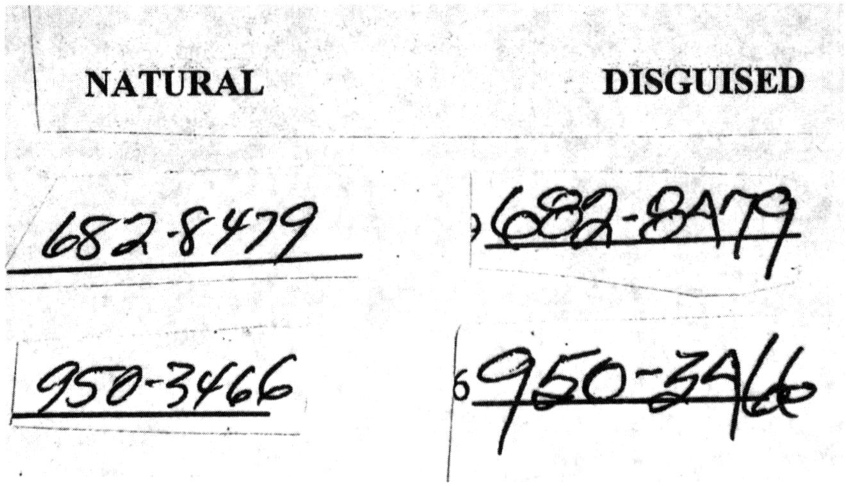

Figure 11.3 Naturally written numbers on the left and disguised numerals on the right. The writer chose to use alternate forms of the 4 and 8.

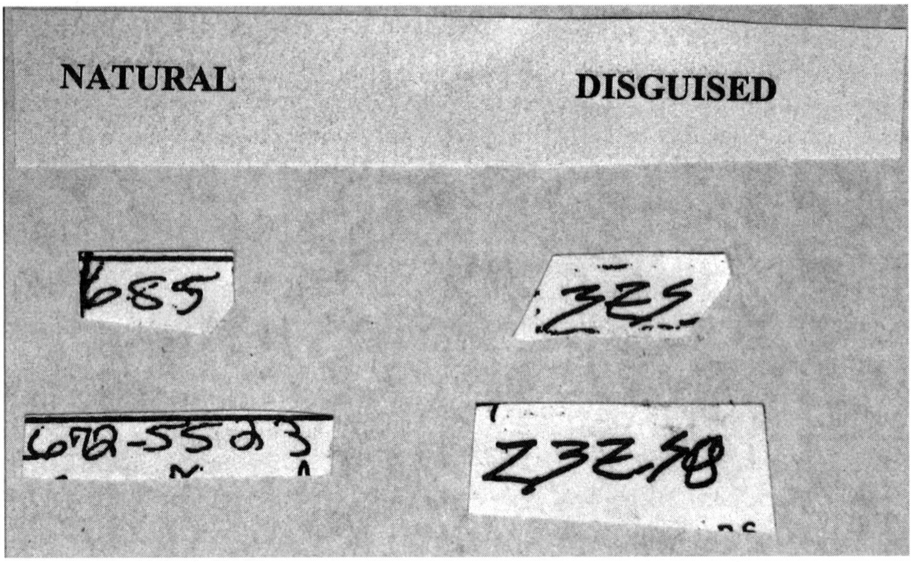

Figure 11.4 Naturally written numbers on the left and disguised numerals on the right. The writer chose to use alternate forms of 2, 5, and 8. Slow execution and angularity of numeral forms are additional characteristics of disguise. The disguised numbers came from the writer's request exemplars. The naturally written numbers were obtained from the writer's employment application. If the FDE suspects he or she has been given disguised hand printing or numerals, an attempt should be made to locate course of business hand printing and numbers. Course of business exemplars may aid in the determination of authorship and demonstrate to the court the writer's failure to comply in providing naturally written request exemplars.

11.2.2 Obtaining Non-Disguised Numeral Exemplars

Examination of disguised numerals presents a challenge to the FDE. If the questioned document reflects numbers comprised of grotesque formations, it may not be possible to determine authorship. In cases where the writer provides grotesque number formations

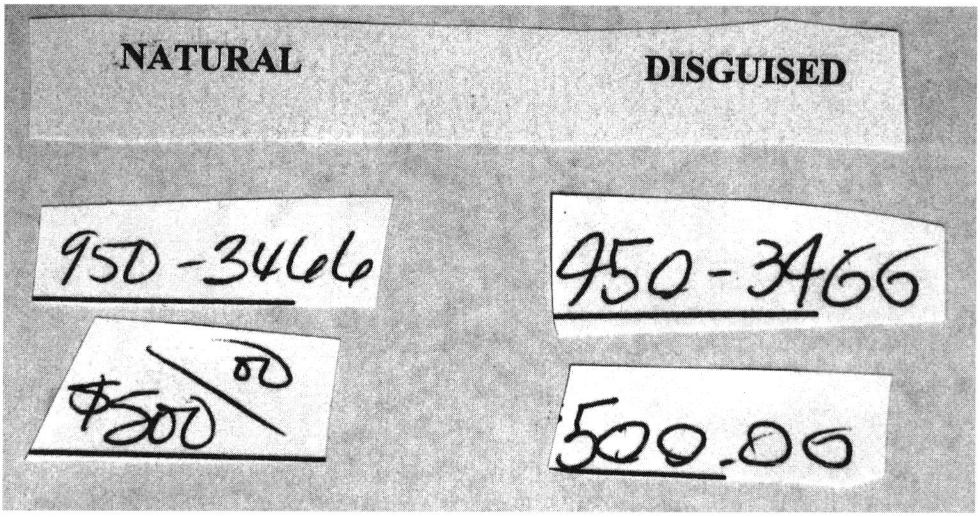

Figure 11.5 Naturally written numbers on the left and disguised numerals on the right. The writer chose to change the pen direction in the formation of the 9 and 0. Use of the tent-shaped 4 was chosen by the writer in his or her attempt to disguise the numbers.

on the request exemplar, the investigator or FDE should continue obtaining dictated exemplars until naturally written numbers appear. If the writer does not provide naturally executed numerals, an attempt to locate course of business exemplars should be made by the investigating officer.

References

1. Purtell, D.J. and Casey, M.A., The Comparison Value of Hand Printing Styles, paper presented at the Annual Meeting of American Society of Questioned Document Examiners, Houston, TX, 1981.
2. Davidson, J.M. and Keckler, J.A., A Study of Disguised Handwriting, Handprinting, and Numerals, paper presented at the Annual Meeting of the American Society of Questioned Document Examiners, Houston, TX, 1981.
3. Hull, J.M., The Relationship between Disguised Handwriting and Years of Formal Education: Final Results, paper presented at the Annual Meeting of the American Academy of Forensic Sciences, Boston, 1993.
4. Dawson, G.A., Brain function and writing with the unaccustomed left hand, *J. Forensic Sci.*, 30, 167–171, 1985.
5. Zimmerman, J., Handwriting Identification Based on an "Unaccustomed Hand" Standard, paper presented at the Annual Meeting of the American Society of Questioned Document Examiners, Orlando, FL, 1991.
6. Dawson, G.A., An identification of handprinting produced with the unaccustomed left hand, *Can. Soc. Forensic Sci. J.*, 26, 61–67, 1985.
7. Seaman Kelly, J., The examination of disguised numbers, *J. Am. Soc. Questioned Document Examiners*, 44, 1027–1028, 1999.

Preparation and Collection of Signatures and Handwriting Standards

12

BRIAN S. LINDBLOM

Contents

The strength of handwriting conclusions depends in part on the quantity and quality of the known writing. Writing standards or samples may be defined as specimens that demonstrate how a person writes. To obtain such writing sounds like a simple task, but actually it may be one of the most complex and difficult steps in the whole writing investigation.

What are the basic requirements for proper writing standards? They must show how the individual writes and especially how he or she would produce the material under similar conditions to those in which the questioned writing was prepared. These specimens certainly must contain enough appropriately prepared material to indicate not only the individual writing habits of the author, but also the usual variation in these habits from one writing to the next. They do not necessarily need to contain all of the identifying characteristics of the person's writing, but rather must contain those that also appear in the handwriting in question.

Two classes of specimens can be relied upon to fulfill these requirements. One consists of writing executed in day-to-day course of business, social, or personal affairs. These specimens may be referred to as *collected standards.* The second class consists of material written at the request of an attorney, investigator, or the FDE for the sole purpose of conducting a comparison with the questioned documents — commonly known as *request* or *dictated standards.* The problems encountered in assembling collected or request standards differ in a number of ways, so each category will be treated separately.

12.1 Collected Standards

The two most important factors in assembling good standards are quantity and contemporaneousness. However, a number of factors influence the usefulness of collected standards. The principal points to consider are

1. Amount of specimen writing available
2. Similarity of subject matter (i.e., letter, numeral, and word combinations)
3. Relative dates of the disputed and standard writings
4. Conditions under which both questioned and known specimens were prepared
5. Type of writing instrument, paper, or form used

12.1.1 Amount of Standard Writing

The importance of an adequate amount of writing has already been emphasized as the keystone of good standards. There is a common belief that a writer can be positively identified from only one or two of his or her signatures. Unfortunately, this is far from the truth and is more the exception than the rule. Normal writing variation alone generally makes this an infrequent occurrence. Only with a reasonable quantity of material can all the writing characteristics of an individual and the variations that usually occur from specimen to specimen be accurately determined.

The fact that no person signs his or her name or writes any combination of words in *exactly* the same way twice is one of the more complicating elements of handwriting identification. This range of natural variation must be revealed in the known writing. Variation is a personal factor — the extent or nature is not the same for each individual. It can be influenced by the conditions under which the writing was done. This is the reason that no hard-and-fast rule can be set for a minimum number of signatures or a minimum amount of extended handwriting. There have been cases where a very limited amount of standard material was available, but due to the narrow range of variation and unique combination of identifying features present, the writer could be identified. Other cases have required a much larger amount of known writing in order to reach a strong opinion respecting the authorship of a contested document.

Ten to 20 signatures should constitute an adequate sample, but there are a certain number of cases that may require 30, 40, or even more in order to accurately reveal the writer's habits, ability, and range of variation. It is poor policy for the investigator or attorney to be satisfied with the bare minimum when just a few additional specimens can greatly fortify the findings.

The minimum amount of writing necessary to identify the author of a handwritten document or an anonymous letter likewise varies. As a working minimum, four or five pages of carefully selected continuous natural writing usually prove satisfactory. Regardless of these suggestions, the party submitting a problem should always strive to obtain as large a quantity of handwriting as possible, rather than merely fulfilling the minimum requirements.

12.1.2 Similarity in Content

The best set of standards not only contains a sufficient quantity of writing, but also includes an ample amount of the same general type of material as that in dispute. Thus, if a signature is questioned, standard signatures should be collected; if a check's face detail is believed to be fraudulent, genuine cancelled checks should be procured; if a handwritten letter is challenged, other correspondence serves as the best standard. The reasons for these choices become obvious following a brief consideration of the various factors involved.

A person's signature, because of both its frequency of use and the nature of its employment (to represent the writer in business and personal affairs), tends to become more individual than any other combination of letters that he or she writes. By way of illustration, the reader undoubtedly recalls several unusual signature styles of acquaintances that differ radically from the remainder of the person's writing. In some countries signatures customarily are so stylized that they bear virtually no relationship to the person's general handwriting and may hardly hint at the spelling of the name. Obviously, in these instances extended handwriting is of little value in considering the authenticity of the signature in question.

When collecting signature standards one must also consider the use for which each specimen was written. Some writers have two or more distinctive styles for particular purposes. For example, a more formal variety may be used on checks, wills, deeds, and other legal documents, while a second style will be reserved for courier receipts, acknowledgments on various forms, and other documents of lesser importance. Some individuals are consistent in their use of a specific signature style, while others may be entirely inconsistent. Therefore, there is a need to obtain numerous signatures employed for the same purpose as the one in question and, whenever possible, it is desirable to include those appearing on other classes of documents. Only with such a set of standards can it be readily determined whether the writer employs more than one style of signature and, if so, what relationship the questioned item bears to any one of them.

Generally, signatures are not suitable for comparing with other types of disputed writing. Principally, this is because they exhibit a limited number of writing features, but also because there are often differences between the signature's letter styles and the remainder of the person's writing. In these latter cases standards made up of letters, reports, and other handwritten documents provide for a more extensive and proper study of the writer's handwriting habits. Similarity in content between the standard and disputed writings greatly assists the comparison process. The FDE needs to compare like items — signatures with signatures, numerals with numerals, hand printing with hand printing, and so on. A general analysis of what is needed to compare with the questioned writing may, for example, reveal a great number of capital (uppercase) letters combined with lowercase letters, suggesting that known material, such as an address book, diary, or recipe book, would produce good comparison writing. These same standards may be of relatively little value in identifying the writer of an anonymous letter or holographic will, just as hastily written notes are poor standards for comparing with formal writing, such as on an insurance or employment application. Thought must be given to the selection of the kind of extended handwriting or printing to be collected.

12.1.3 Relative Date of Preparation

In previous discussions of handwriting identification, it was pointed out that over the course of time a person's writing and signature may undergo gradual changes. The rate and nature of change varies and depends on such factors as how often and how much the individual writes and his or her age, writing skill, and mental and physical conditions. For example, the writing of a middle-aged person in good health may change very little from year to year, but during a severe illness, it may deteriorate sharply only to return to its original vigor.

A signature may have been written immediately after a serious accident or operation, when it was essential to execute the particular document. During a severe illness one is not apt to do much writing, so that there may be only a limited number of comparable signatures, if in fact any exist at all. In such a case the date of preparation is extremely significant. There are circumstances in which appropriate standards do not exist.

Normally, a typical adult's basic writing habits change gradually. Therefore, material written 2 or 3 years before or after the disputed writing serves as satisfactory standards, but as the number of years between the date of standards and questioned material becomes greater, the standards have a tendency to be less representative. Consequently, an effort should always be made to procure some specimens written near in date to the disputed document.

12.1.4 Writing Conditions

The conditions under which the writing was prepared may affect its value for comparison purposes. Haste, lack of care, or unnatural writing position — for example, resting the paper on the knee — introduce variations that may make the specimens entirely unsuited for comparison with carefully written material. In the case of illness, writing in bed in a somewhat awkward position may introduce variables in addition to those produced by the writer's poor physical condition. Other noticeable variations can be caused by writing on a rough or irregular surface, or in a moving vehicle. Illegible receipt signatures and notes scribbled on scrap paper, the backs of envelopes, or message pads are common examples of specimens that reflect qualities typical of the conditions under which they were written.

No writing prepared under such unusual conditions should be depended on exclusively for comparison with writing done under more normal circumstances, although it can often serve as a valuable supplement. A brief investigation into the conditions under which a document was written can indicate its appropriateness as a standard.

12.1.5 Writing Instruments and Paper

Since the identification of handwriting depends on consideration of all its elements, the kind of writing instrument used for a particular sample may have some influence on its usefulness as a standard. Pen and ink writing contains certain identifying characteristics that are not fully revealed in pencil specimens, while a change in the style of pen (for example, from a ballpoint to a broad-tip fluid ink marker) also may introduce writing variations or conceal features that would otherwise be present. Furthermore, the reader must remember that a ballpoint pen produces writing that is not completely comparable to work of a nib or fountain pen, for the latter represents a distinctive class of writing instrument. In addition, the porous-tip pen produces writing strokes that have still other slightly different characteristics. Gel pens represent yet another class of ink line.

The composition, size, shape, and ruling of paper may measurably affect the writing. Many writers adjust the size of their signature to some degree to the space allowed for signing. Standards on ruled forms permit consideration of how the individual habitually arranges his or her writing in relation to the printed baseline. Each element may assist in reaching a more accurate solution to some problems. Standards prepared with comparable writing instruments and on appropriate paper or forms provide the FDE with the most useful material. It may be possible to reach an accurate conclusion, and the document examiner often does, with standards that were not prepared with similar pens or pencils or on paper exactly the same composition, size, shape, and ruling as the questioned document. However, the FDE should look for the very best standards upon which to base an opinion, not just what will do. There are many cases every year in which less than ideal standards seriously restrict the findings. Thoroughness on the part of the investigator or attorney from the start means each limiting factor will be considered and eliminated whenever possible.

12.2 Sources for Standards

Writing is a part of the daily life of almost everyone. Consequently, the potential sources of writing standards are numerous, and those who frequently investigate handwriting cases soon develop a comprehensive list of sources. However, many who are confronted with a handwriting problem have had little or no experience with this type of investigation and will undoubtedly find the following suggestions of assistance.

Among the possible sources of signature standards are

- Cancelled checks
- Traveler's checks
- Signature cards for savings, checking, and charge accounts and safe deposit boxes
- Signed courier receipts or registered letters

- Diaries or agendas
- Business and personal letters
- Credit and loan applications
- Sales and credit card receipts
- Leases
- Deeds and mortgages
- Agreements
- Bills of sale
- Contracts
- Notes
- Stock certificates and shareholder and corporate minutes and transfers, along with other legal or business documents
- Partnership and incorporation forms on file with government agencies
- Court records and affidavits, such as naturalization papers, bankruptcy proceedings, divorce papers, probated wills, and estate files
- Powers of attorney
- Passports
- Marriage license affidavits
- Driver's, automobile, chauffeur, and other types of licenses and license applications
- Applications for natural gas, electricity, water, and telephone services
- Tax returns
- Insurance applications, records, and beneficiary forms
- Employment applications and records
- Records from currency exchanges, check-cashing agencies, and pawn shops
- Voter's registrations, petitions, and polling lists
- Hospital and physician patient records
- Time sheets, payrolls, pay receipts, and personnel forms
- Union and trade association files
- Welfare, unemployment, social security, Medicare, and old-age compensation records
- Signatures for narcotic and poison purchases
- Hotel and motel registrations
- Church, charitable organizations, club, and professional society records
- Veteran and military records
- Fingerprint records
- Board of education or university class cards, exams, and essays (in the case of younger persons)

When specimen checks are being provided for comparison purposes, it may be advantageous to request consecutively numbered examples that bracket the questioned document's date. This reduces the likelihood that the submitted material is self-serving or selected because the signatures appear the most divergent from the contested signature. However, if several of these standards happen to have been signed all on the same day, a larger collection of checks may be desirable.

In investigations of handwriting other than signatures (e.g., an anonymous letter or holographic will), these same sources may disclose writing of the proper class. Other writing specimens very often can be obtained from acquaintances, business associates, or correspondents. Business meeting notes, records, and messages furnish other excellent sources of standards, as do personnel forms and applications for employment and club or association memberships. In fact, any organization or activity, either business or social, in which the person is interested and active is a potential source of specimens.

12.2.1 Verifications of Standards

It is absolutely essential to verify who wrote the standards. No matter how logical the assumption may be, the investigator or attorney should not merely assume that the suspect or litigant prepared the writing. The importance of this is revealed in the following case history. In a problem of ascertaining whether a group of employees individually signed affidavits, several endorsed paychecks were submitted as standards. Some endorsements and affidavit signatures failed to agree, but before the study was completed signatures written in the presence of a personnel officer were requested. These latter samples revealed that one employee had not been endorsing his paychecks, but he signed the affidavit. A serious error might have been made if the FDE had relied on the assumption that the endorsement on the paycheck was genuine, although such an assumption may be true in many cases. In another case involving a denied signature, standards written over a period of several months established that the same writer had executed the contested signature. Further investigation shortly before trial disclosed that an employee, rather than the suspect himself, had signed all the standards as well as the questioned signature. Before any standard can be used in court, it must be proven by competent evidence. It is the responsibility of the submitter of the case to verify that the standards are by a particular writer; this task is not to be undertaken by the FDE. However, the FDE will scrutinize the samples to ensure there are no obvious differences, suggestive of more than one writer's involvement.

Standards should be verified through a detailed field investigation and by obtaining specimens written in the presence of reliable witnesses. Careful questioning of those bringing forth the standards and, if possible, the writer himself or herself can serve as a means of authentication. Writings used for different purposes, prepared at various times and under diverse circumstances, permit technical cross-checking and verification. As mentioned earlier, standards are the cornerstone of disputed writing cases, and no identification can be more accurate than the standards that support it.

12.3 Request Standards

Despite the vast potential sources of handwriting standards, circumstances do arise under which it is difficult or, perhaps impossible, to obtain an adequate set of collected standards at the moment they are needed. This is most common in, but by no means confined exclusively to, criminal investigations. Therefore, if the victim or suspected person is available and willing, a set of request or dictated standards should be secured.

The conditions under which these standards are prepared make it imperative that certain precautions be observed so that their comparison value is not impaired:

1. The material must be dictated to the writer.
2. The dictated text must be carefully selected.
3. An adequate amount of writing must be produced.

4. Each dictation should be executed on a separate sheet or slip of paper.
5. Some portion of the dictation should be repeated, preferably three or more times.
6. Writing instruments and paper should be similar to those used in preparing the disputed document.
7. The dictation should be interrupted at intervals and the dictation content changed from time to time.
8. Normal writing conditions should be arranged.

It must be recognized that the above guidelines can lead to serviceable standards at best — ones that can identify the actual writer and, one hopes, can eliminate all who did not write the material in question. Request specimens normally are not completely representative of the full range of a writer's habits, but they should, if well controlled, tell a great deal about his or her writing habits. Experienced FDEs recognize the shortcomings of these specimens. For this reason, while identification may be possible with a particular writer, a full elimination is generally impossible. Apparent differences can be brought about by disguise, nervousness during preparation, or simply the restricted scope of the specimens written at a particular time and with the knowledge that the results are to be used for comparison purposes. On the other hand, if the standards compare favorably with the disputed material, the fact that they do not fully reveal the writer's habits does not preclude or restrict an identification.

12.3.1 Dictation of Material

Dictation has been found to produce the most representative request standards. Initially the dictated text must be read to the writer without suggestions as to the arrangement of material, spelling, punctuation, capitalization, or other points that may cause him or her to disguise or modify his or her natural writing habits. Once a sufficient number of such samples have been prepared, additional specimens can be requested in which instructions are given to, for instance, write using capitals, write in large letters, etc. These directions should be documented so that the FDE is aware of the special instructions provided to the writer.

The manner of dictation has much to do with the ultimate value of the standards. If the subject matter is continuous, as with a letter, the rate of dictation should be gauged so that the person writes continuously rather than intermittently. The initial speed of dictation should be set so that he or she is not rushed, but subsequent specimens ought to be taken at several different writing rates, with some portions written hurriedly. Only in this manner can his or her normal range of writing variation be even approximately reproduced in the request specimens. Rapid dictation prevents a suspect from furnishing only his or her best or neatest writing, which is often as undesirable as poorly written specimens, since the quality of the disputed material generally lies between these extremes.

Furthermore, if the writer attempts to modify or disguise his or her specimens, a more rapid rate of dictation lessens his or her chances of success. Here it is well to observe that many unskilled persons can write only very slowly, focusing their main attention on the formation of each letter. One who attempts disguise may write in a similar manner. Thus, upon first impression, the natural manner of writing of the unskilled writer may suggest disguise, but throughout extensive specimens his or her writing characteristics will be consistent. On the other hand, a mass of extended specimens prepared with unpracticed disguise lacks such uniformity, and the numerous inconsistencies reveal its true nature.

Request standards obtained by any means other than dictation, for example, by having a person copy typed, handwritten, or printed matter, should be avoided. By allowing the writer to use a prepared copy, the manner of arrangement as well as the correct spelling and punctuation are indicated, and these individual characteristics, which might otherwise be helpful for identification purposes, are lost. Furthermore, a person who must copy from a script alternately reads and writes and produces discontinuous specimens. Since each page of writing contains numerous stops and starts, the standards may lack the writer's personal freedom and rhythm. Obviously, carefully conducted dictation eliminates these pitfalls. In some circumstances it may be desirable to have the FDE dictate and collect the request writings, although this approach must be weighed against any possible suggestion of bias on the part of the document examiner, having had direct contact with the writer.

12.3.2 Selection of Text

The request standards may be based upon any of three types of text:

1. The contents of the disputed document
2. Some similar material, which contains many of the same words, phrases, and numeral and letter combinations
3. A standardized form that includes all the letters of the alphabet and a number of the more commonly used words

Whenever the suspect or litigant has at least partial knowledge of the contents of the questioned document — and this is a common condition — there is no serious objection to dictating it. In fact, there are advantages to this procedure that make it superior to the others. In so doing, the investigator or attorney is relieved of any need to prepare special material for dictation, while the FDE is greatly aided by being able to compare the same combination of words and letters in both the standard and questioned writing.

Occasionally, however, it may be desirable not to disclose the contents of the disputed document to the person who is furnishing the comparison specimens. Experience has shown that in these instances the preferable procedure is to dictate material similar to that disputed, such as a continuous text for comparison with anonymous letters or specimen checks and receipts for standards to compare with disputed checks. As many words and letter and numeral combinations from the disputed matter as possible should be included. Be sure that any misspelled words or unusual uses of capitals or punctuation found in the questioned specimens are included in the dictated material.

Probably the least useful request specimen is standardized or form dictation, such as "The quick brown fox jumps over the lazy dog." More elaborate texts that include all uppercase and lowercase letters of necessity contain unusual and unfamiliar words or names. Consequently, even though the dictation is repeated several times, the resulting standards tend to lack the freedom of the person's normal writing and will not contain the same letter and word combinations present in the questioned document. There are times, however, when form dictations prove valuable. They can be used as the opening series of dictations to be followed up by the production of more valuable request standards.

The reader should not infer from this discussion that only one type of dictation should be used for each writer. On the contrary, more representative standards can often be prepared by using two or even all three classes of text. These combinations can be employed to best advantage when the disputed writing is brief. Otherwise, merely using the disputed subject matter would require extensive repetition in order to secure sufficient material. Yet this technique may have equal advantages with more lengthy disputed writing. For instance, by changing the subject matter — such as from that in question to some unrelated material — the writer may relax and thus introduce more of his or her typical writing variations into the standards.

Under certain conditions, particularly in criminal investigations, request standards may be obtained without resorting to dictation. When a suspected writer is apprehended, he or she generally attempts to explain away any action that might incriminate him or her. The alert investigator will ensure that the suspect writes out his or her statements. While composing such an explanation, he or she generally centers his or her efforts upon furnishing a logical excuse for the suspicious actions and gives little thought to the actual execution of the writing. Thus, these written statements can represent a good sample of normal writing.

A person who is to supervise the preparation of dictated standards should always have full knowledge of the questioned material. If not, he or she will have to restrict the request specimens to a generalized form; even then, there may be the danger of not getting the right kind of standards. Key factors, such as capitalization, spelling, and the importance of certain words within the questioned material, may not be fully apparent until the field investigator has seen or discussed the questioned material with those who know more about the case or about the considerations of the FDE.

12.3.3 Amount of Material

It is extremely difficult with only request standards to obtain a truly representative picture of the usual variation in an individual's writing. Even under the most favorable circumstances it requires the preparation of extensive specimens. The great majority of request specimens are deficient simply because they do not reveal the variation that is part of the suspect's normal handwriting.

Nervousness and deliberate disguise are two factors that frequently curtail normal writing variation. The initial portion of request standards furnished by a writer who has had no part in the preparation of a disputed document may well reflect nervous tension. This nervousness may be instilled, at least in part, simply by the accusation of having prepared the disputed document or merely because the writer knows that the specimen is to be compared with a questioned writing, such as a threatening letter. This anxiety may disappear when the dictation is continued at length, so that subsequent writing may assume a more natural character. On the other hand, many guilty individuals and others who choose to be uncooperative deliberately attempt to disguise their writing in order to avoid detection. Fortunately, only the exceptional writer can continue an unpracticed disguise throughout several pages. Therefore, if we are to ensure that the request specimens portray the natural writing variation of the individual and are free from the effects of nervousness or deliberate disguise, it is necessary to have the writer furnish at least five or six pages of continuous handwriting or 20 or more signatures, each written on a separate sheet of paper.

The most common defect of request specimens is the failure to have the person write enough. At times this is because the investigator does not know how much writing is needed for suitable standards, but often it is merely because the investigator finds this time-consuming (which it is) and believes that other aspects of the investigation are more important. However, if identification or elimination of the writer is to be accomplished, extensive standards should be prepared.

12.3.4 Repetition

Regardless of the type of subject matter employed, the value of the standards is increased by some repetition. In the course of dictation a page of writing should be repeated at least three times. It should preferably be taken from the questioned material. Repetition allows the writer to become familiar with the subject matter and helps him or her to write more freely and naturally. At the same time, it serves as an effective means of discouraging disguise. At times the guilty writer, realizing that he or she cannot continuously reproduce the modifications of his or her first specimen, abandons the attempt altogether. If, on the other hand, he or she continues his or her efforts in this direction, marked variation between successive specimens should appear. Upon detection of these inconsistencies, additional and more extensive writing, including still further repetition of previous material, must be obtained. In this way, attempted disguise could be more readily detected, and with enough request writing, even in extreme cases, some specimens ultimately should be obtained free from disguise.

12.3.5 Writing Instruments and Paper

The influence of the writing instrument and paper can be readily controlled with request standards. It is necessary only to furnish the writer with a pen or pencil similar to the one used in the questioned document. At the same time, paper should be selected with the same physical characteristics and ruling as the disputed sample, trimming it, if necessary, to the proper size and shape. With this preparation personal habits affected by writing materials are less apt to be excluded from the request specimens.

When the questioned writing is in ink, it is important to recognize what kind of pen was used. *Pen* is an inclusive or general term, and, as discussed in another chapter, there are several classes of instruments, each with substantially different writing characteristics. Today ballpoint, gel, and roller ball pens are the most common and popular. Ink rolls onto the paper with a non-flexible ball. Some different writing characteristics are revealed with a gel and fiber pen, especially if there is some flexibility of the point, as can be the case with fiber pens. Fiber-tip pens generally produce wider strokes than other instruments, strokes that can hide details that would otherwise be present. Not every writer can write with equal ease with each instrument, and if one class of pen has been used for all questioned writing, then this kind of pen should be used for at least some of the request standards.

12.3.6 Interruption of Dictation

Breaking dictation up by a rest period or two is good procedure. Writing fatigue brought about by the preparation of extensive specimens may be eliminated, and disguise likewise may be discouraged or rendered ineffective. If all completed specimens are removed from the writer's view during these pauses and his or her attention is focused on other matters, consistent disguise becomes more difficult, since after such a rest period the writer's recollection of details can be less accurate. Finally, interruptions aid in introducing more natural writing variation into request specimens than is to be found in the same amount of continuous writing.

12.3.7 Writing Conditions

The writer should be seated comfortably at a desk or table. Where questioned specimens are known to have been written under less normal conditions, such as while standing or resting the paper against the wall, the suspect should be asked to execute some supplementary standards in a similar position. When the exact writing conditions are unknown but the questioned writing suggests some abnormal writing position, specimens should be taken in several different ways, for example, standing and resting the document on a high counter or bending over a table; writing with the paper on a clipboard held in the hand or on a pad resting on his or her knee. These will supplement the normal seated writing position. An exact record of how each specimen is written should be made. Thus, with a combination of standards prepared under common and unusual writing conditions, the effects of these changes can be fully studied.

12.4 Special Considerations for Request Signatures

By far the most troublesome problem with request writing is obtaining a satisfactory set of signature standards. Since the amount of writing involved is small even when a large number of signatures are prepared, the effects of nervousness or deliberate disguise may not be entirely eliminated. Request signatures, too, have a tendency to be more uniform than signatures written from day to day. For this reason, they may fail to fully show how the person usually signs his or her name.

Several techniques help to overcome these limitations. Increasing the amount of writing through the introduction of different subject matter tends to improve the standards. Instead of executing a set of signatures alone, the writer might fill in the details on 20 or 30 specimen checks or receipts, each of which also requires his or her signature as the issuer, endorser, or both.

Another tactic, especially effective when a person is thought to have written a fictitious name in his or her natural handwriting, a situation common in police investigations of fraudulent checks, is to include among the standards a number of similar names together with the fictitious one. Following such a technique, if the fictitious name was Joseph Martin, several variations, such as Joseph Harts, Rudolph Martin, and Stephen Marvin, could be added.

When circumstances permit, an excellent procedure is to have the writer prepare portions of the standards on different days. In this way, the request signatures tend to be more like those written from day to day. This procedure need not be confined to signatures, as it is equally effective in making any writing more representative and should be employed whenever possible.

The most common problem with request signatures and extended handwriting is their failure to contain a representative amount of writing variation. Every effort must be made to overcome this deficiency. To achieve this end, each specimen should be written on a separate sheet of paper similar to the questioned document. The shift of writing position accompanying each change of paper tends to introduce slight variations that are not generally encountered in a series of signatures executed on a single sheet. In fact, 10 signatures written one after the other on a single sheet often have less value than 2 or 3 signatures written on separate occasions.

A preliminary examination is often made to evaluate the various request specimens submitted. To assist the FDE, each sheet of writing should be numbered and accurate notes should be made on how the specimens were written, when the dictation was interrupted, which sections were written at rapid or slow rate, and, if the specimens were not all prepared on the same date, the date and conditions under which each was compiled. These notes, particularly when disguise has been attempted, greatly assist in segregating the abnormal standards.

12.5 Combination of Request and Collected Standards

There is no reason that request and collected standards cannot be used simultaneously. In fact, under certain conditions the two classes of specimens are desirable in order to obtain an adequate and accurate set of standards. This is often the situation where anonymous letters are in question. Whenever there is a suspicion that the effects of deliberate disguise have not been and cannot be completely eliminated from request standards, the standards should be supplemented by collected specimens. The combination of the two classes of standards is often the best material for comparison purposes, even when disguise is not an issue. The standards must, of course, be by one writer. The FDE will review the material to ensure that the specimens are not contaminated.

In some instances even the most thorough and exhaustive search for material results in inadequate collected standards. To supplement these effectively with request writing usually requires that a complete set of dictated standards be prepared. Only by such a procedure can one be sure that the latter specimens are representative.

12.6 Standards for Special Problems

From time to time the FDE is called upon to provide an opinion about documents bearing abnormal or atypical signatures. Problems of this type include signatures written under the influence of alcohol or drugs (medical and illegal usage), during serious illness or near death, and, in some instances, by elderly persons. Appropriate standards are generally needed to answer these questions since each suspect signature prepared under any one of these situations may have abnormal qualities.

12.6.1 Intoxication

Excessive consumption of alcohol produces different effects on different writers. Some individuals have far greater tolerance than others and can consume large quantities of alcohol before their signature is affected. In the earlier stages of alcohol consumption, the signatures of an individual may be only slightly affected, generally being somewhat larger and slightly less accurately written, but as more alcohol is consumed, there may be further deterioration in the writing. Lack of coordination may lead to inferior design and poor writing alignment. The overall writing skill may decline even though the writer is still striving to execute a normal signature. The writing may begin to take on the appearance of carelessness and may ultimately assume "a drunken stagger." Successive signatures might wander away from the normal design and even from the design of the immediately preceding signature in somewhat unpredictable ways. In other words, variations become great. Successive specimens written during the same night of drinking may vary greatly in many identifying elements. What we are concerned with in this discussion is not the extent that alcohol affects writing, but how to obtain comparable specimens that can be used effectively in accurately identifying a signature that clearly reflects the influence of alcohol.

In these problems the task of obtaining known specimens may be less difficult than with other classes of abnormal signatures. A writer who may have signed a questioned document under the influence of alcohol probably has signed others on similar occasions. In most cases, especially those involving heavy drinkers, proper investigation uncovers a number of signatures showing various indications of intoxication. Again, the wider range of variation requires more authentic specimens and as many specimens as possible should be intoxicated versions rather than sober variations.

Standards for these problems consist almost exclusively of material collected from various sources. Request specimens (or any specimens, for that matter, written when the writer is not drinking) may have limited supplementary value. Of course, if the writer is apprehended while intoxicated and can be persuaded to cooperate, specimens so prepared should be of value. Seldom is it possible to obtain request specimens with the suspect intoxicated.

In passing, it should be observed that physical disability or the use of drugs can produce symptoms suggesting intoxication, and this may be reflected in the writing. On the other hand, the writing of sufferers from palsy or Parkinson's disease may be improved by properly prescribed medication.

12.6.2 Deterioration in Writing of the Aged

Very difficult problems may be encountered with signatures that have seriously deteriorated due to the writer's age or to terminal illness. Of these, the deathbed signature is particularly perplexing since the deterioration may have been rapid, with little or no forewarning in earlier signatures. With decrepit signatures of an aged writer, the decline normally occurs gradually over a period of months or years. More known specimens revealing writing weaknesses are therefore available than with deathbed signatures.

Writings of this nature are characterized by a lack of fluency in execution and inconsistency in details of form. They exhibit poorer writing quality than earlier signatures by the same person. Study of a series of signatures by an infirm writer reveals much greater variation from one signature to another than was typical of vigorously written signatures of earlier years. This inconsistency, which in some cases is very prominent, complicates the problem. In order to reach the most accurate conclusions, two or three times the normal number of exemplars may be needed, and they must be closer in date to the signature in question than is the usual case. Although a sufficient number of signatures may have been executed, it is often hard to locate them.

We are confronted with the same identification problem with the deathbed signature, since a dying person signs his or her name only for the most urgent reasons. Available genuine signatures are often almost as suspect as deathbed signatures, since circumstances surrounding their preparation are so like those surrounding the questioned signature; for example, at the time all specimens were written, the writer was being cared for by the one member of the family who is strongly favored by a disputed will. Because of the very small number of specimens comparable to the deathbed signature, it is essential that all be studied.

Adequate standards for these problems must include as many signatures written during the period of decline as can be obtained. There is advantage in supplementing them with earlier, more normally written signatures, even though they are not representative of the questioned period. These specimens do give the FDE some idea of the basic model the writer is attempting to duplicate. When a large group of contemporaneous signatures, say at least 25, can be obtained, reasonably meaningful solutions can be reached. Unfortunately, far fewer signatures are usually located. Then there must be serious leaning on experience to judge whether the questioned signature would digress from the standards in the way that it seems to. The fewer contemporaneous signatures there are for study, the more unanswered divergencies there are that will remain between the known and questioned signatures. Any opinion rendered must reflect the problems encountered when comparing the questioned signatures with inadequate specimens.

12.6.3 Sickbed Signatures

Signatures of an infirm writer represent badly deteriorated productions by an individual who may subsequently regain some or all of his or her writing skill. A signature of this kind may have been written immediately after a serious accident or operation. Periods of illness, especially severe illness, are not times when one is apt to do much writing, and so there may be only a limited number of signatures written.

The question of a writer's weakened condition is normally complicated by being propped up in a bed or at times under worse conditions. The writing position, as much as the physical weakness, leads to a low-quality production. These factors must be considered in collecting standards and examining the case.

Sickbed signatures, like most of these special cases we are considering, can most effectively be solved with more signatures than in the general run of cases. Unfortunately, the writer may never have executed a sufficient number of signatures while sick to satisfy the true needs of the FDE. In fact, the questioned signature may be the only example written during the illness. Writing produced after does not help much if the person has completely regained his or her writing abilities. In some instances, the person may have gone through a long recovery period during which he or she did some writing and the quality of the signatures are below his or her pre-illness or post-recovery strength. These specimens can be of help in giving some idea of how his or her signature declined and are certainly better than having to depend exclusively on normal specimens.

12.7 Hand Printed Standards

The procedures and principles involved in the collection of hand printed standards closely parallel those for handwriting. Since the requirements for adequate standards are the same as those for handwriting, the various steps in the preparation of both collected and request standards just discussed apply.

Hand lettering is used more frequently in daily life than is generally realized. Few people, except those who letter professionally (draftsmen, architects, engineers, and illustrators, for example), prepare any extended specimens in this manner, but many applications, questionnaires, and other forms require portions to be printed. Printing is used from time to time to prepare legible addresses on envelopes and to write postcards or even entire letters. While it may seem difficult to locate the equivalent of three or four pages of continuous lettering, this may be accomplished with the exercise of a little diligence, particularly with individuals active in the business world.

It is true, nevertheless, that with many individuals the potential supply of collected standards is more frequently limited in hand printing problems than in cases involving disputed handwriting or signatures. Consequently, we must turn to request standards that are naturally written and reflect the style of printing habitually used by the writer. Clearly, no instructions should be given to print only capitals, print small letters and caps, or the like. If it is suspected that the style of lettering being employed is different from the writer's usual habit, he or she should then be instructed to use a specific type of lettering. This fact should be noted on the specimen. The absence of a particular individual letterform, however, does not mean that the writer should be instructed to make a certain type of *e*, for example. This procedure will only destroy the value of the standards and prove nothing about how the writer usually letters.

Each of the factors discussed in connection with handwritten request standards must be closely observed, but the amount of request lettering necessary for adequate standards becomes by far the most important consideration. In all instances sufficient material should be procured to ensure that the specimens are free from disguise or nervous restraint, and at the same time contain those usual variations so typical of the individual's normal lettering. On average, five or six pages of request material should produce satisfactory specimens.

12.8 Conclusions

To build an adequate set of standards requires perseverance and painstaking care. It is not sufficient simply to include enough material to show what the individual's writing looks like. The standards must represent as fully as possible his or her writing at different times and under varying conditions, but they definitely should show how he or she wrote at the time of the questioned writing.

To obtain request writing one cannot merely furnish a suspect with paper and pen and ask him or her to write a few lines. One must thoughtfully select the material to be dictated, supervise the suspect or litigant while he or she writes in order to minimize disguise, and simultaneously keep in mind all the various factors discussed above.

Rather than pick up the first two cancelled checks that are at hand, the best collected standards are gathered only after exhaustive searches, checking all possible sources of writing while being constantly aware of those conditions that enhance or detract from the usefulness of any piece of writing. The quality of the standards determines largely the strength and accuracy of the ultimate opinion. It is only through diligence on the part of the person collecting them that they fully serve their intended purpose.

The importance of accurate standards cannot be overemphasized. Every year a number of indefinite conclusions are rendered in cases where the acquisition of better standards would have meant a positive identification or elimination of a particular writer. This situation is illustrated time and again where the original standards are inadequate and the preliminary results are inconclusive. This often prompts the field investigator to search out and provide the FDE with additional specimen material for a supplementary examination, and the consequence is oftentimes a report with a more definitive conclusion.

Section V

Mechanical and Electronic Impression Examinations and Comparisons

Pens and Pencils

13

BRIAN S. LINDBLOM

Contents

Since ancient times, various writing instruments have been used to communicate the written word. Today quill and nib pens have given way to other classes of writing instruments, namely, the ballpoint, porous-tip, roller ball, and gel pens (Figure 13.1). A number of special types of pens, including those used for drawing, lettering, and artwork, are also currently marketed. Each class has its own identifying characteristics potentially distinguishing it from the others (Figure 13.2). The use of a particular pen or ink class often has importance in the investigation of a disputed document, particularly when there are issues of backdating, insertions, or other alterations.

13.1 Ballpoint Pen

The ballpoint pen was introduced to the American market in 1945, at the close of World War II.* Its popularity grew rapidly despite early faults. With it, the nibs of the fountain

* The ballpoint pen was actually first sold in Europe as early as 1935, but by 1939 sales totaled only about 25,000 pens (*Fortune*, July 1946, p. 144). L. Biro began manufacture of these pens in Argentina in 1943 and sold them in limited numbers in South America (*Readers Digest*, December 1946, p. 60). A few found their way to the U.S. by 1945. It was his pen that lead to the introduction by Reynolds of the domestically made ballpoint pen in the fall of 1945, and the accompanying promotional program quickly jumped sales into the millions. Thomas Whitside, in *The New Yorker* (February 15, 1951), tells of the pen's early history and promotion.

Figure 13.1 Examples of four modern pen classes: ballpoint, porous tip, roller ball, and gel.

pen were replaced by a small steel ball. The ball is held in place at the tip of the pen by a housing that allows it to rotate freely during the writing process. As the ball rolls on the paper, ink that was picked up from the reservoir is transferred to the paper surface.[1–3] Ballpoint pen ink is a paste-like substance of high viscosity, which contrasts with the fluid ink used in roller ball and porous-tip pens. Most ballpoint pen inks have three components: dyes, resins, and volatiles. Certain manufacturers, such as Parker, produce refill cartridges, while others, including Bic, sell disposable pens.

The work of the ballpoint pen can be recognized and distinguished from other writing instruments both by its ink and by its line morphology.[4] In lighter strokes the transfer of the paste-like ink occurs only on the edges of the higher fibers, while in heavier strokes the pressure on the ball forms a groove or compression in the center of the stroke. Ballpoint pen inks dry rapidly once deposited on the paper. Other characteristics of the ballpoint

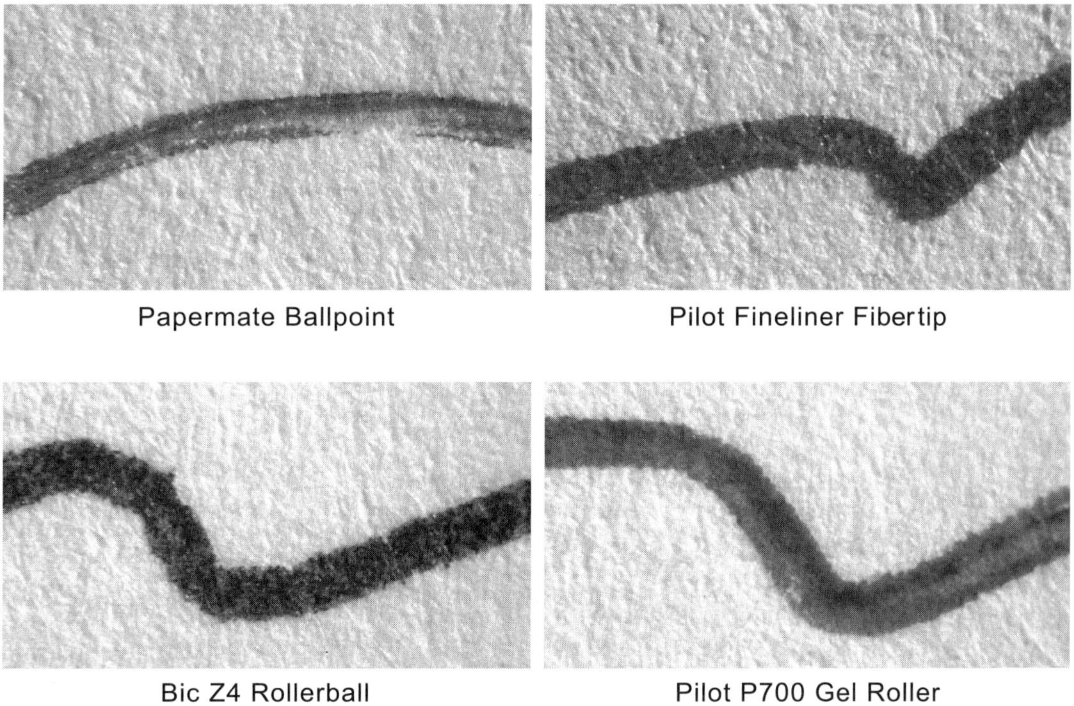

Figure 13.2 Microphotographic enlargements of ink lines created using four modern pen classes.

pen are derived from the ball, which controls the width and quality of the writing stroke produced.

Ballpoint pens can be differentiated from one another by their ink color, ball size (medium, fine, or extra fine), and defects that may be present in the writing line. These defects include gooping (a small dot-like deposit of ink, especially at points of redirection, such as in loops and connecting strokes), skipping or short gaps (Figure 13.3), and striations within the stroke (Figure 13.4). Very fine burr striations may occur at times in a pattern distinctive enough to individualize the pen.[5–7]

13.2 Porous-Tip Pens

The porous- (fiber- or felt-) tip pen has been a popular writing instrument for several decades, having been introduced in the mid-1960s. The writing point of this pen is porous and allows a supply of aqueous (water-soluble) ink to be spread on the paper. While broad tips predominated in the early pens, fine and very fine tips are now commonplace. Some porous-tip pens utilize a hard, perforated plastic tip. The hard-tip pens produce a stroke very similar to the pure fiber-tip pen, except that under some conditions they can create a slight furrow or trough. With all, the ink tends toward intense colors and the pens deliver a heavy, though quick-drying, line.[8] The ink lines produced with a porous-tipped pen can be contrasted with the product of a ballpoint pen. When properly functioning, the former creates a broad, solid, and ribbon-like stroke. With light pen pressure and rapid execution, ink lines will sometimes have a streaked or brush stroke appearance. A particular pen rarely

Figure 13.3 Portion of a signature written with a ballpoint pen showing skips and gooping characteristic of this class of pen.

Figure 13.4 Ink line morphology defects in an example of ballpoint pen writing. The *ug* combination shown above exhibits both a striation pattern and an instance of ink gooping.

develops enough individuality to distinguish it from others containing similar ink; however, as with ballpoint pens, many tip widths are available and can be a distinguishing factor. Pens of this class are factory filled and are generally disposable.

Due to the flow of the ink and stroke width, subtle line quality characteristics, such as hesitations and inconspicuous pen lifts, are sometimes harder to detect when examining handwriting executed with this class of pen.

13.3 Roller Ball Pens

It was only a few years after the introduction of the ballpoint pen that the first attempt was made to adapt this design to use fluid ink. It was not until the late 1960s that such a pen was perfected to write with sufficient reliability to ensure general acceptance. This class of writing instrument represents a hybrid of the fountain and ballpoint pens.[9] Like the porous-tip and fountain pens, the ink used is water based. The pens are sold as either disposable items or refillable with a replaceable ink cartridge that includes the ball. These pens produce a stroke not unlike that of a porous-tip pen, except that the ball tends to emboss the paper, creating a trough much like that seen in the work of a ballpoint pen. Differences in ink distinguish this product from that of the conventional ballpoint pen. The ink generally saturates the paper with slight bleeding into adjacent fibers, and ink flow-back can often be seen at the end of a stroke. Unlike ballpoint pens, they do not exhibit the characteristics of skipping and gooping. Some stroke defects do appear with use, but a particular pen typically does not develop sufficient individuality to distinguish it from others of the same class.

13.4 Gel Pens

This newest class of pen has become a very popular writing instrument throughout the world. It was first developed by Sakura Color Products Corporation of Japan in 1984.[10] Sakura was interested in producing a writing instrument that was environmentally friendly and did not contain volatile organic components such as those found in roller ball pens. Initially this class of writing instrument used a high-viscosity, pigment-based opaque ink, which does not tend to bleed into the paper fibers as much as is found with water-based roller ball or porous-tipped pens. Modern black gel inks incorporate both dye-based and pigment coloring. A broad range of ink colors are available and can generally be seen through the semitransparent barrel.[11] Some varieties are advertised as suitable for arts and graphics, including those that use an ink that is metallic in appearance. Given the vast array of colors available, these pens are very popular among children and artists. Like ballpoint and roller ball pens, the ink is delivered using a housed metal ball. Pilot, Zebra, and Pentel are common brands. One characteristic that can be used to differentiate its ink line from that of other writing instruments is a tracking effect along the outer edges of the ink stroke (Figure 13.5). While this effect is not necessarily seen throughout an entire writing, it is nevertheless a common occurrence and suggests that the ink is being pushed by the ball to the outer edges of the stroke.

13.5 Fountain or Nib Pens

Today the fountain and nib pens are more a sign of prestige than a commonplace writing instrument. An individual may select this class as his or her official signing pen, while using other writing instruments for everyday use. The writing characteristics vary. The nib point width and its flexibility are two characteristics of this pen. Occasionally, a steel-pointed dip pen may be encountered, especially where very old documents are concerned.

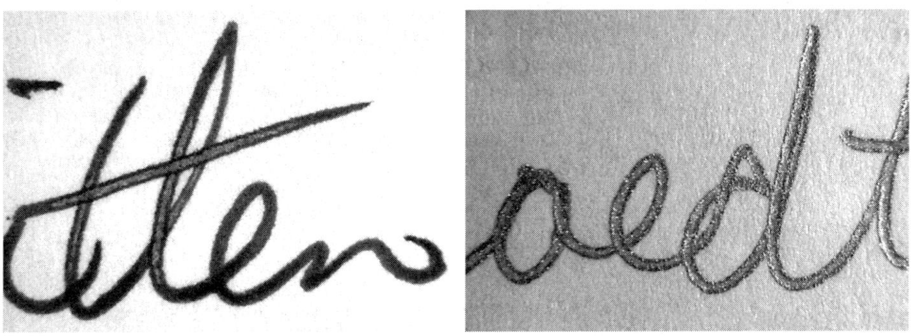

Figure 13.5 The example on the left was written with a Pilot P-700 gel pen, while that on the right was produced using a Staedtler Gel Roller with metallic silver ink. The ink line created by gel pens can take on a track-like appearance, as seen in the examples above. When this occurs, it is a straightforward process to determine the sequence of lines as the last stroke, and its tracks break underlying strokes, as seen in the *t*-crossing at the left.

Normally, a particular nib pen cannot be identified as the source of a specific writing, as it lacks individuality in its ink line morphology. Still, not all of these pens produce the same writing strokes,[12] which makes it possible to say in some situations that a certain pen was not used to execute a questioned document or that more than one pen was used.

Nib pens produce a stroke different from those of other types of pens. The nibs can make a distinctive, darker double track within the stroke, but this double track is not as pronounced with modern, stiff-pointed fountain pens as it is with more flexible points.

The ability of the fountain pen to provide sufficient ink to execute lengthy documents with comparable ink densities from page to page can distinguish it from the steel dip pen. The latter produces more intense writing each time it is filled, and gradually the intensity of the strokes diminishes. If the point of a fountain pen is flexible, definite shading can be a part of the writing, which is recognized by the gradual increase in the width of the stroke due to the pressure of the pen, particularly on downstrokes.

13.6 Writing Inks

The history of ink dates back to ancient times.[13–16] Among the earliest materials used was carbon, which produces a very permanent ink. Today it is still employed in certain drawing inks, often referred to as India inks. During the Middle Ages iron gall ink was developed. A characteristic of writing produced with iron gall ink is its oxidation and resulting fade from black to brown. The ink's corrosive nature also has the effect of eating through the paper on which it is applied, a phenomenon observed in many historic documents. It was ultimately improved to become blue-black ink and was widely used until the ballpoint pen became the most popular and common writing instrument.

Various natural dyes were initially adopted for colored inks. With the introduction of aniline (discovered in 1856) and other synthetic dyes (such as synthetic indigo, 1861) these natural dyes were gradually replaced.[17] The changeover came in the late 1800s and early 1900s. During the 1930s dyes were developed that required the use of strong alkaline solutions instead of the mild acid ink. These newer inks enjoyed limited commercial popularity. With all of the inks discussed to this point, water was the chief solvent.

With the invention of the ballpoint pen in the mid-1940s came a different type of ink. It is a thick, paste-like material using organic chemical solvents rather than water.[18] In this respect it resembles inks used in old-style typewriter ribbons and in the printing industry, although ballpoint pen ink has its own special properties.

Since the 1970s fluid water-based inks have found expanded use, first in the porous-tip or fiber-tip pen, and more recently in roller ball pens. The fiber-tip pen marked a modification of the felt marking pen, which had had some specialized use and employed a non-water-based ink. Synthetic dyes and pigments form the coloring matter in almost all present-day inks.

One can see that the variety of pens that have been invented require a corresponding assortment of inks. As a result, analytical methods have been developed by document examiners and chemists to distinguish between the different classes of inks and pens. Document examiners are called upon not only to differentiate between inks, but also to identify the source of a particular ink. Very similar, but different, inks can be distinguished by proper tests. However, a positive identification of a pen-and-ink combination is rare. Typically the best that can be stated is that the questioned ink is consistent with that found

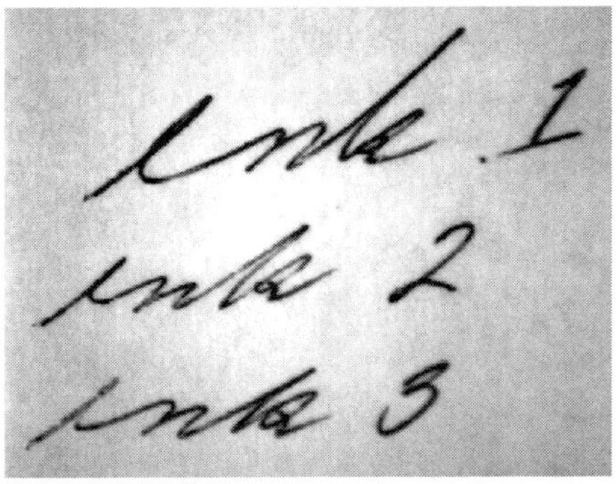

(1) Unfiltered

(2) Infrared luminescence using a light filter of 530-700 nm, a camera filter of 715 nm and an integration setting of 0.3 seconds.

(3) Infrared reflectance using a camera filter of 715 nm.

Figure 13.6 Three black ballpoint pen inks that appear similar to one another under normal lighting can be distinguished with the aid of optical filtering. Entry 1 is from a pen manufactured by Papermate, 2 by Bic, and 3 by an unknown pen manufacturer.

in the suspected pen and could have originated there or in any other pen having an ink of the same formula.

Non-destructive tests, including visual, microscopic, infrared, and ultraviolet examinations, are almost universally employed as the first step in any ink study and may permit one to distinguish between the work of two pen classes or two ink formulas. Testing starts with an examination of the ink for color and the appearance of the stroke on the paper using hand magnifiers or a stereoscopic microscope under various types of illumination. Relative ink color qualities may be detected with a change of illumination from daylight to fluorescent or tungsten light.

Studies of the ink strokes using diachronic filters[19] or infrared luminescence[20–24] are further means of distinguishing between similar but not identical inks (Figure 13.6 and Figure 13.7). None of these tests alter or damage the document.

A microscopic spot test — the reaction of a minute drop of chemical reagent on a portion of the ink stroke viewed under the microscope — helps to determine that the document was written with a particular class of fluid ink, e.g., iron base, synthetic dye, carbon ink, or pigment. When properly performed, these tests make only a microscopic change in the document. Although not utilized as much today as formerly, they are of particular value in demonstrating that two different inks were employed on a single

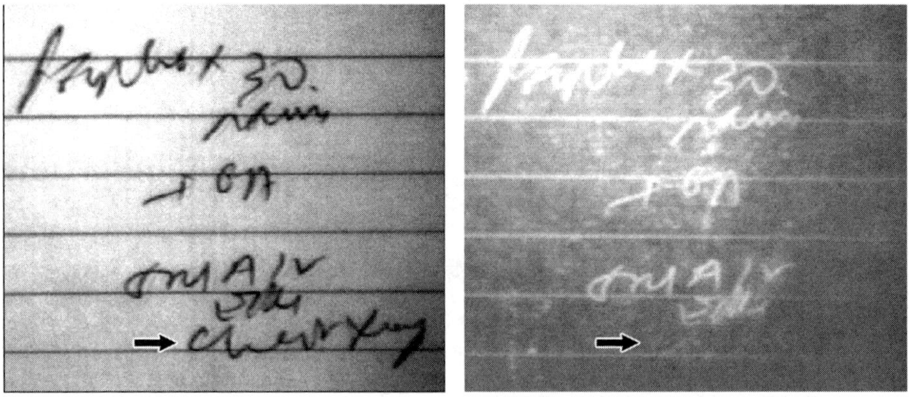

Unfiltered Infrared Luminescence

Figure 13.7 Infrared analysis was used in a case where it was suspected that a doctor had altered a patient's medical records. The entry in question is seen under normal lighting and under infrared luminescence. Note that the words *Chest Xray* become infrared transparent, while the remainder of the appointment note is luminescent.

document. Unfortunately, spot testing does not necessarily distinguish between brands of the same class of ink.

It is possible for different inks to react similarly when examined using the nondestructive techniques described above. In this case, further discriminatory testing using a technique called thin-layer chromatography (TLC) may be appropriate.[25–28] Because TLC results in a change to the document's condition, it should be considered destructive (or at best semidestructive). The method requires removing very small samples of ink from the paper, a step that may require a court order or an agreement among all parties to the dispute. The removal technique typically employs a microdisc hole punch or a scalpel to remove small quantities of ink and paper. TLC is the most popular of chemical tests currently available (see color Figure 13.8 following p. 366).

Other instrumental analyses are now available that allow further discrimination of inks, such as gas chromatography/mass spectroscopy (GCMS), Fourier transform infrared spectrometry (FTIR), and Raman spectroscopy.

Chemical tests are an important part of the identification of inks. Chemical composition can been determined by a combination of tests and demonstrated in legal proceedings in order to show that the document has been backdated or altered.

In general, it may be said that ink examinations offer more opportunities to disclose fraud than to clearly confirm genuineness. The inconsistencies between various parts of the document can at times, although not always, of course, lead to the former conclusions, but consistent results do not conclusively establish that the document was necessarily prepared at one time.

13.7 Pencils

Like pens, pencils can be grouped into classes based on their physical properties. There are three general categories: encased graphite, encased color, and mechanical.[29] In the course of manufacture, the graphite forming the marking substance is mixed with clay and waxes and then baked. The amount of graphite and clay and the period of baking are factors that determine the hardness of the lead.* Pencils designed for drawing range from very hard to very soft — capable of making only a light stroke, even with heavy pressure, or of making relatively dark lines with only slight pressure. There are also several grades of common writing pencils.

These instruments can be classified in a general way by their relative grade or firmness. Writing pencils have numerical designations, with 1 being the softest and 4 being the hardest. Artists' drawing pencils are distinguished separately, with 8B being the softest and 2H the hardest.[29]

Inexpensive pencils are made with less refined graphite and clay, while the top-grade pencils use fine materials and are manufactured under excellent quality controls. Microscopic study of pencil strokes helps to differentiate between the cheaper and more expensive writing leads. A small percentage of pencils use resin as a substitute for the clay base. Their appearance is not substantially different. Mechanical pencils can use a lead that is either clay or resin based. The polymer leads are most often used by draftsmen. Lead sizes in this latter category include, but are not limited to, 0.3, 0.5, 0.7, and 0.9 mm and vary in hardness.

When a specimen of pencil writing is examined, it may be very difficult to establish the particular hardness of the pencil used, although broad groupings, such as soft, medium, or hard, may be recognized. The differences between two adjacent grades prepared by the same manufacturer are slight. Two writers using the same grade of pencil may produce slightly different effects because of individual habits of pressure and emphasis. Furthermore, a change of writing background from a hard table top to a softer writing pad impacts on the density of the stroke produced, as does the relative sharpness of the point. Softer points in wood-encased pencils may wear down appreciably during the writing of several pages, whereas a harder lead holds its point much longer.

On the other hand, the question does arise as to whether all the writing on a page was prepared at one time and with a single pencil. A sharp change in width, clarity, embossing, or intensity of the stroke may be evidence of two separate writings. A study of the pencil stroke may indicate execution on more than one kind of writing surface or with more than one pencil. These conditions are more consistent with preparation of the parts on separate occasions than with continuous writing. Complete uniformity strongly suggests continuous writing, but it is not absolute proof of the latter. If the writing was prepared with a mechanical pencil on several different occasions but with the same writing background throughout, there may be little evidence of the interruptions.

It can seldom be established that a particular pencil was used to write a document. Identifications of this nature must be based upon microscopic flaws in the writing stroke, resulting from impurities in the graphite, and involves extensive microscopic examination combined with controlled lighting and photography. Even so, a pencil does not usually have sufficient individuality to permit a definitive conclusion.

A copy pencil, often referred to as an indelible pencil, forms a special subclass. In its manufacture, an aniline dye is combined with the graphite. The dye has little color until moistened, then the stroke assumes a pronounced blue or purple color and cannot be readily erased with an ordinary rubber pencil eraser. Identification and segregation of these

*Early pencils were a pressed mixture of clay and graphite. The lead of these pencils was not baked and, consequently, was very soft, giving only a broad stroke. Modern pencils, in which the clay and graphite mixture is baked to give added hardness, came into use about 1810 (Woodward, W.E., *The Way Our People Live*, Dutton, New York, 1944, p. 120 footnote).

pencils parallel very closely the pattern in the case of black leads, except that the dye allows some additional group classification and there is less variation in the hardness of the leads.

Colored pencils, especially red and blue, have widespread use and their work may become questioned. Within each color group there are a variety of tints. Some pencils are water soluble, some are not. A drop of water will resolve this question. Some contain high-quality leads producing smooth strokes, while poorer grades create a flawed stroke. All these factors, plus the basic lead width, distinguish different pencils.[30] Besides visual and microscopic examinations, ultraviolet, infrared, and infrared luminescence examinations can be used to distinguish between the work of different pencils (Figure 13.9). Identifying a particular make by these factors is a risky undertaking. Materials change from time to time with all manufacturers. Certain manufacturers sell to a group of wholesalers, each of whom brands the same pencil with his own trade name. Uniformity of all identifying factors suggests that one pencil of a particular color was used throughout a document, but it does not establish this with a high degree of probability.

Examinations of non-colored pencils are basically restricted to visual study, but can be supplemented with the use of the stereoscopic microscope and hand magnifiers. It is important to consider the degree of embossing, condition of the point, and intensity of the strokes. Spotlight illumination at different, especially low, angles can reveal a secondary

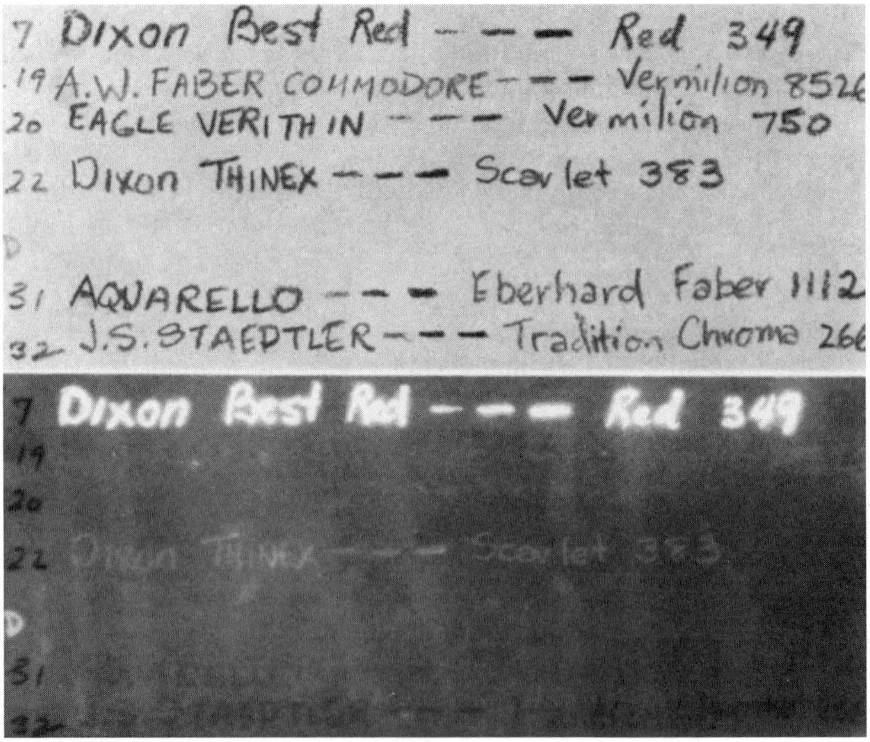

Figure 13.9 The upper sections show six red pencils of similar tints. The lower section shows their reaction in an infrared luminescence test. Note the different intensity in the bright luminescence of pencils 7 and 22 and the dark recording of pencil 32. The others are semi- or completely transparent. If these writings were the work of unknown pencils, the test would permit differentiation of 7, 22, and 32 from each other and from the other writings.

color sheen. Except with colored pencils, infrared and ultraviolet examinations are not of particular assistance. Since black pencils all contain a basic carbon ingredient, they react in a similar manner under infrared study. With colored pencils, some absorb infrared radiation, while others do not. An ultraviolet study helps to differentiate between some dyestuffs. Infrared luminescence also can separate some similarly colored strokes.

A number of chemical and instrument analyses have been attempted, including thin-layer chromatography (TLC), mass spectrometry (MS), and x-ray fluorescence spectrometry. However, there have been problems both with extracting the lead from the paper surface and with contamination.[31,32] More promising is the use of scanning electron microscopy–energy dispersive x-ray (SEM-EDX) and laser ablation inductively coupled plasma/mass spectrometry (ICP-MS). The latter approach is semidestructive in that small spots are burned on the document.

Evidence derived from the examination of pencil strokes may not necessarily form a definite solution to the problem at hand. Instead, this information, combined with other physical document evidence, may assist in reconstructing the circumstances surrounding the preparation of a questioned document.

References

1. Making ball point pens, *Machinery*, London, 135 , 1951.
2. Ball point pens, Bulletin of the Bureau of Criminal Investigation, 15, 7, 1950.
3. Lindsly, C.H., Schmidt, R.B., and Casey, R.S., Ball point pens and inks, *Chem. Ind.*, 50 , 1948.
4. Hilton, O., Characteristics of the ball point pen and its influence on handwriting identification, *J. Criminal Law Criminol. Police Sci.*, 47, 606, 1957.
5. Poulin, G., The Ball Point Pen Striation Patterns as Evidence in Document Examination, paper presented at CSFS/ASQDE meeting, Montreal, 1985.
6. Black, D.A., Identifying ball pens by the burr striations, *J. Criminal Law Criminol. Police Sci.*, 61, 280, 1970.
7. Lewis, J., Striation Patterns in New and Used Ball Point Pens, paper presented at the American Society of Questioned Document Examiners meeting, San Diego, 2002.
8. Black, D.A., Fibre tipped pens, *J. Criminal Law Criminol. Police Sci.*, 57, 521, 1966.
9. Dick, R.M., Rolling Marker Pens, paper presented at the American Society of Questioned Document Examiners meeting, 1973.
10. Wilson, J.D., LaPorte, G.M., and Cantu, A.A., Differentiation of black gel inks using optical and chemical techniques, *J. Forensic Sci.*, 49, 364, 2004.
11. Gerandt, M.N. and Urlaub, J.J., An Introduction to the Gel Pen, paper presented at the American Academy of Forensic Sciences meeting, Seattle, 1995.
12. Osborn, A.S., *Questioned Documents*, 2nd ed., Boyd, Albany, NY, 1929, chap. 11, p. 151.
13. Carvalho, D.N., *Forty Centuries of Ink*, Banks Law Publishing Co., New York, 1904.
14. Grant, J., *Books and Documents*, Chemical Publishing Co., New York, 1937, p. 41.
15. Mitchell, C.A. and Hepworth, T.C., *Inks: Their Composition and Manufacture*, 4th ed., Griffin, London, 1937.
16. Lucas, A., *Forensic Chemistry and Scientific Criminal Investigation*, 4th ed., Arnold, London, 1946, p. 79.
17. Nickell, J., *Pen, Ink and Evidence*, University Press of Kentucky, Lexington, 1990, p. 34.

18. Harrison, W.R., *Suspect Documents: Their Scientific Examination*, Praeger, New York, 1958, p. 217.
19. Dick, R.M., A comparative analysis of diachroic filter viewing, reflective infrared, and infrared luminescence applied to ink differentiation problems, *J. Forensic Sci.*, 15, 557, 1970.
20. Pfefferli, P.W., Application of an Infrared-Comparison-Projector in Document Examination, paper presented at the International Association of Forensic Sciences meeting, Oxford, 1984.
21. Ellen, D., *The Scientific Examination of Documents: Methods and Techniques*, Taylor & Francis Ltd., London, 1997, p. 99.
22. Dalrymple, B., Visible and infrared luminescence in documents: excitation by laser, *J. Forensic Sci.*, 28, 692, 1983.
23. Tanaka, T., A Review of the Spectrometer and Chromaticity Capabilities of the VSC 2000, paper presented at the American Society of Forensic Document Examiners meeting, Indianapolis, 1998.
24. Lyter, A., A Comparative Differentiation of Ball Pen Ink by Infrared Reflectance and Luminescence, Raman Spectroscopy and Thin Layer Chromatography, paper presented at the American Society of Questioned Document Examiners meeting, Ottawa, 2000.
25. Tholl, J., Applied thin-layer chromatography in documented examination, *Police*, 14, 6, 1970.
26. Crown, D., Brunelle, R., and Cantu, A., The parameters of ballpen ink examinations, *J. Forensic Sci.*, 21, 917–922, 1976.
27. Lewis, J.A., Thin-layer chromatography of writing inks: quality control considerations, *J. Forensic Sci.*, 41, 874, 1996.
28. Vargas, R., The Search for a Universal Solvent Solution for the TLC of Fluid Inks, paper presented at the Southeastern Association of Forensic Document Examiners meeting, Atlanta, 1999.
29. Masson, J.F., Pencils, paper presented at the American Academy of Forensic Sciences meeting, New Orleans, LA, 1986.
30. Hilton, O., Identification and differentiation between colored pencils, *J. Forensic Sci.*, 6, 221, 1975.
31. Cantu, A.A. et al., A scientific study of pencil lead components, *J. Forensic Sci.*, 23, 662, 1978.
32. Zoro, J.A. and Totty, R.N., The application of mass spectrometry to the study of pencil marks, *J. Forensic Sci.*, 25, 675, 1980.

Typography 14

WILLIAM J. FLYNN

Contents

14.1 The Anatomy of a Font

14.1.1 Parts of Type

Many of the terms that are used by typographers to depict the parts of a printed character have been used by FDEs to describe typewritten letters. Examples of such terms might be *ascender*, *descender*, and *serif*. There are other words commonly used in the typographic world, such as *stress* and *stem*, that have equivalent meanings for document examiners. Rather than using the word *stress*, document examiners have often referred to the transition from a thin line to a thicker line as *shading*, and the vertical backbone of a letter as its *staff* rather than its *stem*. When it comes to describing the parts of a typographic character, however, the vocabulary of the typographer is much richer than that of the FDE. This is probably to be expected, since printers have had about a 500-year head start.

Because modern computer-generated documents utilize typographic fonts almost exclusively, it is certainly time for the FDE to adopt the accepted typographic terminology. Figure 14.1 details the most commonly used terms. Note that the stem of the letter *h* (the

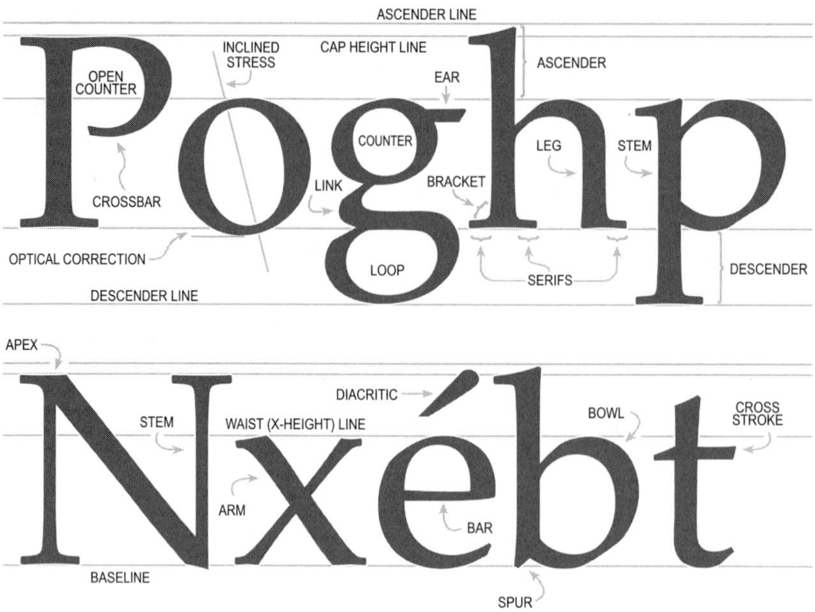

Figure 14.1 The anatomy of a font.

portion above the *x* height, also called the ascender) is actually taller than the cap height. The fact that lowercase ascenders extend beyond the height of capital letters is more often the rule rather than the exception in Roman fonts. Another important point to note is the area marked "optical correction." Typographers will almost always position letters with round bases to actually fall slightly *below* the baseline. This is one reason why baseline measurements should not include round letters.

14.1.2 Type Classification Systems

Typographers utilize several different systems to classify typeface. Some classification systems, for instance, use the physical appearance of the type to differentiate the faces. The broadest classification, Serif and Sans Serif, is in fact based on the physical attributes of the characters (whether they have serifs or not). Other classification systems attempt to use the time period in which the typeface was designed to differentiate them. Examples would be found in terms like *Old Face*, *Modern*, and *Transitional*. Still other systems use the type of bracket* to subclassify the chronological classifications. Slab-serif is one example of this type of subclassification that is used to describe a special class of Modern fonts.

It has been the experience of the author that efforts to classify a font by the time period in which it was created can be somewhat difficult at best, and often highly ambiguous. This is true because font designers are artists, not engineers. If a font designer believes that a typeface with an Old Style inclined stress would look good with a Transitional bracket and Slab-serifs, then so be it. Let the document examiners worry about how to classify the "Goudendonlubarock"** font. On a more practical note, it can sometimes be very difficult

*The bracket is the transition between the vertical stem and the horizontal serif on a Roman font. See Figure 14.1.

**The author has taken some liberty with the English language here to merge the names of three well-known Old Style, Transitional, and Slab-serif fonts.

	Slab Serif P (Boton)	Old Style P (Goudy)	Transitional P (Times)	Modern P (Bodoni)
Stress	Minimal	Oblique	Vertical	Vertical
Brackets	None	Full	Medium	Minimal
Serifs	Thick	Medium	Medium	Fine

Figure 14.2 Type classification by time period.

Figure 14.3 Font groupings.

to determine whether a typeface exhibits oblique, moderate, minimal, or vertical stress. Figure 14.2 illustrates a chronological classification system that uses stress, bracket type, and serif width. Please note that like all classification systems, these are the idealized examples — in the real world, the choices become much fuzzier.

A classification system that is far more consistent, easy to use, and preferred by the author contains six distinct types of characters: Serif, Sans Serif, Display (or decorative), Blackletter, Script, and Dingbats. Examples of these are shown in Figure 14.3.

Because the vast numbers of business documents are prepared in either Serif or Sans Serif typefaces, these will be the two that are seen most frequently in forensic laboratories. Since few people, outside the world of professional typographers, take the time to install and use the (literally) thousands of TrueType and PostScript fonts that are available, the default typefaces that come with various word processing programs, office suites, and operating systems are the ones that will be most often encountered. That is not to say that the FDE should ignore all of the other available fonts — just the opposite is true. Like

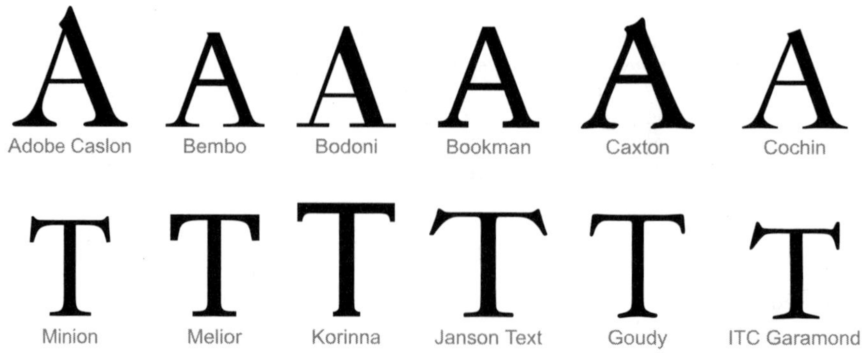

Figure 14.4 Capital letter differences in various Roman fonts.

Figure 14.5 Differences in the design of *Q*'s and ampersands.

everything in forensics, it is the oddball occurrence that is the most significant. Figure 14.4 depicts some of the differences in two of the capital letters found in various Roman fonts.

The most unique characters in nearly all Roman fonts, unfortunately, do not occur as frequently as other letters in many texts — these are the capital letter Q and the ampersand, &. For some reason, font designers have taken more artistic license with these two characters than with any of the others. Examples of the letter Q and ampersand from readily available fonts are shown in Figure 14.5.

The ampersand is actually a ligature — two letters combined to form a single character. The two merged letters are a capital *E* and a lowercase *t*. Together they spell the Latin word *Et*, which means "and." The merger of the *Et* combination is much easier to see in some fonts than in others.

Another common ligature is the symbol known as the business "at," @, which is a stylized form of the letters *at*. Other not so common ligatures found in most Unicode[*] fonts are Æ, æ, Œ, œ, *fi*, and *fl*. Please note that the last two are a single symbol for an *fi* combination and an *fl* combination (Figure 14.6).

14.1.3 Measuring Font Size and Line Spacing (Leading)

Unlike monospaced typewriter text — which was almost always based on characters of fixed widths and a unit line spacing of one sixth of an inch — the vast majority of (though

[*]A character code that defines every character in most of the world's alphabets.

fi vs. ligature ﬁ

fl vs. ligature ﬂ

Figure 14.6 Examples of ligatures.

This is a **12 point** Arial **Black** text: **The Quick Brown Fox**

This is a **16 point** Arial **Narrow** text: The Quick Brown Fox

Figure 14.7 Comparison of Arial Black vs. Narrow.

by no means all) electronic fonts are created in proportional widths and variable line spacings. Proportional fonts have widths predicated on two factors: the natural size of each character (rather than each letter being constrained or expanded to fit a fixed horizontal measurement) and the aesthetic whims of the font designer. The size of typographic fonts is always given in the vertical axis and always in units of measure known as points (now defined as 1/72nd of an inch). A 16-point font, therefore, is always taller than the same font at a lesser point size. The width of characters, however, is quite a different matter. In most typeface families it is very possible for a 12-point Black variant to be much wider than the same letters in a 16-point Narrow version (Figure 14.7).

Another important consideration in font measurement is that individual letters (even capital letters) will almost always measure less than their true point size. This is true because the actual font size is based on a rectangle that corresponds to a traditional printer's type slug. For every character in the font, the slugs must be large enough to accommodate the highest vertical ascender, the lowest descender, plus a small amount of space above and below these features. This is known as internal leading (see color Figure 14.8 following p. 366 and Figure 14.9).*

Because typographic font sizes are only close approximations rather than absolute measurements, the best way to determine the size of a font on a printed page is to use a typographic *E* ruler. Typographic rulers are available at most art supply stores and can also be ordered online. These rulers will typically contain a series of Serif and San Serif capital letters *E* that are printed on a translucent plastic in sizes ranging from 6 to 8 points up to about 60 to 72 points. If the font to be measured is a Serif font like Times New Roman, the graduated Roman letters *E* on the ruler are overlaid with a capital letter

*Internal leading is usually used to keep consecutive lines of type from clashing (overlapping). Modern OpenType fonts contain glyphs for multi-language support. The resulting accents, diacritics, etc., can result in the total measured height of the characters actually *exceeding* their declared point size.

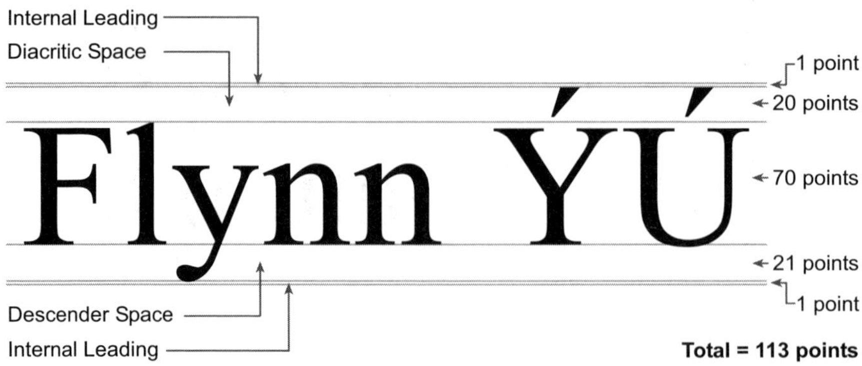

Figure 14.9 One example of the various measurements of a 100-point TrueType font.

(obviously a capital *E* is best if available) on the font to be measured (Figure 14.10). The size of the text can be read directly from the ruler when the height of the *E* on the ruler matches the size of a capital letter on the questioned document. One additional thing to keep in mind when measuring fonts is that most fonts are part of a font family that contains multiple variants of the typeface. For instance, nearly all fonts come with normal, bold, and italic variants. Italic variants are not just the normal version leaned over to the right. Italic fonts are designs unto themselves and may or may not measure the same size as their Roman (normal) counterparts (Figure 14.11).

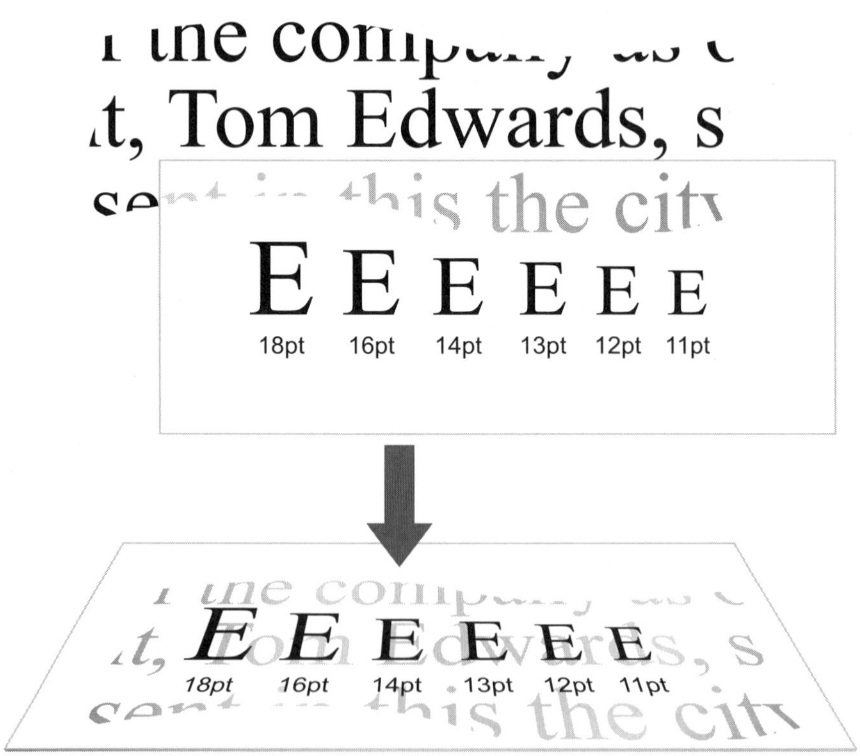

Figure 14.10 Using a typographic *E* ruler to determine font size.

Figure 14.11 An example of a Times New Roman *m* and *i* in normal and italic styles. Note the differences in vertical size and letter shapes.

One of the other potential problems involved in font measurement is the ability of some software, especially desktop publishing programs, to produce fonts in fractional sizes — 12.5 points, for example. Since a so-called 12-point font will usually fall within a range of measurements on either side of exactly 12 points anyway, it may be impossible to differentiate a tall 12-point font from a desktop publishing 12.5-point font.

14.1.4 Measuring Fonts in Reinsertion Cases

In actual casework, font measurement is often performed by a document examiner in order to determine if text has been inserted in previously printed text. Except that newer technology is being used, these cases are identical to the classic typewriter reinsertion problems that have been examined by FDEs for 80 years or more. There are, however, some unique situations that can arise in documents that have been prepared using electronic fonts rather than a typewriter.

In the classic typewriter reinsertion examination, a glass or plastic grid was laid over a page of type in order to determine if the suspect text aligns horizontally, vertically, and is parallel to the unquestioned portions of the page. This technique can still be used, to some extent, in typographic cases if the document has been prepared in a monospaced font such as Courier and is set 12/12 (12 on 12).* Because 12 points equals one sixth of an inch, it is even possible to use typewriter measuring plates in this situation. Be aware, however, that a printer's internal bitmapped Courier may use very different line spacing than the TrueType version of Courier — Courier New (Figure 14.12).

Clear plastic measurement grids can still be utilized quite effectively in typographic reinsertion cases if one has access to (or can make) line spacing grids set on typographic units. Fortunately, it is fairly simple to create tables in most modern word processing programs and then set the vertical cell spacing to a given height in point units. For instance, making a one column table with 50 rows and then setting the spacing to 13.6 points will create a typographic grid with a 13.6-point spacing that can then be printed on clear overhead transparency material. In some recent tests conducted by the author, it was found that the finest increments that were being used by modern word processing programs were 0.5 decipoints. Although finer gradations may be possible with desktop publishing programs, it seems likely that a set of grids with point spacings, such as 13.0, 13.05, 13.1,

*When a 12-point font is set on 12 points of leading, typesetters say that the type is set 12 on 12, abbreviated 12/12. Courier, however, is an odd typographic font because it was originally designed as a typewriter typeface. If an *E* ruler is used to measure a 12-point Courier font it will actually show as a 10-point character. In addition, many printers ship from the factory with an internal Courier bit-mapped or scalable character set. The 12-point versions of these fonts are sometimes designed to print on 1/6 inch line spacing. Courier New, however, is a typographic version of Courier and, as such, will generally follow the true typographic line spacing rules that determine the leading as a percentage of the font size.

```
XXXXXXXXXXXXXXXXXXXX
XXXXXXXXXXXXXXXXXXXX
XXXXXXXXXXXXXXXXXXXX
XXXXXXXXXXXXXXXXXXXX
XXXXXXXXXXXXXXXXXXXX
XXXXXXXXXXXXXXXXXXXX
12 point Courier internal
printer font is "hard-
wired" to print 6 lines
per inch.
```

```
XXXXXXXXXXXXXXXXXXXX
XXXXXXXXXXXXXXXXXXXX
XXXXXXXXXXXXXXXXXXXX
XXXXXXXXXXXXXXXXXXXX
XXXXXXXXXXXXXXXXXXXX
XXXXXXXXXXXXXXXXXXXX
12 point Courier New -
uses typographic line
spacing rule.
```

Figure 14.12 Differences in Courier line spacing between an internal printer font and a version using typographic line spacing.

13.15, 13.2, etc., will be sufficient for nearly all typical forensic line spacing, cases. Line spacing will be covered in some detail in the next section (see Figure 14.13).

14.2 Typographic Line Spacing (Leading)

The vertical distance between lines of type (interline spacing) has traditionally been known as *leading*. The term is derived from the thin slivers of lead that were used by printers to adjust the space between lines of type in a hand-set galley. Unlike the fixed line spacing associated with typewritten documents, typographic line spacings are generally adjusted for aesthetics and readability, predicated on the typeface and layout of a page. Since the measurement of line spacings in a document is probably the single most important aspect of a forensic typographic examination, a considerable amount of material will be covered in this section.

Now is the time for all good men to come to the aid of their country.
The quick brown fox jumps over the lazy dog.
Now is the time for all good men to come to the aid of their party.
The quick brown fox jumps over the lazy dog.
13.7 pt line spacing grid

Figure 14.13 The line spacing grid was produced in Microsoft® Word.

14.2.1 Measuring Line Spacing

Line spacing measurements are given in points and picas. There are (by modern typographic convention) 72 points to an inch, 12 points to a pica, and 6 picas to an inch. Most typographic lines that are computer generated, however, will sit on fractional line spacing. As an example, typical line spacings used by modern laser printers when rendering a 12-point Times New Roman font are about 13.6 to 13.8 points. In order to accomplish the fractional settings, most word processing programs actually use a 10th of a point (decipoint) as their incremental unit.* As far as the printer is concerned, however, the line spacings described above are simply locations on the printer's internal grid system, the fineness of which is predicated on the output resolution of the device. For all practical purposes, however, line spacing grids incremented in 0.5 decipoints should suffice for document examiners conducting forensic examinations of computer-generated documents.

There are several reasonably practical ways in which decipoint unit measurements can be made. One of the methods employed by the author involves the use of a microscope measuring reticle that is graduated in 100 evenly spaced lines.** The reticle is inserted into one of the eyepieces of a stereo-zoom microscope and calibrated in the following way:

1. An accurate typographic ruler is placed under the zoom microscope with the pica and point gradations in view through the eyepiece.
2. While looking through the eyepiece, the magnification of the microscope is zoomed so that 100 of the reticle increments occupy the same length as 10 points on the ruler. Each reticle step will then equal 1/10 of a point (a decipoint).

With some practice, the above method can result in very accurate line spacing measurements. In addition, it is easy to replicate the calibration from use to use, and the technique employs equipment already found in many forensic laboratories. On the downside, the above settings can be used to *directly* measure only the decipoints over a linear distance of 10 points. It is still possible, however, to use the calibrated reticle as a highly accurate vernier*** in order to complete a measurement over any distance that can be spanned by your typographic ruler.

Another technique that can be used to show line spacing deviations over the length of a page is using grids printed on transparency material. Using grids that are based on typographic spacings (such as 13.6, 13.75, or 13.8 points) can be a very effective way to determine if text has been reinserted into a document. The same grid can then be used to make a record of any line spacing discrepancies that occurred in the reinsertion process, by photographing or scanning the questioned document through the transparent grid. These grids can be manufactured in one's own laboratory using most modern word processing or desktop publishing programs. In the author's laboratory, Microsoft® Word has been used to create sets of grids with decipoint spacings ranging from 10.0 to 15.0 points in 0.5-decipoint increments.

*As of this writing, the most accurate word processing program found was Microsoft® Word, the newer version of which allowed line spacing increments of 0.5 decipoints, i.e., 13.8, 13.85, 13.9, etc.

**Meiji microscopes make several such reticles for different types of eyepieces.

***Vernier (noun): A small moveable scale that slides along a main scale; the small scale is calibrated to include fractional divisions of the main scale (www.hyperdictionary.com).

Within the past few years, several companies have created computer-based measuring systems for use in a forensic laboratory. These systems generally consist of a high-resolution video camera, a frame grabber, software that superimposes user-defined grids over the captured image, and some means of annotating the images. In addition, many of these software packages can be calibrated to translate mouse movement (in pixels) into other units, such as millimeters, inches, or points. These computerized measuring systems can be extremely helpful in suspected page substitution cases by cross-comparing the line spacings found on different pages of the suspect document. For instance, an on-screen grid can be drawn very quickly using the line spacing found on one of the unquestioned pages. These settings can then be used as the benchmark. Leaving that grid on the screen, the remainder of the document's pages can be examined for line spacing anomalies. The same technique also can be used on single-page documents wherein one block of text is suspected to have been added (reinserted).

One caveat to keep in mind with any computer measuring system is that the ultimate accuracy of these measurements will be predicated on the lowest resolution of any of the components. For instance, it may be possible for a software program to detect mouse movement at a higher resolution than the image on the monitor. In this example, the measurement might change on screen before the corresponding on-screen movement of the grid markers can catch up.

In summary, some means of making highly accurate measurements must be available to a document examiner before beginning a line spacing case. Whether one utilizes microscope reticles, measurement grids, or computer systems, the golden unit of measurement is 0.5 decipoint. Although these measurements are many times finer than those that were typically made in typewriter cases, they are well within the reach of a modern forensic document laboratory.

14.3 What Factors Affect Typographic Line Spacing?

Fortunately for FDEs there are only three types of people who create computer-generated documents: (1) typographic experts, (2) typographic expert wanna-bes, and (3) those who would not know how to change the default settings if their life depended on it. We can give further thanks to the fact that most fall into the latter category.

Below are some of the questions we can ask concerning the variables that affect the line spacings found in a document:

1. If a word processing program's default settings are considered to be the normal line spacings, then what factors affect these normal spacings?
2. How are the default line spacings of typographically set documents predetermined?
3. Should we expect to find that all 12-point fonts on a page would have the same interline spacing?
4. How much control do users typically exert on the line spacing of documents they create vs. adjustments made by the software and hardware without a user's intervention?

14.3.1 Why Can Normal Line Spacing Be Different for Same-Size Fonts?

There are three primary reasons why typographers (and font designers) adjust line spacings: (1) readability, (2) the design of the font requires extra leading to avoid clashing, and

NOW IS THE TIME FOR ALL GOOD MEN TO COME TO THE AID OF THEIR PARTY. NOW IS THE TIME FOR ALL GOOD MEN TO COME TO THE AID OF THEIR PARTY

Enlargement of 10 pt Arial Narrow set to a "negative leading" of 10/9

NOW IS THE TIME FOR ALL GOOD MEN TO COME TO THE AID OF THEIR PARTY. NOW IS THE TIME FOR ALL GOOD MEN TO COME TO THE AID OF THEIR PARTY

The same text set to 10/12 (normal) leading

NOW IS THE TIME FOR ALL GOOD MEN TO COME TO THE AID OF THEIR PARTY. NOW IS THE TIME FOR ALL GOOD MEN TO COME TO THE AID OF THEIR PARTY

The same text set to 10/13 leading

Figure 14.14 Comparison of Arial Narrow using various leading settings.

(3) page aesthetics. Those who are typographically trained look at a printed document differently than most of us. One of the characteristics they see (even if the document is printed in black ink/toner only) is known as color. Color refers to the overall tone or texture of the type. One individual, skilled in desktop publishing, once stated that color is the "harmonious blending of black and white on a page." The author likes to think of color as the gray level one sees if you squint and unfocus your eyes while looking at the text on a page. The point of this discussion is that line spacing is one of several ways that white is added to or subtracted from a page. Adjusting the color, then, is one of those aesthetic considerations (generally used only by graphic artists) that may result in line spacings that are different than the default.

There are two other (often related) reasons why line spacings are adjusted: font design and readability. The elongated, italic characters found in decorative script fonts, for instance, are often more difficult to read and also require more leading due to their long upper and lower extensions. As explained below, the fact that line spacings are usually adjusted automatically to accommodate different fonts should be of considerable interest to FDEs, especially because these adjustments are usually made without the user being aware of them. First, let us look at the four criteria that generally have the most impact on the readability of a document: (1) the size of the typeface, (2) the typeface variant, (3) the line spacing, and (4) the output resolution of the printer.[*] Here is a good rule of thumb: the easier it is for the eye to identify the outline of the characters in a font, the more closely (both horizontally and vertically) the type can be set. In Figure 14.14, a narrow, all-caps Sans Serif font (a combination that is usually difficult to read) is set on various amounts of leading in order to illustrate the importance of line spacing (white space).

[*]Typeface refers to an entire family of letters. Helvetica and Times Roman are typeface families. A variant would be described as Helvetica Condensed Oblique. When a size is appended to the variant, it becomes a font. For example, 12-point Helvetica Condensed Oblique is a font description.

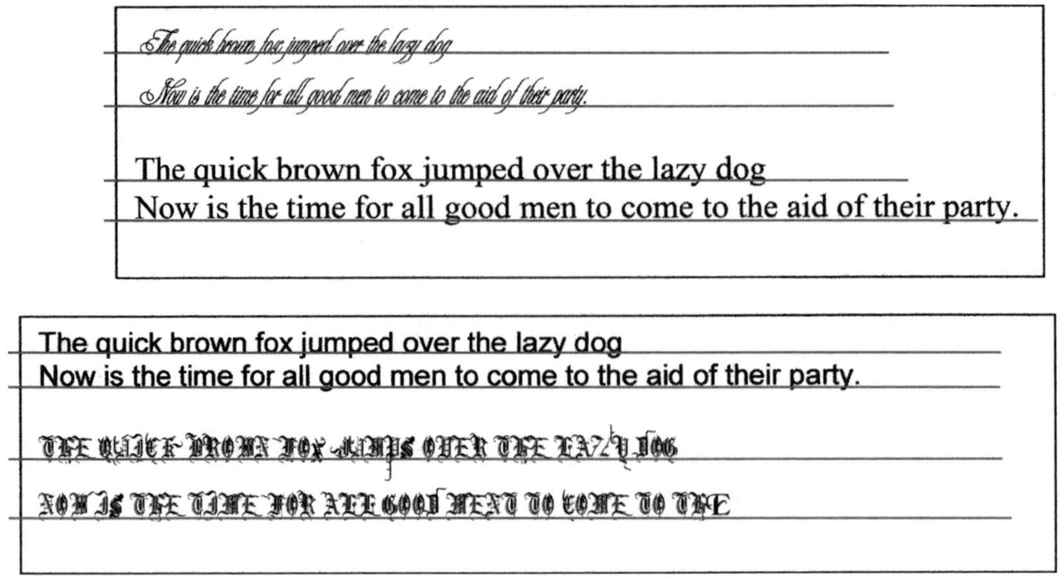

Figure 14.15 Readability comparison.

At normal font sizes, it is easier for the eye to distinguish the outlines of a character printed in a Roman (vertical Serif) or Gothic (vertical Sans Serif) font than any of the other classifications. The designers of a typeface take this into consideration when deciding how much white space they want to appear between lines of type in order to make it more readable. Figure 14.15 shows a striking example of a 14-point Script font and a 14-point Roman font printed on the same printer as consecutive paragraphs with no human adjustments. Note the different levels of difficulty involved in reading the texts and the added leading for the script font needed to prevent the long extensions from clashing between lines.

The font samples in Figure 14.15 also illustrate how additional leading was automatically added to the 12-point Black Letter font compared to the Gothic text. In addition, it shows why an all-cap Black Letter 12-point font should never be used for long passages.

Another factor that influences the readability of a typeface is the resolution (dots per inch, dpi) at which it is rendered. In keeping with the principle described above, typographers (and the software engineers who write the word processing programs) have tended to decrease the default leading (decrease the line spacing) as output resolutions of available printers has improved. In some cases this may be a dating phenomenon. When the typical output resolution of readily available laser printers was limited to 300 dpi (until October 1992), the default recommended line spacing was 120% of the font size. Using this 120% line spacing rule, a 12-point Times Roman font would sit on 14.4 points of leading ($12 \times 1.20 = 14.4$). As the output resolutions have gone from 300 to 600 dpi, and even 1200 dpi, the normal line spacing factor for Times Roman (Times New Roman) has gone down to about 115% of the font size. This results in lines of 12-point Times Roman font typically resting on about 13.8 points of leading.* All things

*As with all things typographic, there are a number of important caveats. The author has noted Times New Roman line spacing changes from the same printer (an HP LaserJet 4) as the printer drivers, word processing program, and operating systems have been upgraded over the years.

being equal, documents created with older software and printed on lower-resolution printers will have greater normal line spacings than newer systems. As always, however, there are some important additional considerations. For example, it was possible at one time (depending on the driver version) to set the output resolution of an HP LaserJet 4 to 150, 300, or 600 dpi and watch the line spacings change with the resolution setting. The author recently attempted to repeat this experiment on a computer with Windows® XP Professional installed. Although it was still possible to change the output resolutions through the printer properties dialog box, the newer driver did *not* adjust the line spacings at all. Because operating systems tend to be upgraded and overwritten (often along with the older printer drivers), it is unknown which HP driver version began to force uniform line spacing, regardless of the output resolution.

14.4 Other Factors That Can Affect Line Spacings

There are many other factors that can have a direct influence on the measured line spacings of text within a document. Some of these factors are operator controlled and others are not. Here is a brief listing of some circumstances that can have a bearing on line spacings:

A. Without conscious user intervention:
  1. An *entire document* is converted from one word processing program to another. For instance, there have been situations wherein a document conversion from WordPerfect (which was a favorite of law firms) to Microsoft Word format has resulted in odd line spacings and formatting anomalies.
  2. *Portions* of an older document (originally prepared in one word processing program) are copied/pasted into a new document being assembled in a different word processor. This sometimes occurs when "boilerplate" legal terminology from an older contract/agreement is used in a new document. FDEs have worked cases where the copied/pasted (inserted) terminology is set on different line spacing than the text that is typed into the new document.
     Copied/pasted text may or may not undergo font substitution. There are many common look-alike fonts that can show up in business documents, for instance, Helvetica/Arial/Swiss, Times Roman/Times New Roman, Times New Roman/CG Times, Courier (internal printer version)/Courier New (TrueType). If a new document is being prepared in Arial and the typist inserts text from an older document prepared in Helvetica, one of two things can happen: (1) if the typist has a Helvetica font on his or her computer, the pasted text will likely remain as Helvetica even though the surrounding text is in Arial; (2) if there is no Helvetica on the machine that is being used to prepare the new document, it is likely that the pasted Helvetica text will be converted to Arial. From a document examiner's standpoint, one must be aware that 12-point Helvetica and 12-point Arial (even though they are very similar looking fonts) may well sit on different line spacing.*
     It is important, therefore, to ensure that all of the line spacing measurements on a page compare text rendered in the same font. This is particularly true in suspected printer reinsertion examinations (see Figure 14.16).

*At least this is true on the author's machine.

Figure 14.16 Many Roman and Gothic fonts commonly found in business documents, such as Times New Roman and Arial, have look-alike font cousins. CG Times is very similar to Times New Roman. Nevertheless, subtle differences can be seen. Note the variation in the design of the percent signs. The superimposed characters at the bottom illustrate some of the subtle differences between the two, such as the lengths of the arms of the *F* and the width of the *C*.

3. Paragraphs that are copied/pasted from one document into another may carry their paragraph line spacings with them even if they are in the same font and prepared in the same word processing program. This can easily give the illusion that text has been reinserted if the line spacing in the earlier document had been forced by the author into an unusual spacing (Figure 14.17).
4. If various parts of a multi-page document are printed from different printers, there may be noticeable differences in the line spacings. Although this may seem far-fetched, just such an occurrence has been directly observed in a forensic laboratory when a network printer went down during the printing of a long document, requiring that the latter pages be printed on a stand-alone laser printer. In a large organization it is also possible that several work groups will produce (and print) their sections of a document on different printers and then later compile them into final form.

B. With conscious user intervention:
1. The numbers of ways that an individual can vary the line spacings in a document are limited only by imagination and the capabilities of the software and hardware. Oddly enough, this is a good thing. As is true in every aspect of forensics, it is the quirky, rather than the normal, that provides the greatest identification

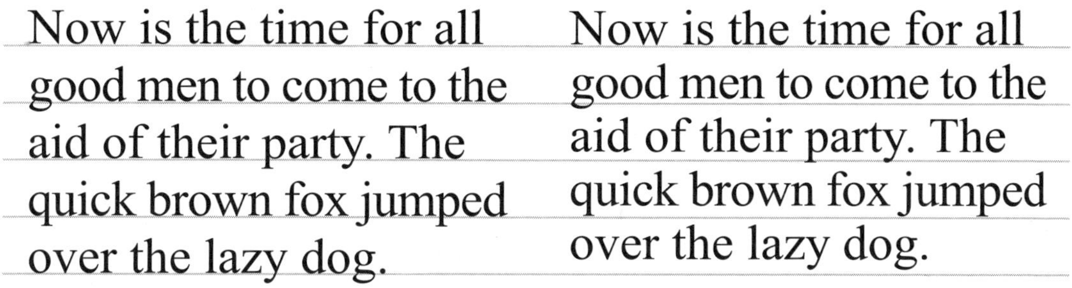

Figure 14.17 The left paragraph is set in the default line spacing for 12-point Times New Roman. The right paragraph was forced to use a line spacing of 13 points.

value. For instance, it may well be that an unusual combination of fonts and formatting found in anonymous letters is also found in a suspect's normally prepared correspondence.

14.5 Horizontal Measurements

Very little has been said in this chapter about horizontal measurements even though at one time (with typewriter-produced documents) these were usually the most important forensic measurements. Most document examiners purchased extensive sets of grids in order to establish horizontal misalignments in a typed text as an aid to the identification of a particular machine. Modern word processing and use of true typographic characters, however, have dramatically changed the way we measure questioned machine-printed text.

There are many factors that can affect the horizontal spacing of a modern proportional, scalable font. Simply changing from a Times New Roman typeface to a very similar looking CG Times face can cause a measurable difference in the horizontal length of a line of text (Figure 14.16). Because there is a slight difference in the stroke widths of Times New Roman and CG Times, they will occupy different horizontal line lengths at the same point size. In addition, two other important factors can have a dramatic impact on horizontal text measurements: fonts can have different weights and can come in condensed and expanded versions.

For web page creation, the TrueType specification explicitly allows for the weights of a font (the stroke widths) to be assigned a numeric value of 100 to 900, with 400 considered regular and 700 designated as bold. Typeface manufacturers, however, are not limited by the specification at all. By combining various permutations of condensed, italic, and weights, Linotype has produced no fewer than 97 different combinations of its popular Avenir typeface. As a result of the subtle differences in the stroke widths of these characters, it could be extremely difficult for the average document examiner to distinguish between a 12-point Avenir semibold and Avenir bold unless there were actual samples of this font at each weight. Table 14.1 lists the names of 15 common weights and 5 levels of Compression/Expansion.

Figure 14.18 illustrates how some of the various weights and expansions can affect text produced in the same font. In addition, the last line of text was prepared in a Garamond typeface from a different type foundry than that which had made the upper four. Note

Table 14.1 Fifteen Common Weights and Five Levels of Compression/Expansion

Weight 1	Weight 2	Weight 3	Condensed/Extended
Ultralight	Regular	Extra bold	Ultracondensed
Light	Medium	Ultrabold	Extra condensed
Thin	Demi	Heavy	Condensed
Roman	Semibold	Black	Compressed
Book	Bold	Ultrablack	Extended

how the different stroke widths of the two normal Garamond fonts have changed their horizontal spacing.

There are a few more factors that need to be considered that can change the horizontal spacing of text. Justification of the right margin, for instance, will attempt to stretch a line of text so that the line lengths are all the same. The effect of full justification is to change both the distance between letters and the distance between words in order to make uniform line lengths. If kerning is turned on, it also can affect the interletter spacing by making microadjustments to particular letter pairs so that they are more aesthetically pleasing. Finally, some word processing software will have the ability to "fit to document." If this feature is available, the operator can have the software attempt to adjust the text and line spacing to fit the document on a set number of pages. This is different than the "scale to fit" option in the printer menu, which simply enlarges or reduces each page of your document to fit on a different size paper during printing. Changing a document that was composed on an 8.5 × 11 inch page to print properly on an A10 sheet of paper during printing would be a classic example of where scale-to-fit might be used.

The quick brown fox jumps over
(NORMAL)

The quick brown fox jumps over
(BOLD)

The quick brown fox jumps over
(CONDENSED)

The quick brown fox jumps over
(EXPANDED)

The quick brown fox jumps over
(NORMAL – FROM DIFFERENT FONT MANUFACTURER)

Figure 14.18 Samples of different weights and expansions using a Garamond typeface.

14.6 Conclusion

The skills that are needed to examine documents prepared on a modern word processing system differ greatly from the days of typewriter identification. In a very real sense, one needs to be typographically skilled in order to reliably draw conclusions from an examination of computer-generated documents. Some of the best training that one can get in this area is by taking courses (or studying texts) on desktop publishing. Virtually every aspect of page layout, font design, and terminology can be learned in desktop publishing classes. In addition, one quickly gets an appreciation for the number of look-alike fonts and their subtle differences.

Today's computer-generated business documents are not typed, but rather are composed and rendered using sophisticated typographic rules that come not from the typewriter but from the typographer. It is apparent that the examination of computer-generated documents has already started to become a subspecialty under the general forensic science of document examination. And even within this subspecialty there are further breakdowns that involve expertise in the software that composes the text and expertise in the machines that render the documents.

There is little doubt that the future of forensic document examination will encompass a number of different computer hardware and software specialties. This will all be in the finest traditions of forensic science — where science and the law merge.

15 Typewriters

MARY W. KELLY

Contents

Although the golden age of the typewriter as a major instrument of document preparation has passed, document examiners can sometimes still be called upon to analyze older documents, forms, or anonymous documents created on a typewriter. Typewriters have been used in the production of everything from travel documents to wills to threatening letters. Even though improved printer technology exists today, the inexpensive nature and portability of typewriters can still make them the instrument of choice for perpetrators of fraud.

Before the FDE can properly examine typewritten documents, he or she must have a sufficient understanding of the technology involved. A working knowledge of the various classes of typewriters and their distinguishing features allows the FDE to differentiate one from another. In addition, there exists a tool, the TYPE classification system,[1] which provides a means by which typestyles can be classified. Classifying a typestyle can lead to the identification of the make and model of the source machine and its possible dates of introduction. The TYPE classification system, an invaluable tool in typewriter examinations, will be discussed in greater detail later in this chapter.

Other examinations associated with typewriting include the reading of certain typewriter ribbons. Reading the ribbon involves removing the ribbon from the typewriter and locating the text in question. An automated ribbon reading device (RAW) has been developed to aid in deciphering typewriter ribbons, but much of this work is still done by hand with the aid of different lighting sources. The most common lighting source used for these types of examinations is transmitted light.

Typewriting examinations also can be conducted to determine whether a particular document had been altered by the addition of text. This examination is aided by the use of typewriter grids, which are measured templates placed over the typewritten text to detect misalignments and spacing irregularities. This is based on the concept that once a paper is removed from the typewriter, it is nearly impossible to reinsert it in the exact same position. Grids contain regularly spaced vertical lines that correspond to the horizontal spacing of a particular model. It follows, then, that typing done in one sitting will be in alignment within the grid, while text added after the paper has been removed from the machine or adjusted may not be. Great caution has to be exercised when making grid measurements, as a certain amount of variation can be found in typewriters.

Many aspects of typewriting technology have tremendous potential for dating significance. As innovations and improvements were made to typewriting, dates of introduction were established.[2] Comparative dating can provide a means by which to gain significant dating information from the examination of a questioned document. Comparative dating is derived from the life history of the individual machine. Dating information can prove critical in certain kinds of typewriter cases, especially with respect to alterations and backdating.

The identification of a typewriter is a function of its operating characteristics. This means that the identification of a typebar machine will involve different characteristics than will a typeball machine or a typewheel machine. Nevertheless, the same kinds of identifying characteristics are sought in all problems, although their frequency and importance will differ with each class of machine. The three classes of typewriters set forth in the following discussion are all impact machines. Impact machines create typewritten impressions through pressure contact with the sheet. They include the typebar, typeball, and typewheel machines. The first step in any typewriter examination is to determine, when possible, which class of machine was used to prepare the questioned document.

15.1 Typebar Machines

The history of typewriters is rich in detail and well documented.[3] Therefore, only a cursory discussion is included here, in order to obtain historical perspective. The first practical typewriter was invented in 1876 by a printer from Milwaukee named Christopher Latham Scholes. He worked in conjunction with Carlos Glidden and Samuel Soule. The men were trying to create a system whereby they could print numbers on the pages of books. Their idea was expanded to include all the letters of the alphabet, and the modern typewriter was born. They marketed their "Type-Writer" to Remington & Sons, a firearms and sewing machine manufacturer. In 1873 the first successfully marketed typewriter, the Remington Model, was launched.

The early typewriters made use of metal typebars, which bore letters and numerals. These typebars are engaged when a corresponding key is depressed. The typebar strikes an inked ribbon and transfers the impression to the paper. Another component of the typebar machine is its moveable carriage. The carriage moves from right to left in order to produce typing across the width of the page. The platen holds the paper in place and provides a striking surface for the keys. Popular makes and models of early typewriters include the Remington No. 1 (1876) and No. 2 (1878), the Calligrapher (1883), and the Underwood No. 1 (1896).

Ribbons most commonly used on a typebar machine are cloth. Cloth ribbons can be made of many different kinds of fabric, but the most common is nylon. Depending on the cloth used, weave patterns and yarn counts can be determined. Under magnification, typebar impressions are not crisp, but rather somewhat fuzzy in appearance, and the fabric pattern of the ribbon can usually be observed (Figure 15.1). The cloth ribbons used on typebar machines are difficult to read for two reasons. First, once a letter is typed, the ink of the cloth ribbon will fill in that void and the letter will no longer be decipherable. Second, subsequent typing using the ribbon will be stacked on top of previously typed text, thus rendering it illegible.

The impact of a typebar machine in creating the typed impression is the strongest of the three classes of machines. The impact results in reverse embossing and is an important characteristic of this class of machine (Figure 15.2). Removing the ribbon to create an impression, as was done in this example, is a recommended way of obtaining comparison standards for typebar machines, as any defect to the typeface is clearly visible and not masked by effects of the ribbon.

From the beginning of the typewriter's development in the 1870s, each manufacturer designed its own typeface, according to its own ideas of legibility and desired appearance. As a result, many of the older machines have type designs that are particular to the given

Figure 15.1 Strike-up of a typebar machine (Royal). Observe the fabric pattern of the ribbon, reverse embossing, and fuzzy quality of the impression. Also note the rebound effect seen on the letter *e*. Rebounding is a common effect seen on typebar machines where two impressions print, instead of one, and they do not quite superimpose.

Figure 15.2 Strike-up of a typebar machine, without the ribbon in place, to demonstrate the depth of the impression. The effect is reverse embossing.

machine. This is another important identifying feature particular to this class of typewriters. The typeface and typewriter are to be considered one entity when conducting examinations and comparisons.

Horizontal spacing, called pitch, is another characteristic to consider when evaluating a typewritten document. Pitch can be measured by use of an ordinary ruler or with the previously mentioned standard typewriter grids. Traditional spacing on the early machines was either pica or elite. Pica spacing is 10 letters per inch or 2.54 mm per character. Elite spacing is 12 letters per inch or 2.12 mm per character. European spacing is based upon the metric system of measurement and includes spacings such as 2.10, 2.20 and 2.30 mm. There was a time when if the horizontal spacing was anything other than 2.12 and 2.54, it could be stated with certainty that that particular machine had been manufactured outside of the U.S. This definitive statement can no longer be made. There came a time when typestyles began being manufactured outside of the U.S. and many styles were copied. Nonetheless, horizontal spacing remains a critical measurement to be made when conducting typewriter examinations.

In 1939 IBM introduced the first proportional-spacing typewriter, marketed as the Executive. This machine produces typewriting that resembles printing in that the letters occupy different units of spacing, ranging from two to five. For example, the capital letter *M* occupies five units, while the small letter *I* occupies two. Proportional spacing cannot be measured with standard grids, wherein all characters are of equal size. Special grids have been developed for proportional measurements (Figure 15.3). Increments for proportional spacing include 1/32, 1/36, and 1/45 inch. The combination of their typeface design and unique spacing distinguishes these types of machines from other typebar machines and provides significant dating potential.

Over time, several important improvements were made to Scholes' original machine. The first was the introduction of the "shift" key that allowed for printing of both upper- and lowercase letters. The second was the introduction of electric power in the early 1920s. In the electric machines, a motor-driven device activated the selected key and resulted in somewhat more uniform typing, as the force imparted by the typist was no longer as critical.

Thus far, only class characteristics of typebar machines have been discussed. The evaluation of these class characteristics is the important first step in any examination. However, the basis for the identification or elimination of a particular machine lies in the

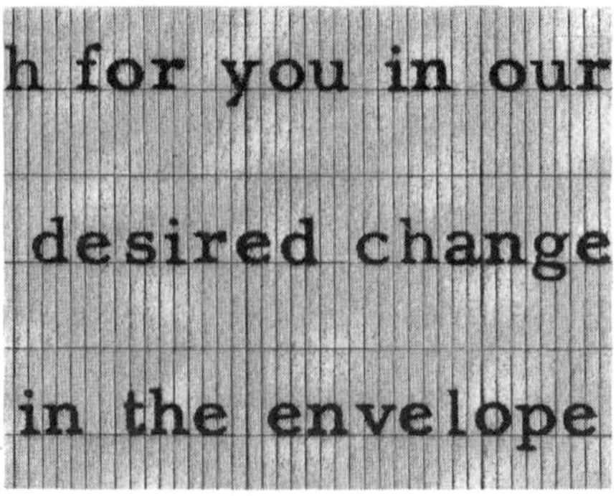

Figure 15.3 Proportional-spacing typewriting under a specially designed grid. Note that the *i* and *l* occupy two units, while the *o*, *r*, *e*, and *s* occupy three.

evaluation of its individual defects or peculiarities. These defects can be divided into two classes: alignment defects and typeface defects.

All machines are ideally designed so that each character prints uniformly on an imaginary baseline. However, the effects of wear and tear often result in defects and cause divergence from this ideal. Defects are especially pronounced in this class of machine due to the collision of the metal typebars. These collisions result in damage to the typeface and skewed alignment. As with any forensic examination, before the proper weight of these individual defects can be assessed, they must be observed to repeat consistently in an adequate sample of known specimens.

In summary, critical features to consider when examining typewritten documents that may have been created on a typebar machine include fabric ribbon, depth of impression, reverse embossing, unique typestyles, and individual defects.

15.2 Typeball Machines

With the introduction of the Selectric typewriter in 1961, the single-element machine was born and the concept of typewriter identification changed. This was the second major class of machine to be developed. The Selectric consists of two distinct typing parts, the machine and the typeball, or element. The typeball contains the letters and characters in rows and columns around the element (Figure 15.4). Depressing a type key causes the ball to rotate and tilt in order to position the proper character and then strike the paper.

The Selectric typewriter replaced typebars and moving carriages with a printing element, the typeball. The ball moves along a slender metal rod, tilting and rotating at a very high speed as it selects the desired character. The ball elements contain characters in a row encircling a ball. All capital letters are on one side and all lowercase on the other. Each key and each letter has an interposer rod that is controlled by latches.

In order to type a particular letter, three rapid movements are necessary. The ball must rotate so that the proper column is centered and it must tilt so that the proper row is centered. This combination of actions moves a particular character into position to type, and then the third action causes the ball to move forward and strike the ribbon to print the letter on the paper. Immediately following the printing action, the typeball returns to its rest position. The entire cycle is completed in the fraction of a second.

Figure 15.4 Typeball element containing four horizontal rows of letters and characters.

These ball elements are interchangeable from one machine to another and have replaced the typebars of the traditional typewriter. The significance of these interchangeable ball elements is that different typestyles can be produced on a single page, without having to remove the paper from the machine. Further, their significance includes that when they become damaged, they are simply removed and replaced and any defect is cured. Other companies that manufactured machines using a ball typing element include Adler, Facit, Olivetti, Remington, and Royal.

By the time of the invention of the Selectric typewriter, fabric ribbons had given way to carbon film ribbons. A carbon ribbon consists of a carbon wax coating on a polyethelene base. The ball element strikes the ribbon causing most of the coating to be deposited on to the paper, leaving behind a clear outline of the letter that was typed. Thus, carbon ribbons, except for specially designed security ribbons, can usually be read and the typed text deciphered. Carbon ribbon impressions are crisper than those of fabric ribbons (Figure 15.5).

The amount of impact generated by typeball machines in creating typed impressions is less than was seen in typebar machines. The impressions are created as the ball is actually in motion, striking the paper. As a result, the impressions are somewhat uneven in depth, being slightly deeper along the edges (Figure 15.5 and Figure 15.6). There is very little reverse embossing, as compared to impressions generated by typebar machines.

Figure 15.5 Strike-up of a ball element (Selectric fitted with Prestige Elite ball). Observe the shiny black surface of the carbon ribbon and crisp quality of the impression.

Figure 15.6 Strike-up of a typeball machine, without the ribbon in place, to demonstrate the depth and unevenness of the impression.

In 1971, with the introduction of the Selectric II, further additions were made to typing technology. It contained a dual-pitch mechanism that allowed for typing at both 10 and 12 characters per inch. As a result, pica-size type at elite spacing (crowded effect) and elite-size type at pica spacing can appear.

Further innovations were made in 1973 with the introduction of the Correcting Selectric. These machines were fitted with two types of ribbons: a special correctible carbon ribbon and a lift-off correcting ribbon. When first typed, the ink of these carbon ribbons does not adhere strongly to the paper. As a result, a second typing of the same letter through a lift-off ribbon, which has a highly tacky coating, can remove the original letter. Only a blank impression of the original letter, which can be observed microscopically using side lighting, remains. With time, this class of ink gradually acts upon the paper and becomes more difficult to remove.

As for defects, damage to typeball elements is infrequent. In cases where it occurs, it is usually the result of the element having been dropped or used to type over a paper clip or staple. Occasionally, defects to a ball element are the result of the manufacturing process, whereby a bead or bump occurs in the plating process (Figure 15.7). Unlike damage that occurs to the typeface of a typebar machine, which remains a permanent part of that typewriter's identity, damage to ball elements is cured with their replacement.

The misalignment defects that occur with ball elements are less pronounced than those in typebar machines, but perhaps are more consistent. In ball elements the misalignments are common to whole rows or columns of characters on the surface of the printing element. Accurate alignment of a character is accomplished by means of two detents or knife edges operating in the center notches in the machine itself for baseline alignment. Thus, four characters in a single vertical column of any given rotation are controlled by the positioning of the detent in the same notch on the typeball, and all four should have the same left-to-right alignment. As a result, there is a direct correlation between alignment defects found among several characters on a typeball, whereas on a typebar machine the only correlation is between the two characters (usually upper- and lowercase) found on a single bar. Thus, alignment depends on the machine and the further combination of the machine and ball element.

Special typestyles were designed for use on the typeball machines. Characters were designed specifically to fit on the ball element. Common typestyles created by IBM for use on the Selectric machines were Courier and Prestige Elite. These typestyles have been copied extensively over time.

Figure 15.7 An IBM Orator typeball bearing a bead defect on the number 6.

In summary, critical features to consider when examining a typewritten document that may have been created on a typeball machine include carbon ribbon (with or without lift-off corrections), depth and unevenness of impressions, unique typestyle, and individual, identifying defects.

15.3 Typewheel Machines

The typewheel machine was introduced in 1974 and represents the third class of typewriters. The typewheel, often referred to as a daisy wheel, is a hub surrounded by a series of spokes or arms, each of which contains a single typeface at the end (Figure 15.8 and Figure 15.9). After a letter is rotated to align the typeface to print, a plunger strikes the back of the typeface, forcing it into the ribbon and onto the paper. Carbon film ribbons are used on this class of machine. The carbon ribbon impressions of a typewheel machine are very crisp and clean (Figure 15.10).

The introduction of the typewheel brought about a series of new typestyles, most of which were very similar to those used previously on typebar and ball element machines, yet differing in subtle details. The typestyles had to be designed so as to fit properly onto a wheel element. As with ball elements, a significant aspect to these interchangeable typewheels is that different typestyles can be produced on a single page without having to remove the paper from the machine.

Figure 15.8 Typewheel element: a hub surrounded by a series of spokes.

Figure 15.9 A close-up of the typewheel spokes, each of which contains a single character.

Figure 15.10 Strike-up of a typewheel machine (Brother). Observe the crisp, clear carbon impression.

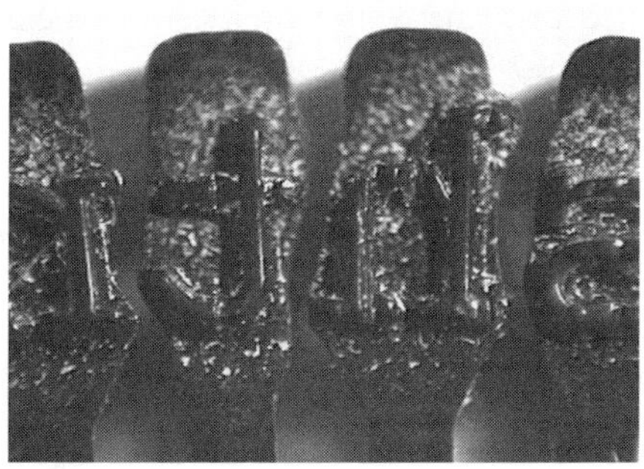

Figure 15.11 A close-up of damaged typeface. It is most probably the result of the wheel having been put on incorrectly and being damaged by the plunger striking the typeface.

The typewheel can develop individual defects when the face becomes worn or the spokes twisted or broken. On certain wheels, damage can be the result of the wheel having been put on the machine incorrectly. In these cases, the plunger strikes the typeface causing damage (Figure 15.11). The metal or plastic coating on the wheel can also flake off causing defects. Each spoke contains a single letter or character and, thus, there is no correlation between the defects of one character to another. As with the ball elements, defects on typewheel are easily cured when the wheel is replaced.

In the typewheel mechanism the scope of misalignments, particularly in the vertical direction, is limited. For the most part, their spacing is very consistent within a given machine. When misalignments do occur, they are small or limited in number. It therefore can be difficult to assess the significance of the misalignments found.

The amount of impact generated by typewheel typewriters in creating impressions is the least of all three classes of machines. The impressions are created as the spoke of the wheel is pressed into the paper by a plunger. It creates a flat, uniform impression (Figure 15.12). There is virtually no reverse embossing, compared to impressions generated by typebar machines and some typeball machines.

Along with the introduction of typewheel machines came the age of electronic typewriters. With electronic typewriters, there are few mechanical actions. Almost all of the typewriters' actions are controlled by the electronic unit. As an example, most electronic typewriters will automatically change the spacing when a wheel is inserted that has a

Figure 15.12 Strike-up of a typewheel machine, without the ribbon in place, to demonstrate the flat, even impression.

different pitch than the previous one used. Because the electronic machines contain fewer mechanical parts, they tend to produce very consistent finished products at high rates of speed.

As a result of this speed and efficiency, along with the electronic typewriter came the advent of word processing. A word processing system reduced to its fundamentals is the combination of a typewheel typewriter and information storage, editing, and search systems. These memory storage devices can be either magnetic cards, tapes, disks, or solid-state memories. Of those typewriters still being manufactured today, the vast majority are electronic typewheel machines.

15.4 Classification System

As previously stated, the classification of a typestyle can be an integral part of the examination of a typewritten document. Early systems by which to classify typestyles were designed by Ordway Hilton,[4] David Crown,[5] Gerry de la Durantaye,[6] and Richard Totty.[7] In the early 1990s a computer-based classification system, TYPE, was designed by Dr. Philip Bouffard.[8] It makes use of up to 40 features by which to classify a typestyle.

The primary specimen sources for this system are the Haas atlases. These atlases are a collection of typestyles compiled by Josef and Bernhard Haas.[9,10] They are written in German and are organized by typeface or type element manufacturer. Also included are dates of introduction of the typefaces, as well as the dates and serial numbers of the typewriters from which the impressions were taken (Figure 15.13). There are tables that cross-reference the manufacturer and typewriter usage of ball and daisy wheel elements.

Differentiating typewriting can be broken down in several ways. There is the size of the typeface, e.g., pica or elite. There is also the basic design, which can be monotone, script, or shaded. Monotone type is the most conventional style of typeface design, wherein all proportions of the characters are of equal weight and, where appropriate, the characters have serifs. Additionally, it can be separated by the mode of spacing — single unit or proportional.

A major concept of TYPE is that different typestyles make use of different letters as the basis for their classification. That is, rather than using the same letters, numerals, etc., for all typestyles, TYPE uses different features for each typestyle. This permits greater flexibility and more efficient classification.

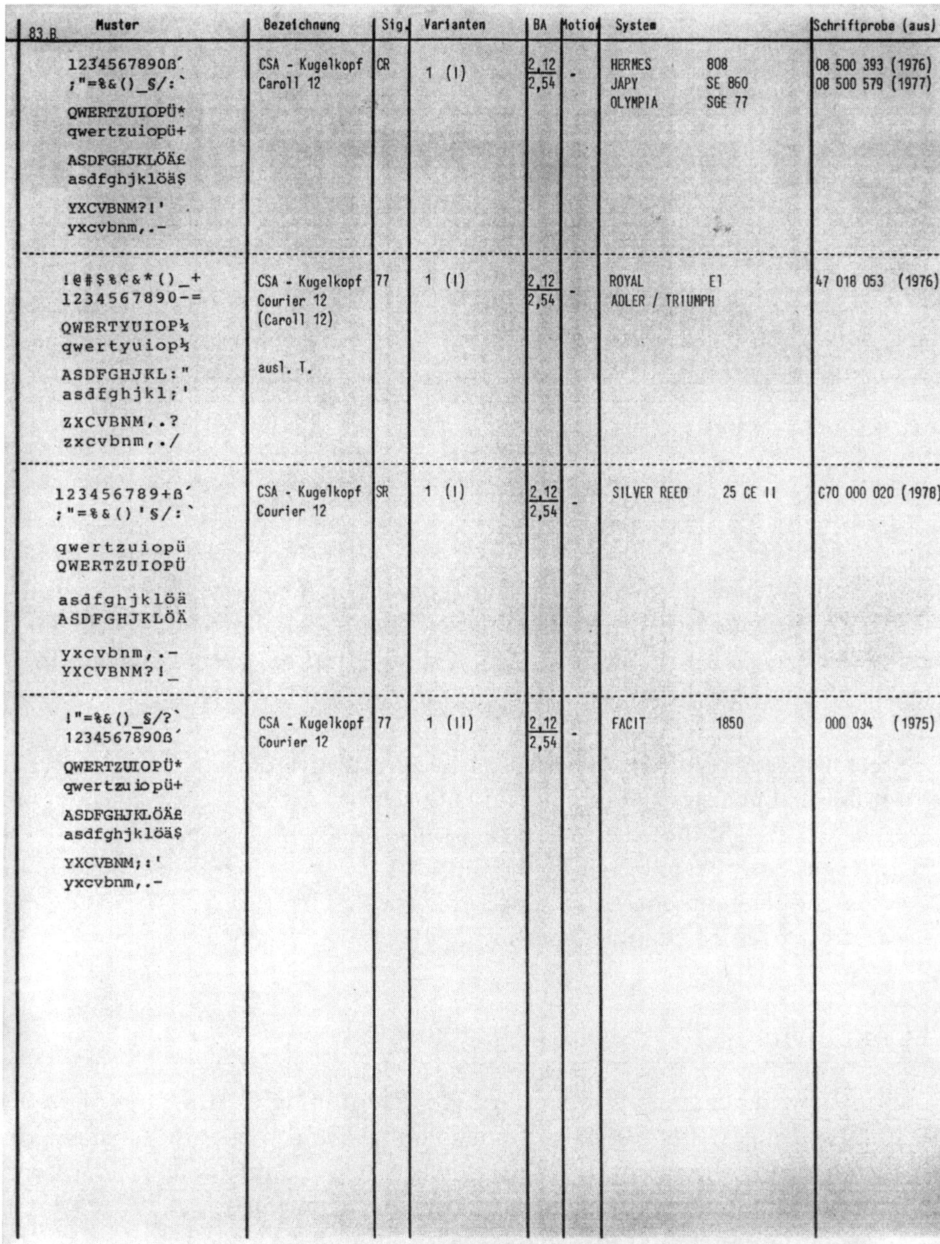

83 B Muster	Bezeichnung	Sig.	Varianten	BA	Motion	System	Schriftprobe (aus)
1234567890ß´ ;"=%&()_§/:` QWERTZUIOPÜ* qwertzuiopü+ ASDFGHJKLÖÄ£ asdfghjklöä$ YXCVBNM?!' yxcvbnm,.-	CSA - Kugelkopf Caroll 12	CR	1 (I)	2,12 2,54	-	HERMES 808 JAPY SE 860 OLYMPIA SGE 77	08 500 393 (1976) 08 500 579 (1977)
!@#$%¢&*()_+ 1234567890-= QWERTYUIOP½ qwertyuiop½ ASDFGHJKL:" asdfghjkl;' ZXCVBNM,.? zxcvbnm,./	CSA - Kugelkopf Courier 12 (Caroll 12) ausl. T.	77	1 (I)	2,12 2,54	-	ROYAL E1 ADLER / TRIUMPH	47 018 053 (1976)
123456789+ß´ ;"=%&()'§/:` qwertzuiopü QWERTZUIOPÜ asdfghjklöä ASDFGHJKLÖÄ yxcvbnm,.- YXCVBNM?!_	CSA - Kugelkopf Courier 12	SR	1 (I)	2,12 2,54	-	SILVER REED 25 CE II	C70 000 020 (1978)
!"=%&()_§/?` 1234567890ß´ QWERTZUIOPÜ* qwertzuiopü+ ASDFGHJKLÖÄ£ asdfghjklöä$ YXCVBNM;:' yxcvbnm,.-	CSA - Kugelkopf Courier 12	77	1 (II)	2,12 2,54	-	FACIT 1850	000 034 (1975)

Figure 15.13 A sample sheet from the Haas Atlas. It sets out the typestyle, spacing, machine type (i.e., Kuglekopt is a ball element), and manufacturing information.

The system relies upon a series of menus. Selections are entered as to major type design and letter spacing and then as to numerals and upper- and lowercase letter designs (Figure 15.14). When sufficient selections have been made, a search of the database is initiated. Once the search is complete, possible matches are displayed. To complete the classification process, the actual strike-ups of the specimen matches located in the Haas Atlas should be compared to the questioned text.

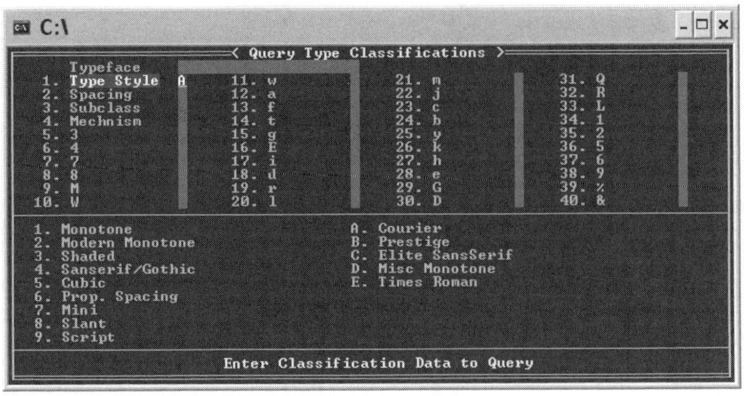

Figure 15.14 The TYPE classification query screen.

15.5 Dating of Typewriting

There are basically two means by which to date typewriting. The first is based upon absolute dates, which are provided by the manufacturers. Two examples of this absolute dating include: (1) the date of introduction of a type design and (2) design changes to certain characters in an established font. This valuable information is located in such resources as the Haas atlases.

The second means of dating typewriting is comparative dating. Comparative dating is derived from the life history of the individual machine. With use, a typewriter gradually succumbs to the effects of normal wear and tear, also known as progressive wear. Certain kinds of deterioration are reflected in the work of the machine. These include broken typeface or damaged alignment. As each occurred for the first time, a date was established. As these defects are remedied, new dates in the machine's history are established.

15.6 Conclusion

A thorough typewriter examination will include determining (1) the class of typewriter, (2) horizontal spacing, (3) typestyle and possible make and model by classification, and (4) any individualizing alignment or typeface defects. The identity of a particular machine rests on a unique combination of class and individual characteristics.

The discussion presented in this chapter is meant only to provide an overview of the subject of typewriting and to reflect the less significant role this technology now plays in the casework of the average FDE.

References

1. Bouffard, P.D., A PC-Based Typewriter Typestyle Classification System Standards, paper presented at the American Academy of Forensic Sciences meeting, New Orleans, LA, 1992.
2. Seaman Kelly, J., *Significant Dates of Modern Typing Methods*, American Board of Forensic Document Examination Monograph, 1993.

3. McCarthy, J.F., A Brief History of the Typewriter and Some Early Exemplars of Early Typeface, paper presented at the American Society of Questioned Document Examiners meeting, 1972.
4. Hilton, O., Identification of the work from an IBM selectric typewriter, *J. Forensic Sci.*, 7, 286–302, 1962.
5. Crown, D.A., The differentiation of pica monotone typewriting, *J. Police Sci. Admin.*, 4, 134–178, 1976.
6. de la Durantaye, G., A Classification System for Typewriting Specimens, paper presented at the American Society of Questioned Document Examiners meeting, 1973.
7. Totty, R.N., Hall, M.G., and Hardcastle, R.A., A computer based system for identification of unknown typestyles, *J. Forensic Sci. Soc.*, 22, 65–73, 1982.
8. Kelly, M.W., The significant contributions of Dr. Philip D. Bouffard to the examination and classification of typewriting, *J. Am. Soc. Questioned Doc. Examiners*, 7, 91–96, 2004.
9. Haas, J., *ATLAS de Schreibmaschinenschrift, Non-PICA*, 1972.
10. Haas, B. and Haas, J., *ATLAS de Schreibmaschinenschrift PICA*, 1985.

The Examination of Computer-Generated Documents 16

WILIAM J. FLYNN

Contents

One of the most extraordinary skills that we as human beings possess is written communication. Unlike a conversation, our written words allow us to transmit ideas not only over distance, but also over time. It would seem that the urge to make a record of our lives is as old as man himself. Even before there was writing, ochre and charcoal drawings in caves show our ancestors recording important daily events, such as a particularly successful bison hunt — creating, perhaps, the earliest form of textbook or, then again, maybe just bragging rights. Our progression from pictographs to graphical user interface has taken place over a remarkably short time. At no time in man's history, however, have the documents we make and print gone through more dramatic technological changes than in the past 30 years. The text of this chapter, for instance, has been created using electrons strung together as binary bits. The ephemeral world of thought has been transformed into something tangible using the equally ephemeral electric charges held in a silicon memory register. Our ancestors would have understood *why* we want to tell a story, but they could never have imagined *how*. To quote the famous science fiction writer Arthur C. Clarke, "Any sufficiently advanced technology is indistinguishable from magic."*

The history of recording our thoughts in printed form is also the story of mankind's changing technology. At almost every major technological advance there has been a corresponding change in the way we record and store our ideas. Certainly some of the earliest attempts at industrial chemistry and technological innovation came from the efforts to make better inks and dyes. The adaptation of moveable, recastable, hot metal type in the late 16th century helped to spread literacy across Europe. During the 17th and 18th centuries metallurgists gradually learned how to increase the strength of iron and steel. The typeface designers quickly took advantage of the harder metals by creating new type fonts with finer detailing and smaller sizes. Because of the greater readability of these newly created alphabets, more letters could be strung together on a page, thus saving paper and ink costs for the printers and publishers. As printing costs were reduced, the distribution of the printed word on pamphlets, flyers, and, of course, books began to spread among the population. The information technology of the day continued to move in lockstep with general technological innovation.

*Taken from Clarke, A.C., *Profiles of the Future: An Inquiry into the Limits of the Possible*, Harper & Row, New York, 1962, which was based on essays written between 1959–1961.

The Industrial Revolution, beginning in the 18th century, saw the first efforts of inventors to put mechanical printing devices in the hands of the public. Although the first English typewriter patents were issued to Henry Mill in 1713, it does not appear that a working model of his design was ever made. Mill is given credit, however, for his attempt in 1713 to write the longest sentence seen at the English Patent Office: "An artificial machine or method for the impressing or transcribing of letters singly or progressively one after another, as in writing, whereby all writing whatever may be engrossed in paper or parchment so neat and exact as not to be distinguished from print."* *Nearly a century later* in 1808, Pellegrino Turri produced at least one functional machine.** The first significant production of typewriters, however, was the Sholes and Glidden machine in 1874.*** The QWERTY keyboard layout (which we use to this day) was designed for these early Sholes–Glidden machines. There are at least two historical reasons given for the unusual key layout. The first reason for the QWERTY design is said to have come about in order to deliberately slow down a typist so that the mechanical linkages of the typewriter could keep pace. The second reason (probably apocryphal) for this layout is rumored to have been for the convenience of the early typewriter salesmen. The layout made it easy to demonstrate the machines by typing the word *typewriter* — all the letters of which fall on the top line of keys.

In all, the history of printing has probably gone through four distinct technological phases:

1. The invention and adaptation of moveable, recastable metal type (ca. 1450 to 1500)
2. The Industrial Revolution: the use of steam-powered mechanical presses, the invention of Merganthaler's Linotype, and the office typewriter (ca. 1870 to 1975)
3. The adaptation of photographic processing by the printing industry — flexography, photogravure, offset, and electrophotography (xerography) (ca. the late 1800s to the present)
4. The digital era: computer typesetting, electronic fonts, and graphical user interface (ca. 1973 to the present)

This chapter deals with the last episode described above: the way modern digital computers and output devices produce data on paper. Specifically, the text will focus on how devices used for computer output place their images and text on documents and what information can be acquired from a forensic examination of these items.

16.1 Computer-Generated Documents and the Forensic Examiner

In the late 1960s it was considered mandatory in the field of questioned document examination to read Ordway Hilton's, Dr. Phillip Bouffard's, and Dr. David Crown's treatises on the examination of typewritten documents (see Section 15.4). At that time the business community was just upgrading their typewriters from manual and electric single-font machines to the new IBM Selectric typewriters with interchangeable type

*From patent 385, which was granted by Queen Anne in 1714 to Henry Mill. Details of Mill's machine are not known.

**Turri's machine was reportedly made for his blind friend, the Countess Carolina da Fivizzono.

***Christopher L. Sholes was a Milwaukee newspaperman, philosopher, and inventor. Glidden was a member of the Kleinstuber Machine Shop gang, where the early typewriter development work started in 1868.

elements. For the first time, the identification of a typewritten text required the identification of a typing system, rather than just the typewriter. Since defects could occur on the type element or in the typewriter itself, it became necessary to contend with the fact that an element could be moved from machine to machine or a machine could use multiple elements. It was, therefore, the combination of typewriter and interchangeable element that was identified as a system.

In the space of just 40 years we have gone from the Selectric — arguably the most technologically advanced typewriter of its day — to the computer age. A document creation system that had just two parts in 1961 (typewriter and element) can now have multiple components — some based in software and some in hardware. The modern FDE must be aware of such things as the operating system, the word processing program, the version of the word processing program, the digital font and version of the font file, the printer driver and its release date as well as the connection to the printer (stand-alone or networked). All can have an impact on the appearance of the printed text, as can the many variables associated with the printer itself.

Make no mistake about it — this is a complex area that requires some knowledge of typography, digital fonts, computer systems, and software. All of these topics could (and have) filled their own textbooks. This chapter provides a boiled-down version of what is believed to be the most important information for those engaged in the forensic examination of computer-generated documents.

We live in the Information Age, a time when information is being generated, published, and stored at an ever-increasing rate, and computers play an integral role in all three of these activities. Perhaps the quote that best illustrates how rapidly things have changed in the past 500 years comes from Martha Beck in the April 2002 edition of *O Magazine*: "Information is King. The Sunday edition of the *New York Times* on average contains more information than was printed in the whole of the 15th century."

16.1.1 The Pre-Examination Evaluation

The examination of typographically prepared documents should begin with the same precautions and care that would be prudent with any type of examination. Obviously, if the documents are to be processed for latent fingerprints, document examiners must preserve the integrity of the evidence. If your laboratory is equipped to perform destructive analyses of the ink, toner, paper, glue, or other parts of the document, then non-destructive tests should be completed first. In this regard, the pre-examination evaluation of typographic documents is no different than document examinations of any type.

16.1.2 Examination Procedures

An examination will often be conducted to help answer the questions that are at issue in the case. Some of these questions are likely to include:

1. Is the document an original or copy?
2. Can the printing technology be identified? Dated?
3. Has more than one technology been used to prepare the document (letterhead printed on a lithographic press — the variable data printed with dry toner)?
4. In a multi-page document, is there evidence that one or more pages are prepared differently than the others?

5. Is there any evidence that text has been altered or added to the document?
6. Are there any obvious (gross) defects in the printing that might lead to the machine being individualized?

In general, the well-established principle that documents should be thoroughly scrutinized on both sides, corner to corner, is just as valid for modern, computer-generated documents as it has been since the dawn of forensic document examination.

16.1.2.1 Is the Document an Original or a Copy?

This is a question that may seem trivial to a new FDE, but daunting to the experienced practitioner. The following examples show why it may be difficult or impossible to determine if the evidence is an original machine-printed document or a machine copy:

Example 1: Modern computer technology can blur how we define an original vs. a copy as well as the physical distinctions between an original and a copy. For example, it is possible for multiple original versions of the same electronic document to be printed on different printers. This occurs daily in the modern world when e-mail attachments are printed out by the recipients. A preparer (say someone in corporate headquarters) sends a policy change to individual employees in branch offices, who then print out the text on their machines. Each can lay claim to having an original document, even though some originals may have been printed on inkjet printers and some on laser printers.

Example 2: Many offices, particularly small businesses, utilize multi-function machines. These devices have become enormously popular because one machine may take the place of a printer, copier, scanner, and facsimile machine. From a forensic standpoint, the problem is that the same machine that was used to print an original document may later be used to copy it. Many of the multi-function devices use inkjet technology, so that copies made on these machines are not traditional dry, xerographic toner photocopies. It can be easy, therefore, to mistake an inkjet copy for an original inkjet-printed document.

Example 3: A document originally printed on a laser printer, then copied on an inkjet multi-function machine, and subsequently copied again on a traditional xerographic photocopier may appear to be a xerographic photocopy of an inkjet original.

Example 4: Original machine printed documents *can* have machine-rendered signatures. Several companies can take one's original signatures and convert them into TrueType fonts. Because these signatures are scalable fonts, they can be smoothly resized, bolded, and italicized to give them visually different appearances. A toner or inkjet signature appearing on a document, therefore, is not necessarily proof that the document itself is a reproduction.

Since the evidence of an original document vs. a copy can be somewhat ambiguous, it is probably best to simply refer to a machine-rendered document in terms of the technology used. For instance, "Item 1 is a six-page contract produced on an office machine that utilizes (1) dry toner, (2) inkjet technology, etc." In addition, it is probably best to avoid terms like *laser-printed* or *bubble jet printed*, as these refer to specific subclasses of toner and inkjet technology.

16.1.2.2 Can the Printing Technology Be Identified?

Conclusions that are drawn from forensic document examinations usually progress from the general to the specific. It is a common practice for document examiners to step through their examinations attempting to first determine class characteristics, followed by efforts to ascertain more individual, identifiable features. Following this formula, the starting point for an examination of a computer-generated document will usually involve a microscopic examination of the printed text in order to determine the most general type of evidence — what technology was used to print the document.

Efforts to identify the printing technology can be thought of as attempts to *classify* the printing process. The classifications that can be made from visual (microscopic) examinations will initially revolve around three basic determinations: Has the document been printed (1) in black and white or color, (2) using an impact or non-impact process, or (3) with toner, wet ink, or other medium? When you use these simple classifications during an initial screening of the document, keep in mind that they are not *mutually exclusive* categories. That is to say, a text can be printed in color, using dry ink (toner) on a non-impact printer. As you work your way through the technology descriptions and sample images below, remember that not all of the printing processes fit neatly into the limited choices above. The dye sublimation process, for instance, utilizes ink that goes on the page as a gas. You also will see that some printers are hybrid combinations utilizing more than one technology.

Table 16.1 is a summary of the major types of computer printing technologies. In addition, the appendix to this chapter provides a summary of some common causes of laser printer output defects. As is always true, there is no substitute for live samples. It is highly recommended that efforts be made to acquire original known samples of the printer technology you are attempting to identify.

16.1.2.3 Has More Than One Technology Been Used to Prepare the Document?

In some instances it may not be possible (without chemical or instrumental analysis) to determine much beyond the technology that was used to produce a computer-generated document. Depending on how the document was allegedly produced, however, this may be all that is necessary to resolve the issue. The two types of cases where this information can be of considerable importance are reinsertion and page substitution. In the first situation, a document is placed back into a printer after the parties have signed and agreed to the terms in the document. If a questioned passage is printed with a different type of printer than the surrounding text, it can be considered proof that the document was changed by reinsertion. An example of this occurred in a patent case, in which all of the unquestioned text had been printed on a dot matrix printer while the questioned assignment of the patent had been inserted with an inkjet printer at the bottom of a page that began with dot matrix printing. In the second type of case (page substitution), one or more pages of the original document are removed and different ones are inserted. This situation arises frequently in probate matters where the signature page of a will is left intact, but the preceding pages are replaced. Depending on what stories the various parties tell concerning the creation of a questioned document, merely being able to determine that more than one printer technology was used may be sufficient to resolve the litigation.

Table 16.1 Major Types of Computer Printing Technologies

Technology	Liquid Ink	Dry Toner	Thermal	Other	Impact
High-speed mainframe	Through ribbon				Yes
Daisy wheel typeball	Through ribbon				Yes
Dot matrix	Through ribbon				Yes
Inkjet	Yes, ballistic (may be more than 4 colors)				No
Laser/LED monochrome		Yes			No
Laser/LED color		Yes, 4 colors			No
Magnetography (similar to electrophotography, but with a magnetic drum)		Yes, ferrous (magnetic)			No
Thermal wax/pigment			Dry wax/pigments melted onto page		No
Dye sublimation			Resinous inks heated to a gaseous state (sublimated)		No
Thermal paper			Heated pins in ceramic head produce characters much like dot matrix		No
Solid-ink/thermal jet Xerox/Tektronix			Dry ink "stick" melted and drops sprayed onto drum	Solid ink sticks, 4 colors	No
Hybrid digital offset HP Indigo and Heidelberg	Yes, liquid toner up to 7 colors			Blanket impression	No

16.1.2.4 Is There Evidence That One or More Pages Are Prepared Differently Than the Others or That Text Has Been Altered?

In the above section we discussed the possibility that different printers were used within the same document. Here, we will deal with situations wherein entire pages have been substituted within a document. The difference in these examples from those above is that the text on substituted pages may actually be printed by the *same* printer as was utilized to prepare the original text. If there is a possibility that page substitution has taken place, such things as font changes, formatting, paper type, etc., must be considered. In general, a forensic document examination that is conducted to determine if entire pages or lines of text have been added or removed from a document is, in a word, an analysis of *consistency.* Although many parts of a document can be checked for consistency (spelling, punctuation, staple holes, formatting, etc.), these analyses typically boil down to three different types of examinations: technological, chronological, and "Does the story fit?" Although it is often true that a forensic scientist should enter an examination without much information about the factual issues of a case, there are exceptions — and potentially altered documents are one of the big exceptions. This is definitely an area where knowing the story about the document's creation, in advance of the examination, can be quite helpful. Following are some reasons why.

Evidence that pages in a multi-page document have been created differently may or may not be evidence of tampering. There are some perfectly logical reasons why pages in a long text are formatted differently, are on different paper, or even printed on a different printer than the surrounding text. For instance, some documents are regularly modified or appended. These may include bylaws, contracts (especially real estate contracts where commission rates, governmental regulations, and fees change often), forms relating to taxes, procedure manuals, and, in general, documents that only need to have sections revised over time (rather than the entire document having to be rewritten).

Another consideration involves the use of boilerplate language. If certain long phrases (such as disclaimers) are used in the creation of, say, new contracts, it is possible that these passages are being electronically cut and pasted from an older document into the one being created. It is not unusual for the original formatting and fonts used in the boilerplate to remain intact after they have been pasted into the new document — the point being that a sudden change in the typeface or spacing characteristics of a page may not necessarily be evidence of alteration or addition. Again, this is another reason why a detailed statement as to how the document was created is such an important piece of information to be obtained before a forensic conclusion is issued. If there is testimony that the document was typed in its entirety (all in one sitting), for instance, being able to prove that part of the document is inconsistent with the other can be highly significant.

Finding chronological inconsistencies can be a powerful tool in discrediting pages within a document or, in fact, the entire document. These examinations are often complex and can require extensive research to establish dates of introduction for the printing technology, paper production (watermarks), business form or letterhead printing dates, etc. On the other hand, discrediting a document may be as simple as showing that an office or individual did not have access to a particular type of printer on the date the document was allegedly produced. As an example, a forensic examination may determine that a four-color toner process was used to render a document, but an investigation proves that an office had not yet purchased its color laser printer at the time the questioned item was supposed to have been printed. It is not, of course, the job of the forensic document examiner to prove access to a particular machine, but rather to identify as thoroughly as possible the machine/process used to prepare the document and then let the case investigators handle the fieldwork.

Virtually every aspect of a document can have some sort of anomaly. FDEs are well aware of the importance of checking staple hole patterns, paper types, watermarks, indentations, printed form numbers, etc., as part of their routine casework. Any of these typical examinations may show evidence of an addition or alteration. In the case of computer printing technology there are several approaches that one can take to help determine if text (or entire pages) have been added, removed, or altered. The first step should be an attempt to determine if the same printing technology was used throughout the document.

In this regard, literally everything that is printed on the document should be examined, including the printing process used to prepare the letterhead.*

Even if only one technology was used to create a document there may be evidence that passages were created on different machines. In some cases it may be possible to make this determination non-destructively; in other instances only destructive testing such as ink and toner analyses will provide definitive evidence in this regard. Infrared examination of the ink on an inkjet-printed document may well show that a questioned passage was printed with an ink that reacts differently than all of the other ink in the document. A word of caution is necessary here concerning infrared examinations of inkjet-printed text. If passages in a document react differently in reflected or infrared luminescent examinations, care must be taken to ensure that both texts are printed in black ink only (rather than composite black, in which combinations of cyan, magenta, and yellow inks are intermingled with the black). Before making a determination that two inkjet inks are different, a thorough microscopic examination should be made to ensure that the comparison is black ink to black ink.

On a more subtle level, the overspray patterns (satellite drops) that occur as an artifact in most inkjet printings may also provide useful evidence of a printer change (Figure 16.1). Most modern inkjet printers are capable of printing the font sizes found in typical documents (about 10 points to 14 points) as multiple lines in a single pass. In other words, in one rightward or leftward pass of the inkjet head, two, three, or even four lines of text may be printed simultaneously.

The rule here is that the printer cannot print more lines in one pass than the vertical size of the nozzle array in the printhead. For example, if a black-ink print nozzle array is 48 points high, then only combinations of text and line spacing of 48 points or less can be executed in one pass. Because the number of inkjet nozzles on printheads has increased over time, the vertical spacing of consecutively printed lines of text may itself be a dating phenomenon (see color Figure 16.2 following p. 366). The micrograph in color Figure 16.3 (following p. 366) shows a Hewlett-Packard 960C black printhead. The nozzles are in groups of three and four, left to right.

The fonts and line spacings should be carefully examined. There are many look-alike fonts that contain only subtle differences — Times New Roman and CG Times are good examples.** In addition, the default line spacings of computer-generated texts have, in general, slowly decreased over time as the output resolutions of the printers have increased.*** What had once been default single-line spacing for a 12-point Times New Roman font (14.4 points of leading) may now be only 13.5 to 13.8 points. The difference in line spacings of just a few 720ths of an inch require careful measurements, but are certainly possible (and probably mandatory) in a modern forensic laboratory.

Of course, the media on which the document is prepared may provide information as to the possibility that pages have been added or substituted. The standard forensic examinations of paper watermarks and other identifying characteristics of the media (such as

*The author has had a backdating case wherein the perpetrator used a Selectric typewriter to prepare the text of a document, but had used an inkjet copier to reproduce a 1960s' letterhead.

**See example at the end of this section.

***As electronic font outlines have been rendered with greater fidelity, it is easier for the eye to distinguish the letters. The higher-output resolutions have allowed the lines of text to be moved closer together and still remain readable.

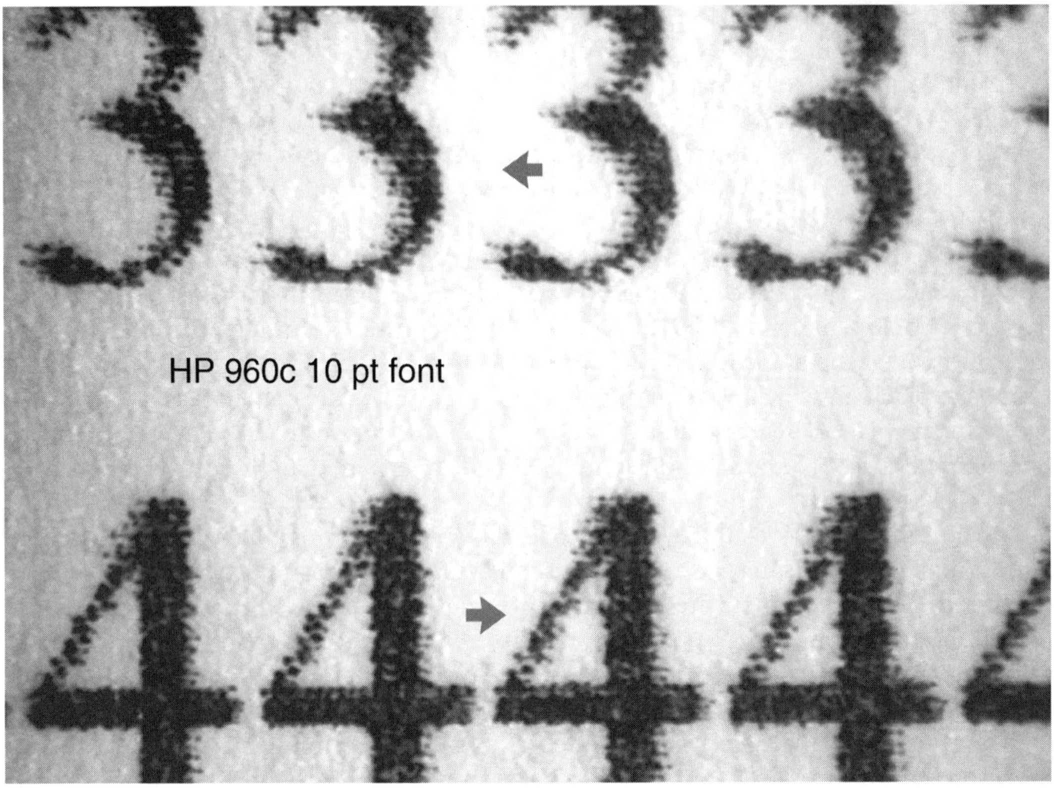

Figure 16.1 Close-up image of the overspray patterns cause by the leftward and rightward movement of an inkjet printhead. This photomicrograph illustrates the alternating overspray pattern.

form numbers, letterhead data, ultraviolet characteristics, etc.) should always be conducted if the possibility of document alteration exists.

In summary, printing anomalies that occur in a document may or may not be proof that it has been altered or appended. There are some examinations, however, that may show conclusively that an entire document or selected pages are bogus. Finding a page printed on a laser printer in Courier New typeface, mixed in with pages typed in Courier on a Selectric typewriter (when the entire document was supposedly prepared in 1969), would certainly prove that portions were backdated. If the computer and printer that allegedly produced the document still exist, standards should be obtained and compared to the questioned material. Although it is sometimes better not to have too much factual information about the documents being examined, it is highly recommended that in these types of cases an FDE obtain as much detailed information about the alleged conditions and equipment surrounding the creation of the questioned document as possible.

16.1.2.5 Is There Any Evidence That Text Has Been Altered or Added to the Document?

In the case of computer-generated documents, alterations or added texts are generally going to be detected by an examination to determine if the document has been reinserted back into a printer. In short, an examination to determine if text has been added or altered will be similar to a test for page substitution in that it is an analysis of the document's consistency. Any discrepancies in the line spacing, printer technology, fonts, or layout in the questioned text may be proof that it was added by printer reinsertion.

16.1.2.6 *Are There Any Obvious (Gross) Defects in the Printing That Might Lead to the Machine Being Individualized?*

Gross printer defects are probably a fairly rare occurrence, but they do happen. One of the problems with some of these defects is that they can be specific to a particular ink cartridge, toner cartridge, imaging drum, or other replaceable component. A defect or set of defects, therefore, may occur on a machine for some period and then disappear when the expendable component is replaced. It is also possible for a defect such as a clogged inkjet nozzle to clear itself during operation. For this reason, it is probably best to get samples of the printer's output during the same period as the questioned document rather than rely on machine standards taken at some later period.

Some manufacturers, like Hewlett-Packard, make their printers' repair manuals available for general purchase. These manuals provide a wealth of information, including common defects in the printing output and the steps that are necessary to correct them. If a manual for a particular printer is unavailable, another potential source for information is the local printer repair shops. The technicians who regularly work on these printers can often provide their direct, hands-on experience with all types of defects and typical printing characteristics associated with various manufacturers' machines.

Lastly, the manufacturers themselves undoubtedly have the best information about their products, but are often hesitant to get involved in a criminal or civil case or to talk freely about the defects in their own printers.

16.2 Descriptions and Histories of the Technologies

16.2.1 High-Speed Mainframe Line Printers: Drum and Chain Printheads

This class of printer was first developed in the early 1950s for use with the huge mainframe computers that were just beginning to come online. There were two main types of line printers: drum printers and chain printers.

The IBM model 716 that sold from May 1952 through July 1969 was the most widely distributed of the drum printers. In this class of printer, the print mechanism consisted of 120 type wheels (disks) arranged together as a cylinder (drum) behind the ribbon. These wheels spun into the strike position and printed an entire line at a time. At full speed these printers could produce about 150 lines per minute. The model 716 print wheels had 26 (all capital) letters, the numerals 0 to 9, and nine special characters, such as the dollar symbol, period, and slash. The most widely used line printer, however, was the IBM model 1403 introduced in October 1959 and discontinued in 1971.* This printer utilized a rotating chain with characters affixed to one side. There were 240 characters affixed to the chain, usually arranged in 5 groups of the same 48 characters. The impact mechanism consisted

*Line printers were big and very noisy. After spending long hours listening to the 1403s print huge batch jobs, some data processing engineers realized that printing certain character strings produced musical pitches. By laboriously keypunching data cards in the proper sequences and for the right duration, the 1403 printer could actually play recognizable songs. For samples of 1403 music recorded at the Richmond California Unified School District in 1970, go to http://www.computerhistory.org/exhibits/highlights

of 132 hammers arranged much like the hammers over the strings in a piano. The hammers were used to strike the back of the paper against an inked ribbon and the spinning chain at exactly the right moment to impress the characters onto the page. An examination of the text should show the typical pattern made by a metal character striking a fabric ribbon. The 1403 was typically fed 11 × 14-inch fanfold (green-bar) paper and could print up to 1400 132-column lines per minute. All of the chains that were used by the 1403 contained monospaced characters. At some point in time during the 1960s, however, special chains became available that utilized a Courier font in a full set of upper- and lowercase letters, numbers, and symbols. These special chains could be ordered with as many as 120 different typographic characters arranged on the chain in two groups. These special chains and the 1403 printer became a particularly popular way for companies to produce their own technical/policy manuals. As of 2004, the Lexmark Division of IBM was still selling nylon ribbons for 1403 printers.

16.2.2 Typewriter Technology Printers: Typeball and Daisy Wheel

The historical record seems to be somewhat unclear as to when the first IBM typewriter was modified to function as an external computer printer. It is certain, however, that as far back as 1964 IBM had released a machine called the Magnetic Tape Selectric™ that was a dedicated word processing system. This system utilized technology that IBM had purchased in 1955 from An Wang (founder of Wang Laboratories). In this system the Selectric typewriter functioned as both the input device (the keystrokes being stored on a magnetic tape) and the printer when the tape was played back.

In 1969 David S. Lee and a small group of engineers at Diablo Systems developed the first efficient daisy wheel printer. In 1972 Diablo was sold to Xerox and Lee formed a company that would produce one of the best-known letter-quality printers of its day — Qume. In 1985, IBM released its version of the daisy wheel computer printer, the ProPrinter I. Although IBM went on to manufacture several later versions of the ProPrinter, the technology had already become obsolete by that time because of the growing use of both laser and inkjet printers.

16.2.3 Dot Matrix Printers

Dot matrix printers are technically described as impact matrix printers. This distinguishes them from non-impact matrix printers such as inkjet, thermal printhead, and even laser — all of which create text and images as a series of closely spaced dots. The earliest development work on wire printers appears to be that of Reynold B. Johnson at IBM in 1949. By the mid 1950s both Burroughs and IBM had developed high-speed wire printers for use with mainframe computers. These machines turned out to be highly unreliable and were never a commercial success. In 1968 Seiko Electronics manufactured a small dot matrix printhead. This unit was used to print out hard copies of timed athletic events recorded by Seiko's electronic timers at the 1972 Olympic Games in Tokyo. This small printer was known as the EP-101. Ten years later, in 1978, EPSON (SON of Electronic Printer) used the EP-101 dot matrix technology to market the first (under $2000) consumer dot matrix printer — the EPSON TX-80. The TX-80 never sold well, however, and the first commercial success for dot matrix technology came 2 years later in 1980, with Epson's MX-80 (the 80 referred to the number of columns it could print). Because of the shear number of MX-80 printers that Epson sold, the MX-80 printer control language, Epson

ESC Codes, set the standard for dot matrix printer control language as well. The Epson ESC codes were the *de facto* language for sending ASCII characters and formatting codes to dot matrix printers until graphical user interfaces and scalable fonts became practical.

The original Epson dot matrix printers used a nine-pin vertical array for their output. Several short-lived models had 18-pin configurations that were used in near-letter-quality (NLQ) mode. These printheads had two closely spaced columns of nine pins. The close spacing of the pins tended to fill in the white space between the dots and gave the characters a more solid quality look. By 1982 Epson had released a 24-pin dot matrix printer utilizing a head with two columns of 12 pins. Both 9- and 24-pin dot matrix printers continue to be sold today and represent about 6 to 8% of the overall printer market in the U.S. Dot matrix printers are used primarily in settings where carbon or NCR copies require an impact process, or where very low output costs are a factor.

A second class of dot matrix printer, developed in 1974 by Printronix for use with large computers and very high speed output, is known as line dot matrix printers. The fastest currently available line matrix printers (the Tally T6218 and Printronix P5220) claim to have an output capacity of 1800 to 2400 lines per minute. The line matrix printers utilize an entire row of magnetically fired pins (or, in some cases, several independent heads that shuttle side to side). The characters in the print run are formed as the paper advances vertically while the pins fire horizontally.

Line matrix printers are often used in applications such as bulk mailings and where it is necessary to fill in preprinted forms. The cost per page is the least of all printers — about a 10th of that of a laser printer.

16.2.4 Inkjet Printing Technology

There are two distinct types of inkjet technology: drop-on-demand and continuous drop. Although the continuous-drop inkjet printers were the first to be used commercially, the drop-on-demand printers are by far the more common today. Within each of the two main categories of inkjet technologies, there are many variations and alternate ways in which liquid drops are created and shot onto a printable surface. Inkjet printers can be tiny, portable, battery-powered devices or room-size commercial printers like those made by Inca™, Scitex™, and Heidelberg™. The very large inkjet printers are typically used for printing billboards, truck-side graphics, and fabrics. These behemoths of the inkjet world are rated not in characters per second, but in their coverage — 100s of square meters per hour.

Although they are rarely heard of outside of the industries that use them, there is another type of continuous-drop technology known as valve-jet printers. As the name implies, these are not much more than a series of in-line nozzles that are controlled by precisely timed valves. Valve-jet printers are usually used to apply precision coatings, rather than to function as printers in the classical sense.

Large, continuous-drop inkjet printers are typically used in high-speed, high-volume environments, rather than on standard correspondence. A document examiner might encounter their output on boxes or containers in the form of the printed product manufacturer and product information (including the bar coding on the boxes), bulk mailing labels, and instant winner lottery tickets.

At the other end of the spectrum, inkjet technology has become so precise that tiny inkjet heads are used in the production of microcircuits to print the components on a circuit board. Development work is also under way to utilize inkjet nozzles for medical drug delivery, fuel cell applications, and propellant-free aerosols.

The core technology for continuous inkjet printers was first patented in 1948 by R. Elmqvist of Siemens Elema in Sweden. Elmqvist was issued U.S. patent 2566443 in 1951 for this same process. Siemens went on to produce the first devices to use continuous inkjet printing — medical strip-chart recorders. Work done in the early 1960s that eventually resulted in the first use of continuous-drop technology to produce text is attributed to Dr. R.G. Sweet of Stanford University. His experiments demonstrated how a pressure wave at an orifice can break up a continuous ink stream into droplets. Sweet's discoveries led to the first commercial-size continuous-drop printers, the A.B. Dick, VideoJet, and Mead DIJIT, which were released in the late 1960s. In 1976, IBM licensed the above technology and introduced the first word processing hard copy-output peripheral, the IBM 4640 inkjet printer. Except in large commercial applications, the continuous-drop inkjet printers never became a success.

Hewlett-Packard (HP) claims to have invented thermal (drop-on-demand) technology in 1979. Canon, however, has taken some issue with this and may, in fact, have produced the world's first working thermal inkjet in 1981. (Canon calls its technology bubble jet.) In 1984, HP released its first inkjet model dubbed the Think Jet. A few months later (in 1985), Canon began marketing its first inkjet printer, the Bubble Jet 80 or just BJ-80. Regardless of who actually discovered the thermal drop-on-demand principle, HP and Canon formed a close technology-sharing partnership in the mid-1980s that would later involve licensing rights to both inkjet and laser printer technology.

Because HP and Canon had quickly acquired patents for the generation and ejection of ink droplets using a heating element (see color Figure 16.4 following p. 366), Epson was forced to either pay the steep licensing fees to build its own inkjet printers or use a completely different method for generating and ejecting ink drops. Epson's solution came in the form of piezoelectric inkjet technology.

In piezoelectric inkjet systems, an electric current is applied to a material (such as piezoelectric crystals) that deforms along some axis. By pulsing the current, piezoelectric crystals can be made to flex very rapidly inside an inkjet chamber, thus forcing droplets of ink out through the orifice (see color Figure 16.5 following p. 366).

16.2.5 Monochrome Laser/LED Printers

All modern laser printers can trace their heritage back to 1939, when the young inventor and patent attorney Chester Carlson received a patent for the dry electrostatic printing process known as electrophotography (or more commonly known today as xerography.) Beginning in 1969, Xerox engineer Gary Starkweather adapted his company's xerographic copier technology to a pulsing laser imaging system and the idea for a laser printer was born. Like many rapidly evolving technologies, there is some dispute, however, concerning who went to market first with a fully functional laser printer. Siemens claims that it introduced a laser printer in 1975, while IBM believes that its 3800-1, which sold in 1976, was the first true laser printer. There is no doubt at all, however, that the Xerox 9700 laser printer that sold in 1977 was not only a fully functional laser printer, but also, at 120 pages per minute, the fastest laser printer ever built. In addition to its incredible speed, the Xerox 9700 was likely the first laser printer to work with cut paper rather than continuous-feed, fanfold paper.

Although all laser printers utilize some form of electrostatic imaging to transfer toner to the page, there are several different variations on this theme. By far the most common method that is used to place images and text on a page is known as paint-black. In this scenario the laser paints the images and text onto the imaging drum and an electrostatic charge holds the toner with just enough "cling" to transfer the data to a page (Figure 16.6). There are also, however, paint-white laser printers in which the drum is fully charged and by default will be covered in toner except where it is discharged by a laser. This latter type of laser printer is utilized most often in graphic arts applications and advertising, where large black areas are more likely to be found. Finally, not all laser printers actually use a laser to form the images. Light-emitting diodes (LEDs) and liquid crystal displays (LCDs) have been employed to form the images on the photoconductor in an electrostatic printer. Lastly, Olympus pioneered a so-called ion deposition electrostatic process in 1994.

16.2.6 Color Laser Printers

The first sub-$50,000 color laser printer was introduced in June 1993 by QMS and was known as the Colorscript Laser 1000. About a year later Xerox introduced its own model, the Color Laser 4900, that, like the QMS printer, utilized an engine made by Hitachi. The following year both Apple and Lexmark introduced their own color laser printers using engines manufactured by Canon.

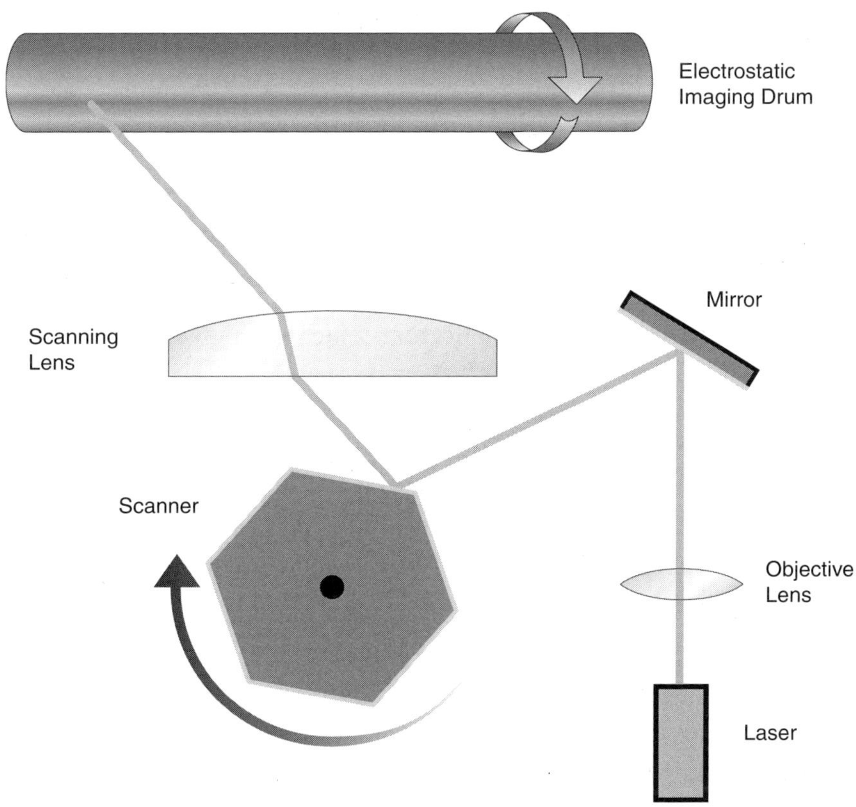

Figure 16.6 Schematic of an imaging unit in a typical laser printer. The scanner is a spinning mirror or prism that plays the laser beam over the surface of the electrostatic drum.

The imaging technology used in color laser printers is identical to that in monochrome printers, except for the addition of cyan, yellow, and magenta toners (see color Figure 16.7 following p. 366). By far, most laser printers utilize print-black technology wherein the laser "paints" the text/image onto the imaging drum. Some specialty lasers, however, using the so-called print-white process (described above), transfer toner except where it has been struck by the laser beam. The basic steps involved in a typical electrophotographic printing process are as follows:

1. An imaging drum or belt (which has the ability to hold an electrostatic charge but then have this charge varied by exposure to light) is first charged by a charging roller.
2. A laser beam or the light from an LED array is focused onto the imaging drum — painting the text or images onto its surface. The light beam either discharges the imaging drum or reduces the electrostatic charge on its surface.
3. Toner is electrostatically charged and rolled over the surface of the imaging drum. Typically, the areas on the drum that are still fully charged repel the toner, while the areas that have been discharged attract the toner.
4. As blank paper is fed from the paper tray, it is given a strong enough electrostatic charge to pull the toner from the imaging drum, thus transferring the image painted by the laser beam onto a sheet of paper.
5. A combination of heat and pressure are utilized in the fusing section of the printer to permanently fix the image onto the paper.
6. The imaging drum is recharged and the process repeats.

16.2.7 Magnetography

Magnetographic printing was first commercially introduced in 1984. In this process a magnetic imager places a latent magnetic image on a metallic imaging drum. An iron-rich (magnetic) toner is then spread across the drum forming the actual image. Finally, the image is transferred to the paper and fused to make it permanent. These printers are typically found only in commercial applications, such as high-speed bar coding, bulk mailings, and labels.

16.2.8 Thermal Wax/Pigment Printers

Thermal wax transfer printers function by melting a wax-based resinous ink onto paper. As the paper and wax-coated ribbon move beneath the thermal printhead, heated resistors melt the wax onto the paper. When cool, the wax fixes to the paper. Monochrome printers have black-only panels for each page to be printed, while color printers have either three (CMY) or four (CMYK) colored panels for each page. Like all process color printers, thermal wax machines rely on a dithering* algorithm to produce the appearance of full color. One problem associated with these printers is the high cost of full-color prints. This is due to the fact that an entire set of color panels is used for each page even if the image only requires the use of a single color (see color Figure 16.8 following p. 366).

*Dithering is the digital equivalent of halftoning, wherein the appearance of a full-color image is produced by using only four colors and the white paper background. Unlike halftoning, however, dithering patterns may not have regular dot patterns.

Thermal wax printers are in general very reliable, thanks to the small number of moving parts. Because of their reliability and the fact that the waxy inks are waterproof, they are often used to print mailing labels, bar codes, and bulk mailing envelopes. Printouts from these printers are sensitive to abrasion, however, as the waxy ink can be scraped or rubbed off or smeared. Some plain-paper facsimile machines also rely on thermal wax transfer technology. The most common industrial thermal printers utilize a 203-dpi (dots per inch) printhead, but heads are also available in 152 to 400 dpi. Thermal wax and thermal pigment printers are often used to produce the "photograph" of an individual on security documents such as passports and driver's licenses.

The earliest thermal wax transfer device was not actually a printer, but rather a typewriter — the IBM Quietwriter 7, which IBM released in 1984. The following year IBM marketed the Quietwriter Printer, which was designed to work with the IBM PC. The original Quietwriter Printer had a 40-electrode head that was capable of printing at a resolution of 240 × 360 dpi.

16.2.9 Dye Sublimation Printers

A dye sublimation printer is a computer printer that uses a thermal heating head to transfer dye to a medium, such as a plastic card, printer paper, or poster paper (see color Figure 16.9 following p. 366). Unlike the thermal wax process, however, the dyes on the plastic panels transfer to the paper as a gas. The term *sublimation* refers to the process whereby a solid (when heated) goes directly to a gaseous state without the intervening liquid phase. Most dye sublimation printers use CMYK colors. Many consumer and professional dye sublimation printers are designed and used for producing photographic prints. Because the amount of dye material transferred to the paper can be controlled to a much greater extent than in the thermal wax process, the intensity of colors can be regulated. In addition, the colors can actually be blended rather than dithered into much more subtle hues than the dithered images in thermal wax printing. In some commercial applications a process known as secondary dye sublimation transfer is used. In this type of printing the dyes are heated once to transfer an image to an intermediate carrier, and then heated a second time to move the image from the carrier to the paper. In some high-resolution applications, such as medical imaging, lasers rather than thermal printheads are used to sublimate the resinous dyes.

Because the dyes used in these printers can be made to adhere to plastic as well as paper, the dye sublimation printing process is often found on identity documents that require a photograph. A document examiner should keep in mind that the used CMYK panels in these printers and in the thermal wax printers retain the images of the material they have printed. If one or more of these printers are suspected of having played a part in a crime, the color panels that are spooled onto the take-up roll should be carefully unrolled and images made of the dye panels.

Discovery of the dye sublimation printing process is generally credited to Noël de Plasse while working for the company Lainière de Roubaix in the north of France in 1957. It was not until the late 1980s and early 1990s, however, that companies such as Mitsubishi, Seiko, and Tektronix began selling commercial dye sublimation printers based on the PostScript page description language.

16.2.10 Thermal Paper Printers (Also Known as Direct Thermal Printing)

Thermal paper printers have seen a dramatic rise in use in recent years due to the proliferation of point-of-sale (POS) terminals that use this form of printing. Once used primarily as facsimile paper, the major sales markets for thermal paper printers have expanded to grocery outlets, large drug chains, banking, healthcare, gasoline outlets, and gaming (Figure 16.10). A recent development in thermal paper printing is the new two-color paper that can print both black and red images. Point-of-sale terminals such as the Epson TM-6000II are made to take advantage of these dual-color papers. Some thermal papers, such as that made by Appleton, can have a UV security tag built in. These tags are invisible in normal light, but can be seen clearly in ultraviolet light. The security tag papers are beginning to be used as anti-counterfeiting measures for product identification packaging and may show up on airline tickets that are printed at the ticket counter.

All direct thermal paper printers utilize a series of heated pins in the printhead that come in contact with the paper. The heat causes a chemical reaction in the paper, resulting in the formation of an image. As of the year 2000, thermal printers dominated the point-of-sale, bar code, label, and ticket markets because of their low cost, excellent print quality, lack of noise, and ease of integration with existing POS terminals.

16.2.11 Solid-Ink/Inkjet Printers

Tektronix, a company that has recently been purchased by Xerox, introduced the first solid-ink printer in 1991, the Phaser. In 2004 the Xerox 8400 series solid-ink printers were marketed for under $1000 in the U.S. The current crop of solid-ink color printers utilizes four sticks of resinous ink that are melted and sprayed onto an imaging drum to form the page. The placement of the drops is controlled by a series of inkjet nozzles on the printhead. The Xerox printers wait until all four colors are transferred onto the drum before the paper begins to traverse through the printer (see color Figure 16.11 following p. 366). This allows the dots to be placed with very fine precision and also allows a very short paper path compared to most four-color laser printers. All of the Xerox 8400 solid-ink printers can be set up to print duplexed (two-sided) pages.

16.2.12 Hybrid (Digital Offset Printers)

There are many new combination technologies that utilize old and new printing techniques. For instance, the printing plates for standard offset lithography may now be burned by a laser scanning device, rather than be exposed to light through a lithographic negative. In this way, text and images (including the typical four-color images needed for full-color processing) can be sent directly from a computer to a laser plate burner without the need for photographic color stripping (see color Figure 16.12 following p. 366).

Another recent entry into the digital arena is digital offset printing (DOP). Digital offset printers utilize an electrostatic imaging drum and special electrostatically sensitive liquid digital inks. Two of the largest makers of the new digital offset presses are Hewlett-Packard, with its Indigo presses, and Heidelberg (Figure 16.13). These types of printers form images on an electrostatic drum and then have the various inks deposited over the drum surface, in much the same way that toner would be laid down in standard laser printing. The ink is then transferred to an offset cylinder (the blanket), where the ink is printed onto the paper.

Figure 16.10 EPSON TM-88iii point-of-sale thermal printer.

Figure 16.13 Heidelberg digital offset press.

Up to seven colors can be laid down in the newest versions of the HP Indigo presses at an effective resolution of 812 × 812 dpi using line screens up to 230 lines per inch. From a forensic standpoint, it can be very difficult to differentiate traditional offset printing from digital offset printing without conducting some form of destructive ink analysis (see Chapter 17).

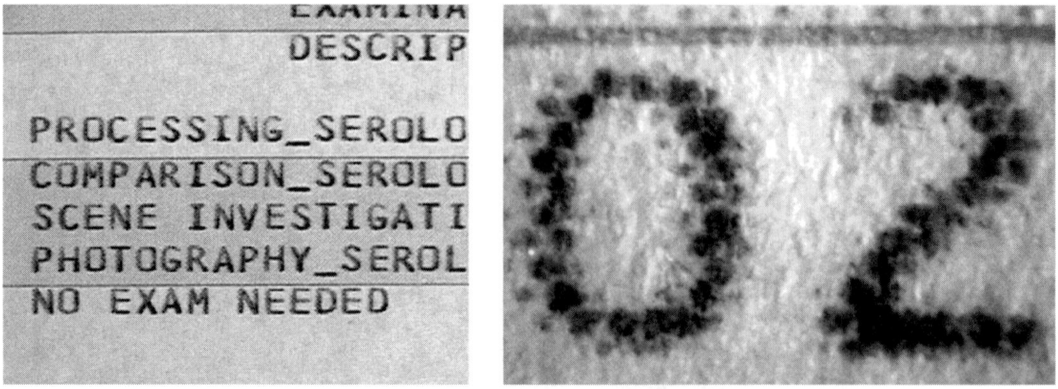

Figure 16.14 Sample of line printer output shown at low magnification on the left and at high magnification on the right.

16.3 Identification Features of the Printing Processes

16.3.1 High-Speed Mainframe Line, Drum, and Chain Printers

Output from these printers is more likely to be found in historical cases rather than on more modern documents. If the document itself is printed on tractor-feed green-bar paper, there is a good chance that it was printed on an impact printer such as dot matrix or one of these high-speed printers. The typical line printer output is monospaced, 10 characters to the inch, with the older printers utilizing all capital letters. The newer line printers, however, could use a full upper- and lowercase alphabet, with Courier being one of the font options.

The letters themselves are solid formed (like typewriter characters) and, also like typewriters, were pressed through an inked ribbon in order to form the text on paper. Although the characters may appear to be made up of tiny dots or dashes, this is an illusion caused by the weave of the inked fabric ribbon. A classic example of line printer output from a mainframe computer is shown in Figure 16.14.

16.3.2 Typewriter Technology Printers: Typeball and Daisy Wheel

It can be impossible to differentiate a document prepared on a standard daisy wheel or typeball typewriter from their computer printer cousins. In fact, some memory typewriters are actually hybrid typewriters–printers that could be used to input data through the keyboard, record the strokes, and then play them back using the typewriter as the printer.

A lack of corrections in a lengthy document, typewriter line spacing, and full justification may be indications that a document was prepared on a daisy wheel or element printer, but by no means is this rule hard and fast. All of the information contained in Chapter 15 can be utilized in making an examination of this class of printer. Figure 16.15 shows a Smith Corona L-1000 daisy wheel computer printer.

16.3.3 Dot Matrix Printers

Because dot matrix printers utilize an impact printing process, they are often used today in settings where carbon or carbonless (NCR) forms are printed. Hospital records, cash

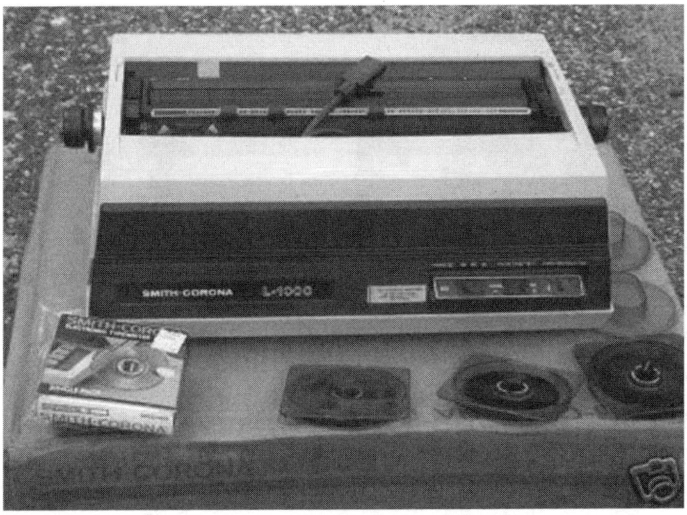

Figure 16.15 Smith Corona L-1000 daisy wheel printer.

Figure 16.16 Dot matrix print sample on NCR paper.

register receipts, and tax forms are typical kinds of documents created on dot matrix printers. The most common pin arrays found in dot matrix printers are 9 vertical, 18 vertical (as 2 columns of 9), and 24 pin (as 2 vertical columns of 12). Figure 16.16 shows dot matrix output from the second copy of an NCR form.

16.3.4 Inkjet Printing Technology

Inkjet printers come in two varieties, continuous drop and drop-on-demand. All of the modern desktop inkjet printers are of the second type — drop-on-demand. It is likely that most of the questioned documents that are encountered in a forensic laboratory will be from this type of inkjet printer. In commercial high-speed applications, however, the continuous-drop printers are more likely to be found. Packaging, bar coding, bulk mailings, and lottery tickets are some of the more common applications for the high-speed, continuous-drop printers. Color Figure 16.17 (following p. 366) shows output from a drop-on-demand printer in monochrome mode (a), a four-color inkjet graphic (b), and a Scitex continuous-drop inkjet printer (Figure 16.17c).

16.3.5 Monochrome and Color Laser/LED Printers

Both black-and-white and color laser/LED printers utilize dry toner as their ink (see Figure 16.18 and color Figure 16.19 following p. 366). Keep in mind, however, that black-and-

Figure 16.17(c) Hi-speed continuous-drop inkjet print sample.

white and color copiers also utilize toner. Although some older xerographic machines used pressure rollers to "crush" the toner into the paper fiber, modern machines use heated fuser rollers to melt a thermoplastic binder mixed with the toner. As a result, some xerographic printers will produce images/texts that have a glassy, melted appearance. Many of the newer microfine toners, however, have a much flatter appearance.

16.3.6 Thermal Wax Printers

Thermal wax printers produce a relatively low resolution output, but have very saturated colors. As can be seen in color Figure 16.20 (following p. 366), a pin or scalpel can be used to scrape the waxy ink from the paper surface. Some plain-paper fax machines also use this process.

16.3.7 Dye Sublimation Printers

Dye sublimation printing is often used to produce photographic images from digital capture devices. Although the actual output resolution is fairly low, the eye is tricked into seeing greater detail because of the fine color tones that are possible with this process. Unlike the other CMYK devices, dye sublimation printers can actually produce shades of each color by varying the amount of heat used to sublimate the dyes. Color Figure 16.21 (following p. 366) illustrates the fine color–tonal changes as the image shades from red to orange to yellow. Dye sublimation printers are often utilized to produce identity documents (both genuine and counterfeit). Keep in mind that the rolls of imaging film that remain in the printer retain the images that they have produced.

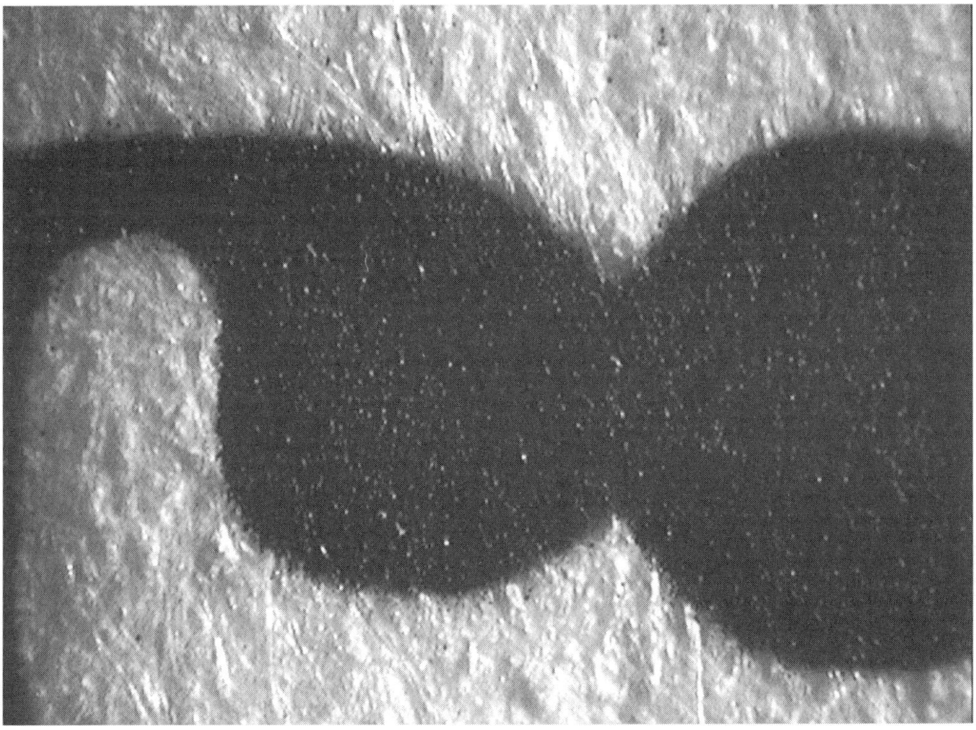

Figure 16.18 Monochrome laser print sample.

16.3.8 Thermal Paper

Thermal paper is generally found in two different formats. Printouts on the first type resemble the older thermal facsimile copies with special coated paper that is slick on the reverse side and somewhat chalky appearing on the imaging side. The other type of thermal paper more closely resembles ordinary heavy paper stock with a thermally sensitive coating. Many of the "Pick"-type lottery machines use this second type of paper. Some newer thermally coated papers have dual-temperature coatings that can print in both black and red. Under medium magnification, the irregular density of the printing elements can often be seen (Figure 16.22).

16.3.9 Solid-Ink/Inkjet Printing (Tektronix/Xerox)

Tiny droplets of melted ink are sprayed onto a transfer drum and then cold-fused (pressed) onto the surface of the paper (see color Figure 16.23 following p. 366). On early models this method of fusing typically resulted in the drops of solid ink (that were round on the transfer drum) becoming elongated along the direction of the paper path. It is reported, however, that under high magnification these droplets tend to remain much more rounded than the earlier Tektronix/Phaser models.

16.4 Standards for Comparison

No matter how familiar one is with the visual appearance of toners and inks, there is really no substitute for live comparison standards. In this regard, a trip to a local computer retailer

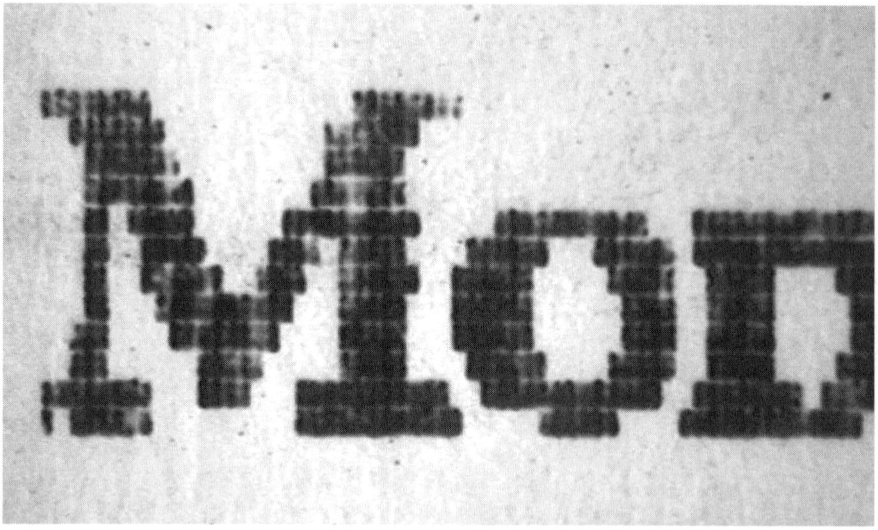

Figure 16.22 Output from thermal paper printers can typically be found at point-of-sale terminals, such as self-serve gas pumps, thermal paper facsimile machines, and cash register receipts. The low-resolution output and uneven density of the character elements are typical indicators of a thermal paper printer.

can often provide samples of several manufacturers' printers as well as different models by the same manufacturer. Remember that the samples obtained from these sources are produced by machines set on demo mode. For the FDE, this means the specimen has been printed at optimum quality and may not reflect images entirely consistent with items being examined. Often the retailer will have printers that utilize different technologies as well.

16.5 Conclusion

This has been a long and complex chapter that has attempted to touch on many different aspects of the forensic examination of computer-generated documents. Even with efforts to cull all but the most important information, we have had to discuss such diverse topics as font design, printer history, typography, and, of course, the ways these topics affect a forensic examination of electronically prepared documents.

It is safe to say that just as the technological methods used to create modern documents continue to change, so the forensic examination of computer-generated documents will continue to evolve. For instance, many laboratories around the world are currently doing research to determine if an analysis of toners, inkjet inks, and other media can be individualized to a particular manufacturer. Devices used by the electronic printing industry itself to determine output quality and image banding are beginning to see their way into the forensic arena as identification tools. All in all, modern document examiners are adapting to computer-generated documents just as FDEs 80 years ago were forced to deal with emerging mechanical (rather than handwritten) documents. The next generation of document examiners will have grown up in a world that has always had computers and computer-generated documents.

APPENDIX: Summary of Some Common Laser Printer Output Defects and Their Causes

In its published laser printer service guide, Hewlett-Packard lists 23 different types of common defects that affect the output quality of a laser printer.[1] Although some of the causes are specific to particular Hewlett-Packard models, many are generally applicable to all laser printers. The following table is a summary of these defects and their most common causes:

Defect	Possible Causes
Blank pages	The toner cartridge is empty or defective; sealing tape has been left on in the toner cartridge; high-voltage power supply is bad; the transfer roller is defective
Black pages	The toner cartridge is improperly installed or is defective; the high-voltage contacts are dirty or the high-voltage power supply is defective; light is leaking into the printer; the laser/scanner cable or assembly is defective
Thin, dark vertical lines	The toner cartridge is defective; the primary charging roller is dirty or defective; the static eliminator teeth are contaminated or defective; the fuser cleaning pad or fusing assembly is defective
Image skew	The paper is loaded incorrectly; the rollers or separation pads are worn; the pickup and separation rollers are worn; the drive gears in the drive train assembly are defective; the registration assembly is dirty or defective; the laser/scanner unit is misaligned
Faulty registration	The paper does not meet specifications; the leading edge of the paper is curled; the paper is loaded incorrectly; the paper tray is overloaded; the feed rollers are worn; the registration solenoid is stuck; worn or broken drive gears
Light print/faded print	The paper does not meet specifications; toner cartridge is low or empty; the print density is improperly set; econo mode is turned on; the transfer roller is defective; the transfer corona assembly is defective
Horizontal lines/repeating defects	The toner cartridge is not seated properly or is defective; the laser/scanner assembly is defective; a defect has occurred on one of the drums or rollers that come in contact with the paper; the static teeth are dirty; the gear train is defective
Poor fusing/character voids	The paper does not meet specifications; printing is on the wrong side of the paper; the fuser is defective; the transfer roller is defective; the high-voltage power supply or DC controller is defective; the print density is improperly set
Background scatter	The print density is improperly set; the toner cartridge is defective; the paper does not meet specifications; the paper is too wet or too dry (optimum 40 to 60% relative humidity); the transfer roller or corona is dirty or defective; when printing an envelope, the print is on the seams; toner buildup in the fuser inlet guide
Back of page dirty	Toner has leaked from cartridge; the inside of the printer is dirty; the transfer roller is dirty; the feed rollers are dirty; the fuser is contaminated; the corona wire is stuck "on;" the input feed roller is dirty; there is toner buildup in the fuser assembly

Thin vertical white lines/stripes	The toner cartridge is defective or nearly empty; foreign obstacles are in the laser beam path; the focusing lenses in the scanner assembly may be dirty; the beam-to-drum mirror is contaminated; the fuser cleaning pad is dirty
Smudged band with overprint	The main motor gear train assembly is not engaging with the toner cartridge drum gears; the fusing assembly is dirty or defective; the static teeth are dirty; the toner cartridge is defective
Toner smear on right or left side of page	The separation belt is dirty; the separation roller is dirty; the pinch roller is dirty; the toner cartridge is defective; the primary corona wire in the toner cartridge is dirty
Vertical fogged stripes	The conductive surface of the guide is worn, causing static charge; the feed guide assembly is defective; the primary corona wire is dirty; the toner cartridge is defective; the inside of the printer is dirty
Right- or left-hand text missing/distorted	A foreign object is in the laser beam path; the toner cartridge is empty; the instruction label on the beam mirror shutter has come loose; the beam-to-drum mirror is bent or misaligned; the top cover assembly hinge brackets are bent
Distortion	The toner cartridge is defective; the gear train is damaged; there is a problem with the paper transport; the main motor is defective; the DC power supply/main motor driver PCA is defective
Bubble print	The photosensitive drum inside the toner cartridge is not grounded; the fuser is defective; the toner cartridge is leaking; the paper does not meet specifications; the high-voltage power supply is defective
Leading edge scatter	The paper does not meet specifications; the fuser inlet guide is improperly set; the high-voltage power supply or transfer roller/wire is defective; the print density needs adjustment; the rollers and high-voltage contact points require cleaning
Compressed print	The toner cartridge is defective; the drum drive gear is broken or worn; the drum drive gear assembly is defective
Portion of the page is blank	Memory overrun — the page is too complex; printing on legal paper when the software is set to letter size
White horizontal line	The toner cartridge is defective
Horizontal fogged stripes	The laser scanner cable is defective; the toner cartridge is defective; the high-voltage power controller is defective; the ground contact to the toner cartridge is defective
Wavy print	The scanner unit is defective; the laser scanner cable is defective; the DC controller power control assembly is defective

References

1. Hewlett-Packard, *Hewlett-Packard LaserJet Family Quick Reference Service Guide*, Part 5021-8908, Hewlett-Packard, October 1997.

Photocopiers

17

BONNIE L. BEAL
SUSAN E. MORTON

Contents

The ability of contemporary photocopiers to create fast, inexpensive document reproductions has lead to their widespread use at every level of business. Today the process used most in stand-alone photocopiers and multi-purpose machines is indirect electrostatic, also referred to as the plain-paper system. For color copiers, the operation has moved away from the silver halide principle to electrostatic and inkjet technologies.

Chester A. Carlson is given credit for the invention of photocopying. Carlson worked for Xerox Corporation, which introduced the indirect electrostatic process in 1960. The machines have come a long way from the older processes, including Photostat®, diffusion transfer, gelatin transfer (Verifax), thermographics, direct electrostatic (Electrofax), and dual-spectrum.

The indirect electrostatic process, also referred to as electrostatic plain-paper, has evolved from analog to digital imaging systems. The difference between the analog and digital processes lies in the imaging and exposure steps of reproduction. The seven steps in the copier's operation are charging, imaging, exposing, developing, transfer, fusing, and cleaning. In the analog process the original document is imaged directly to the photoconductor, whereas in a digital system, the machine scans the document and stores the image, or a modified version of it, that is then transferred to the photoconductor drum via a

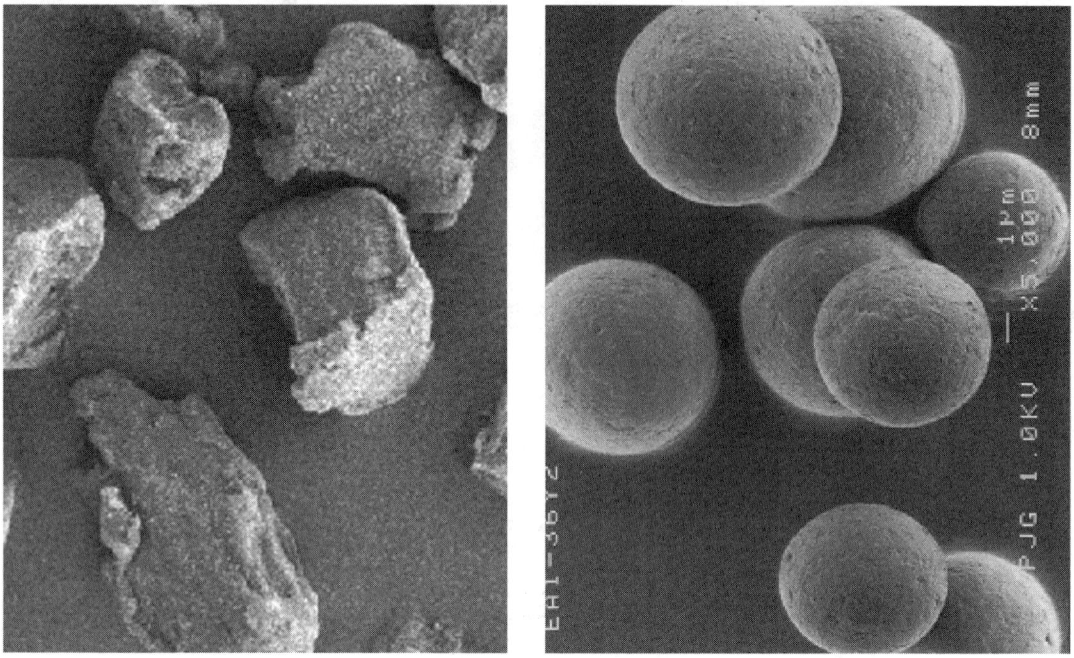

Figure 17.1 This image depicts on the left the irregular shape of the mechanical toner and on the right the spherical-shaped chemical toner.

laser.[1] The output from these two operations can be difficult, if not impossible, to differentiate where the resolution of the digital copier is high (e.g., 600 dpi or greater).

With some digital copiers the user is able to edit the scanned image by cutting, cropping, rotating, or repositioning selected areas before printing the image. Different models afford the user various options in this regard.

Along with changes in the imaging system have come developments in the toners used. Of the three toner classes — liquid, dry, and chemical — the latter two are the most commonly used in copier machines. Chemical toner was created to improve performance and achieve a more even charge over the toner.[2] This was achieved by producing a consistently spherical toner of less than 7 microns in size. The older, mechanically made toner is irregular in shape and has a tendency to clump together. Figure 17.1, a scanning electron microscope (SEM) image, shows a side by side view of mechanical and chemical toner in pure form.[3]

The original liquid toner, although not prevalent today, is available through some retailers for certain models of Savin copiers still in use. When viewed with slight magnification, the older liquid toner appears to dye the paper fibers.[4] Newer liquid toners (sometimes referred to as Electroink) are now being used in high-end printers, and may eventually be seen in photocopiers. According to manufacturers of contemporary liquid toner, the print is comparable to that seen in offset lithography. As with offset lithography, liquid toner can be used to create halftone images.[3] Although the resulting image is very clean and sharp, microscopic examination will reveal fine toner particles (less than 2 microns in size) scattered between the halftone dots (see color Figure 17.2 following p. 366).

The toner can be bonded to the substrate in different ways, including: radiant heat, heat and pressure, cold pressure, hot soft roller, hot hard roller, and hard-soft roller methods. Depending on how it is fused to the paper different characteristics will be seen.[4]

Figure 17.3 Illustrated is a counterfeit protection system (CPS) code enhanced utilizing Adobe Photoshop®.

The development of color copiers brought concerns about their use for fraudulent purposes. To help alleviate these concerns, some manufacturers place a counterfeit protection code onto every copy. This code contains information about the color copier used to reproduce the document. There are reference papers available describing ways to enhance these codes for easier viewing[5,6] (Figure 17.3). In addition, some machines employ a currency lockout system that prevents the user from reproducing paper currency.

Information on new color copier processes such as cylithography, thermal transfer, and cibachrome can been referenced in articles published by James[4] and Doherty.[7] Although most color copiers today use toner, some utilize an inkjet process. Inkjet copiers fall into two categories: stand-alone and multi-functional (copier, fax, scanner, and printer).[8] Due to the higher cost, stand-alone devices are less common.

The printing mechanisms of inkjet printers and copiers exhibit the same characteristics and therefore require the same method of forensic analysis. The FDE will need to be careful when examining a document produced by inkjet technology as it may, in fact, have not been generated by a computer printer, but rather a copier.

17.1 Analysis of Photocopies

The analysis of a photocopy (monochrome or color) can run the entire gamut of instruments in a document examination laboratory, including microscopes, measuring devices, IR imaging systems, and electrostatic detection devices (EDDs).[9] Every case is unique and the entire document needs to be examined thoroughly.

17.2 Identifying Individual Photocopiers

This section will focus on electrostatic plain-paper copiers. An article by James M. Davidson[10] discusses three photocopier examinations: determining whether or not two or more copies come from the same machine, determining whether or not a particular machine produced the copies, and determining the approximate date of the copies. Another article by James Gerhart[11] deals with using fusing roller defects to help identify a photocopy with its source machine. He states that fusing roller defects should be considered a key component in the identification of photocopiers.

Like any mechanical device, a copier will develop its own idiosyncrasies with time and wear, which often will result in the introduction of a unique constellation of marks onto a copy. Knowing how to find and interpret these marks may enable a document examiner to associate a copy with the machine that made it.

Copy machine mechanisms afford several areas where defects cause individualizing features to be transferred to copies (Figure 17.4). Some are ephemeral and require samples made close in time; others are fairly stable, but may be eliminated by repair or service on the machine. It might be necessary to study the repair record of a particular copier before observed features can be properly interpreted.

The first area where unique markings can occur is on the glass imaging window or platen. The glass can become soiled and smudged with fingerprints, adhesives from tape or labels, opaquing solutions, and general debris. Such materials, in sufficient quantity, can produce a constellation of unique marks on each copy. Cleaning of the glass window

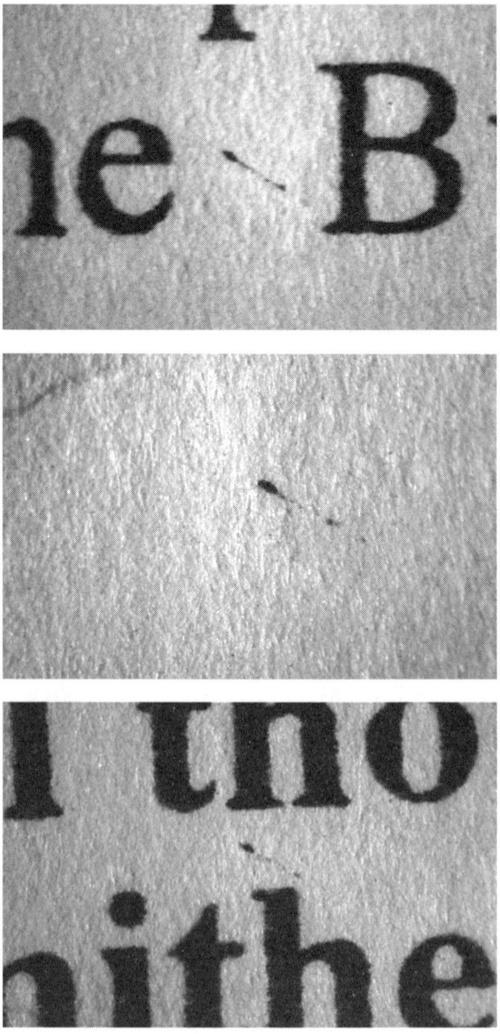

Figure 17.4 An example of trash marks found on each sheet of a multi-page document. (Courtesy of Jan Seaman Kelly.)

may remove these marks. To rule out this possibility it is prudent to obtain, for comparison, known samples prepared on the machine close to the same time as the questioned copy. The glass window may acquire more permanent marks in the form of scratches or pits. Provided these are not so obvious as to annoy the user and lead to replacement of the glass, they will be present for the life of the copier.

The rubber-backing blanket that holds the original document in place on the glass window can also be a source of unique features. It can become soiled or damaged through prolonged use. If the entire imaging area is not covered, a portion of the blanket, with its unique features, will be reproduced. However, with a standard- or legal-size original, the blanket is not usually within the imaging area.

Occasionally a speck of dust or other matter will adhere to the imaging lens and produce a characteristic mark on copies. This type of defect is rare and may be so noticeable as to lead to a service call.

The light-sensitive imaging drum can be a source of unique features as well. Pits, scratches, or marks on the imaging drum will produce characteristic marks and dots on copies. Unlike other imaging defects, drum defects will not appear in the same place on each copy. The drum makes revolutions with each copy and does not land in the same rest position after each rotation. Signs of a drum defect will always be the same distance from the top and bottom (or from the left and right) margins depending on the paper feed direction but will vary along the axis of feed and may not appear on all copies. High-speed copiers have large drums that take three different images at a time. Such machines may only reproduce a drum defect on every third copy.

Paper-feed grippers and rollers can also leave distinctive impressions on copies. This generally occurs when the paper is not seated properly in the paper tray or the tension on the paper-feed device is out of adjustment. These marks are frequently in the form of indented impressions in the paper rather than a visible toner mark. The impressions can be observed with oblique light and may be recoverable with an electrostatic detection device.

17.3 Comparison Samples

An FDE rarely has the luxury of access to an actual suspect copier. It is necessary to base examinations on samples taken from the machine, or machines, in question. Several samples from a copier help to establish which of the tiny marks present on nearly all copies are repeating individual features and which are random electrical misfires. Extraneous marks are nearly always present. In order to capture every significant defect and to eliminate meaningless random marks, an FDE must intercompare several sample pages. Adequate samples can generally be obtained by following these instructions:

1. With the rubber-backing blanket pressed in place against the glass imaging window, activate the machine and make at least six consecutive copies. Number them in the order they were made.
2. Set the machine on its darkest setting and repeat.
3. Set the machine on its lightest setting and repeat.
4. Place a piece of blank white paper on the imaging window and repeat the first three steps.

It also is very helpful to research the repair and cleaning history of the copier. If it has undergone significant repair and cleaning since the alleged time of the preparation of the questioned copies, it may be necessary to locate copies made on the machine prior to the cleaning/service. Knowing the paper's feed direction, long or short side, can also be of assistance.

17.4 Comparisons

In conducting a comparison of a questioned copy to samples from a known machine or machines, the FDE may start with either the samples or the questioned document(s). It is useful to overlay the document under study with a sheet of clear plastic film and create a template by recording the non-image marks with ink dots on the film.[12] The template can then be placed over other copies to establish whether the same set of accidentals is present and to identify any different marks that appear. The documents can also be scanned, overlaid, and compared using the layers function of a graphics software program.

In considering similarities and differences in a copier comparison, the examiner must take into account that copy machines nearly always make random marks on their products. Therefore, the questioned documents and known samples will almost always have some differences, the presence of which does not mean the copies were produced on two copiers. If a sufficient number of consistently repeating marks are present on both questioned and known documents, the document examiner may still conclude they have a common source.

An absence of any similarities between the questioned and known copies does not necessarily warrant an elimination of the suspect copier. The machine may have been cleaned or repaired between the time that the questioned copies were made and the known samples were taken. While such a lack of similarities may preclude an association, it may not justify elimination of that particular copier. Knowledge of maintenance to the machine is always helpful.

There is no set number of similarities needed to effect an identification of a copy machine because it is not necessarily the number that determines the matter. It is the type of defect that carries the most weight. The determination of significance must be in the judgment of the document examiner, and is based on study and experience.

17.5 Dating Copies

It is sometimes possible for the FDE to determine the approximate date a copy was made. To do so, he or she will need copies known to have been made during the suspected time frame. By studying the evolution of marks made by dirt and debris and removed by cleaning, the FDE might be able to narrow the window of time during which a copy could have been created. In addition, knowing the machine's introduction date may help in determining if a copy was made at the time suggested.[13]

17.6 Iteration

Most copy machines do not reproduce documents in an exact 1:1 ratio. Instead, the manufacturers set the machines to enlarge very slightly to make a visual compensation for the slight degradation in resolution. This enlargement factor is usually around 1%. The enlargement factor of a particular copier can be determined by measuring an original document against a first-generation copy. By knowing the enlargement factor of a machine and having the original document, it is possible to determine the number of iterations (generations) the copy is removed from the original. Is it a copy of the original? Copy of a copy? Copy of a copy of a copy?

In order to calculate the enlargement factor for a particular machine, the exact size of a line of text or a figure and the comparable item on the copy is measured. A precise typewriter grid is useful for this analysis. The larger the item, the more accurate the measurements are likely to be. As an example, let us suppose that a figure on the original measures 10 mm and the same figure on the copy measures 10.1 mm. That machine's enlargement factor is 1%. A questioned document with a figure that measures 10.2 mm is most likely a second iteration, or copy of a copy. This assumes all copies were made on the same copier with no adjustment to the image size options on the machine. Caution should also be exercised regarding the possibility of non-linear distortion in which one area of the copy reproduces at a different scale than another.

Extreme care must be used in making and interpreting these measurements. The dimensions of the paper itself can vary slightly with changes in atmospheric conditions; it can expand in humid, and contract in drier, conditions. Precise measurements are very difficult to make and 0.1 mm is a tiny amount.

17.7 Fabrications

Copiers can be used to create fabrications of non-existent original documents. Signatures can be placed on contracts and agreements, dates or amounts altered, or paragraphs added. Usually these cut-and-paste efforts are quite detectable with careful scrutiny. The first red flag is that a copy, rather than an original document is proffered. Since no original in fact exists, the proponent must explain why only a copy of a valuable or otherwise important document can be found.

The most decisive way to demonstrate a fabrication is to locate the source item from which it was made. No single signature can exist on more than one document. If a particular signature is found on one document, a replication of that signature cannot validate a different document.

Often physical evidence on the fabrication will establish its nature and indications of its method of creation even when the source document cannot be located. There may be shadow lines, slight cutoff of ascenders or descenders of letters, or other signs of the cutout. Printed/typed text may not be in proper alignment or may even be in a different type style and font size. If there is a typed signature block that can be measured accurately, it may be possible to prove that the typing underneath the signature is a different copy generation than other text on the document.

17.8 Performing Handwriting and Other Examinations on Photocopies

With the proliferation of copies, FDE's are often asked to perform handwriting or other examinations on non-original documents. No one disputes that original materials are better than copies for such analyses; however they are sometimes not available. For example, many account holders do not receive cancelled checks back from the bank because they are scanned and stored electronically. By the time an allegation of forgery is made, the only evidence that still exists is in electronic form. In other cases, the original documents do exist but are unavailable. They may be in court records in distant jurisdictions or otherwise not available.[14]

The imaging systems of modern copiers have improved greatly with a corresponding increase in the quality of photocopies. Research by Dawson and Lindblom[15] shows that FDE's can assess line quality characteristics when analyzing non-original signatures on good quality reproductions. Found, Rogers, and Herkt[16] further confirmed this. The latter authors compared FDE analyses of questioned documents as copies and in their original form and found that the participants were able to render accurate opinions respecting both line quality features and structural characteristics.

What limitations do copies impose? If the reproduction process is high contrast it may delete faint lines such as pen drags and tick marks. Line quality may be harder to interpret as the third dimension of pen pressure — the depth of the ink line into the paper — cannot be seen. Ink lines can seem more uniform in appearance than they are in the original writing. Some copiers are not panchromatic — they do not copy blue ink as clearly as other colors. Since blue is a very popular color for writing ink, the signature may appear broken up and fragmented. The copying process may mask the presence of tiny hesitation points or it may introduce artifacts that look like hesitation points.

Each copy must be evaluated on its own merits given the differences in image quality and resolution. Obviously, a third generation reproduction will lack the detail and accuracy that is seen in an earlier iteration.

Early models of digital copiers produced a stair-step effect or "jaggies" in the image. This made it difficult to evaluate line quality and subtle structural features in the copies. The same problem occurs with facsimile reproductions that are produced at relatively low resolutions. Fortunately, the resolution in modern digital copiers has increased significantly thereby minimizing some of the obvious effects of pixelation.

Because of the limitations imposed by photocopies, some document examiners will not render definitive results as to handwriting or signature identification. Other FDE's will consider each copy and decide what level of assurance the reproduction quality can support.[17]

Other types of examinations may not even be possible. The minute typeface defects that allow an impact typewriter to be identified with its work are likely to be lost in the copying process. Similarly testing for indented impressions, ink analysis, and paper examinations can only be performed on original documents.

The possibility of a fabrication must always be considered when examining a non-original document. With modern copiers, scanners, and publishing software, such manipulations are easy to create. Sometimes the context of a case precludes that possibility, and the FDE can examine the copy on its own merits.

17.9 Conclusion

The examination suggestions in this chapter are not exhaustive. It is recommended that the document examiner read all of the current literature on the subject because the field is forever changing and evolving particularly where office technologies are concerned.

References

1. Canon, Inc., *The Copier Basic Series, Part 1 and Part 2*, Canon, Inc., Norcross, GA, 2000.
2. Tuvesson, E., The soul of a new machine, *Future Graphics LLC*, 26, 2001.
3. Gibson, G., Xerox Corporation, personal communication, 2004.
4. James, E.L., The classification of office copy machines from physical characteristics, *J. Forensic Sci.*, 32, 1293, 1987.
5. Tweedy, J.S., Class characteristics of counterfeit protection system codes of color laser copiers, *J. Am. Soc. Questioned Doc. Examiners*, 4, 53, 2001.
6. Li, C.K. and Leung, S.C., The identification of color photocopiers: a case study, *J. Am. Soc. Questioned Doc. Examiners*, 4, 8, 1998.
7. Doherty, P., Differentiating Full-Color Photocopiers by Class Characteristics, paper presented at SWAFDE, Long Beach, CA, October 1990.
8. *Color Copier Guide*, Guide 159, 2003, www.betterbuys.com.
9. Shiver, F.C. and Nelson, L.K., Non-destructive differentiation of full-color photocopies, *J. Forensic Sci.*, 36, 145, 1991.
10. Davidson, J.M., The Document Examination of Machine Copies, paper presented at ASQDE, Montreal, September 23-27, 1985.
11. Gerhart, F.J., Identification of photocopiers from fusing roller defects, *JFSCA*, 37, 130, 1992.
12. Tytell, P.V., Transparencies and Acetate Overlays: Their Use in the Investigation and Demonstration of Questioned Document Problems, paper presented at ASQDE, Crystal City, D.C., August 26-30, 1989.
13. Crown, D., Dates of introduction of specific photocopy machines, *J. Forensic Sci.*, 34, 110, 1989.
14. Morton, S.E., A look at newer photocopiers, *J. Forensic Sci.*, 34, 461, 1989.
15. Dawson, G.A. and Lindblom, B.S., An evaluation of line quality in photocopied signatures, *J. Forensic Sci. Soc.*, 38, 189-194, 1998.
16. Found, B., Rogers, D., and Herkt, A., Comparison of document examiners' opinions on original and photocopied signatures, *J. Forensic Doc. Exam.*, 14, 15-30, 2001.
17. Beck, J., Detecting Photocopied Forgeries Using Genuine Signatures, paper presented at ASQDE, Seattle, WA, 1985.

18 Facsimile Machines

BRIAN S. LINDBLOM

Contents

Over the past 10 to 15 years the popularity of fax machines has increased considerably, to the point that most businesses and many households now have them. As a result, disputes involving faxed documents often find their way to forensic laboratories. Concerns may be raised about whether a reproduced signature is genuine, whether information has been added to or deleted from the body of the text, or whether the fax was actually transmitted on a particular date.

Most of the major manufacturers market both multi-function and stand-alone machines; the former incorporate scanning, printing, copying, and faxing capabilities. The printing function of both types can be thermal, thermal transfer, inkjet, bubble jet, or laser, with the latter three being by far the most popular. In addition, fax modems are integrated into many computers or are available as peripheral devices and may be driven by a wide selection of software. Although sheet-fed machines represent the majority of models now marketed, roll-fed mechanisms are still encountered from time to time and typically use coated paper that discolors and fades with time.

Those who have had the opportunity to examine faxes will have noted that the resolution in such documents is low compared to photocopied or scanned and printed material. This is because transmissions are most often performed at a resolution of approximately 100 to 200 dots per inch (dpi), in contrast to the at least 300 to 600 dpi of photocopiers and computer printers (Figure 18.1).

Figure 18.1 Comparison of a signature reproduced on an inkjet printer (1), a digital photocopier (2), a facsimile machine using the fine setting (3), and a facsimile machine using the standard setting (4). Enlargements are seen to the right of each signature. Note the differences in image resolution.

18.1 Faxed Handwriting

An examination of a faxed signature or handwritten entry must be approached with considerable caution, as an assessment of line quality characteristics cannot be fully undertaken. Subtle detail, such as inconspicuous pen lifts, hesitation, retouching, and minor tremor, are unlikely to have been reproduced.[1] Another concern is that transmission noise may create what appears to be tremor in the signature and other handwritten lines. This is not to say that faxes should not be analyzed, but rather that the document examiner must be cognizant of the reproduction limitations. Structural characteristics can be evaluated and may allow for a qualified opinion as to the authenticity of the signature or handwriting. Clearly, determining that there are dissimilarities consistent with non-genuine writing is much easier than the alternative.

Unlike those from computer printers and photocopiers, facsimile transmitted documents may undergo substantial dimensional changes that can result in significant reductions in size and even distortion to an irregular trapezoid shape. Research has documented changes in size ranging from a 2.9% increase to a 12% reduction.[2]

CAMPRO VIDEO 613 728 0013 P.01

JAN 25 '99 17:21 613 728 0013 PAGE.01

Figure 18.2 Sample of a TTI (top) and RTI (bottom) combination taken from a single fax page received (some white space has been removed from between strings for illustrative purposes). In this instance, the RTI appeared along the bottom of the page. There is a very noticeable difference in formatting and typestyle between the two, resulting from the fact that the TTI is created by the transmitting machine, whereas the RTI is added by the receiving machine.

18.2 Transmit Terminal Identifier

Any examination of a fax document should include a consideration of the Transmit Terminal Identifier (TTI) that appears along the top of the document. This header line is generated by the transmitting machine and may include the sender's fax number and name, transmission date and time, the recipient's telephone number, and a page number. All of these are programmable features that the operator may deactivate or modify, either entirely or in part. Different makes and models use a variety of formats and typestyles, allowing one model to be distinguished from another. This information is often very helpful in identifying the possible sender involved and often serves as a mechanism to potentially date a transmission. Unlike the transmitted document, the TTI is not scanned and, therefore its image is sharper than that observed in the remainder of the fax. Also appearing on some transmissions is a Receive Terminal Identifier (RTI), which is generated by the receiving machine. This function is not incorporated into all makes and models, but when present, the RTI will appear as a second line either across the top or at the bottom of the page.[3] An example of a TTI/RTI combination is seen in Figure 18.2. Note the different formatting and typestyles between the TTI and RTI.

In cases where the origin of the faxed document is at issue and no samples are available from the sending machine, the use of a computerized database of TTI samples can be a valuable aid. The American Society of Questioned Document Examiners (ASQDE) has developed one such database.[4,5] The questioned TTI can be compared against an example within the reference collection by using the purported make and model as a starting point. In the alternative, a search can be conducted for all makes and models showing similar TTI formatting and typestyle characteristics to those on the questioned document. Figure 18.3 illustrates a possible match with a questioned TTI using such parameters as the arrangement, date format, page terminology, text design, and text font combination. Data retrieved from this search may reveal that the questioned TTI is consistent with that seen in a different make of fax machine than suggested by the originator of the fax. While the database is not exhaustive, it does provide a good starting point from which to make inquiries with the manufacturer.

Particular caution is needed when studying the TTI data. It must be remembered that much of the information is programmable, so dates, times, and telephone numbers should not necessarily be relied upon as being accurate. In some models, the user also can change certain formatting characteristics.[6] Two examples are the page indicator symbol and date format. Makes and models that connect to a computer often allow the configuration to be modified using accompanying software.

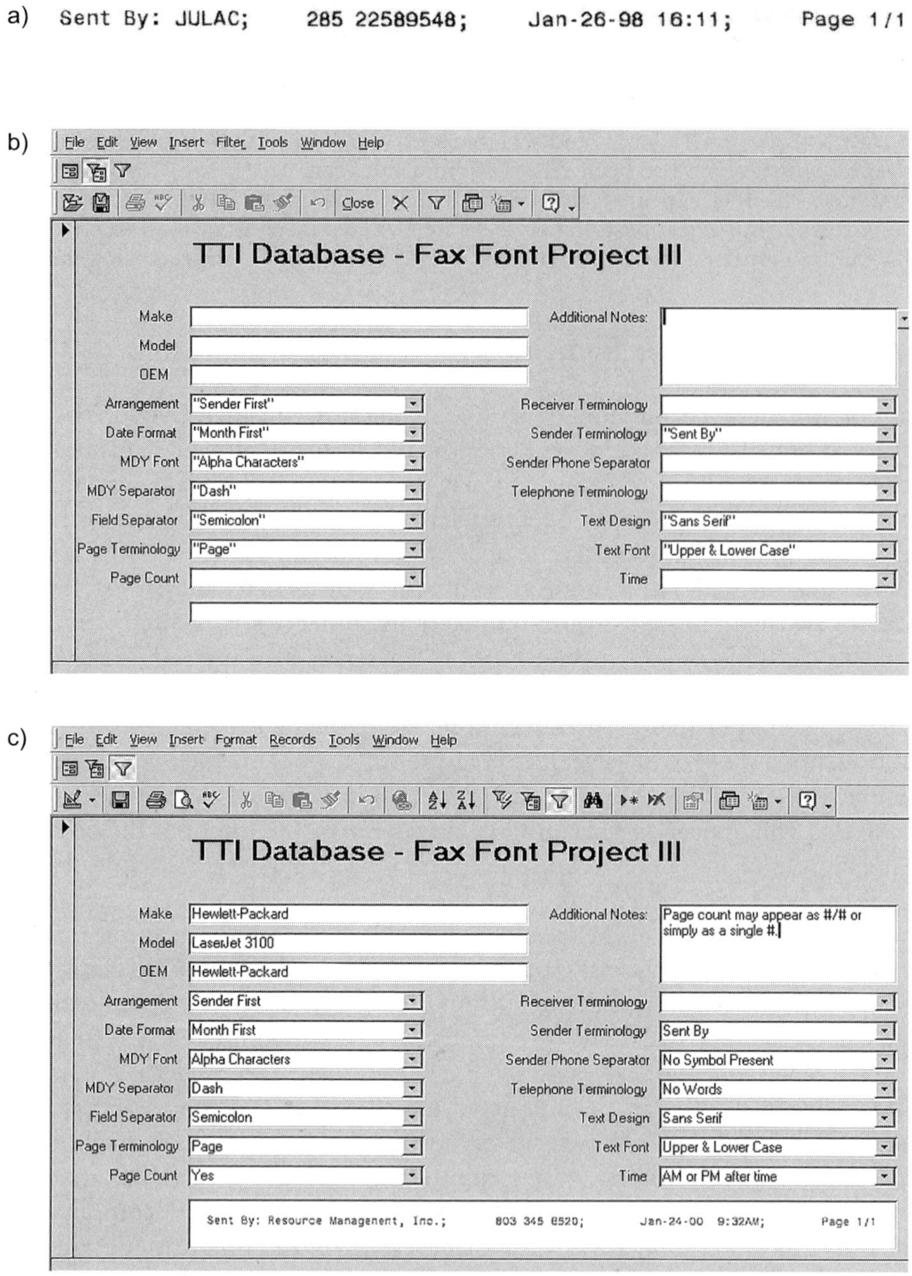

Figure 18.3 When the formatting information of the questioned TTI header (a) is entered into the TTI database Fax Font Project III filter (b), the results of the search (c) provide only one machine that matches: the Hewlett-Packard LaserJet 3100. While the database is not considered an exhaustive reference for TTIs, a unique match does allow a qualified identification of the transmitting model.

18.3 Manipulated Faxes

The poor quality of faxes provides an attractive medium to those hoping to conceal the manipulation of a document, be it in the form of cutting and pasting a signature, adding text, or altering the TTI. Cases have been reviewed in which dates within the TTI have been replaced or entire strings substituted. Without careful scrutiny, fraudulent changes to the document may go undetected. Discrepancies in the degree of pixelation between one area of the document and another should alert the document examiner to alterations. Figure 18.4 illustrates an obvious difference between signatures supposedly placed on the document before it was faxed. While the Haughey signature shows the classic stair-step effect associated with low-resolution reproduction, the Fendor signature displays no such attribute. The inconsistency between the two signatures is an indication that portions of the document have not been faxed. Further support for this position is demonstrated in Figure 18.5, where sample TTIs from the same machine are compared with that on the questioned document. While individual pixels can be seen in the two samples, a solid, blurred image is apparent in the questioned header. The filling-in effect is often seen when faxes have been photocopied multiple times.

18.4 Fax Machine Defects

Fax machines are no different than other office products in that they develop defects over time. One must recognize, however, that imperfections may originate from either the sending or receiving machine, or both. Furthermore, noise resulting from transmission anomalies must be distinguished from actual fax machine defects. For instance, vertical marks or lines found on a transmitted document may be absent from all other faxes involving the same two machines. This could result from a transitory problem or may involve communication line noise.

One of the defects that may be observed in faxed documents is a vertical black line, or printing void, resulting from a malfunction of the sending machine's scanning head. A similar feature may be the consequence of a malfunction in the receiving machine's printing system. Another example is an incomplete image resulting from crumpling of the original by a faulty transmitting machine paper-feed system. A related defect is the consequence of the receiving machine's feed mechanism failure to properly scroll the sheet, causing a horizontal portion of the image to be lost, obscured, or elongated. The printing apparatus of most modern fax machines utilizes the technology found in inkjet, bubble jet, or laser printers. As a result, many of the defects that manifest themselves in these printers will also be found in the product of facsimile machines, particularly when they are multi-function devices.

The condition of the TTI or RTI, as well as the nature and positioning of the defect, provides insight into whether the characteristic observed originates with the transmitting or sending machine (Figure 18.6). For example, a vertical void breaking all printing on the page except the TTI must result from a defect in the sending machine's scanning system. Similarly, a vertical black line running the length of the image area is likely the result of an obstructed scan element, such as might occur when opaquing fluid is transferred to the sensor during scanning. By contrast, if the same vertical line extends the full distance of

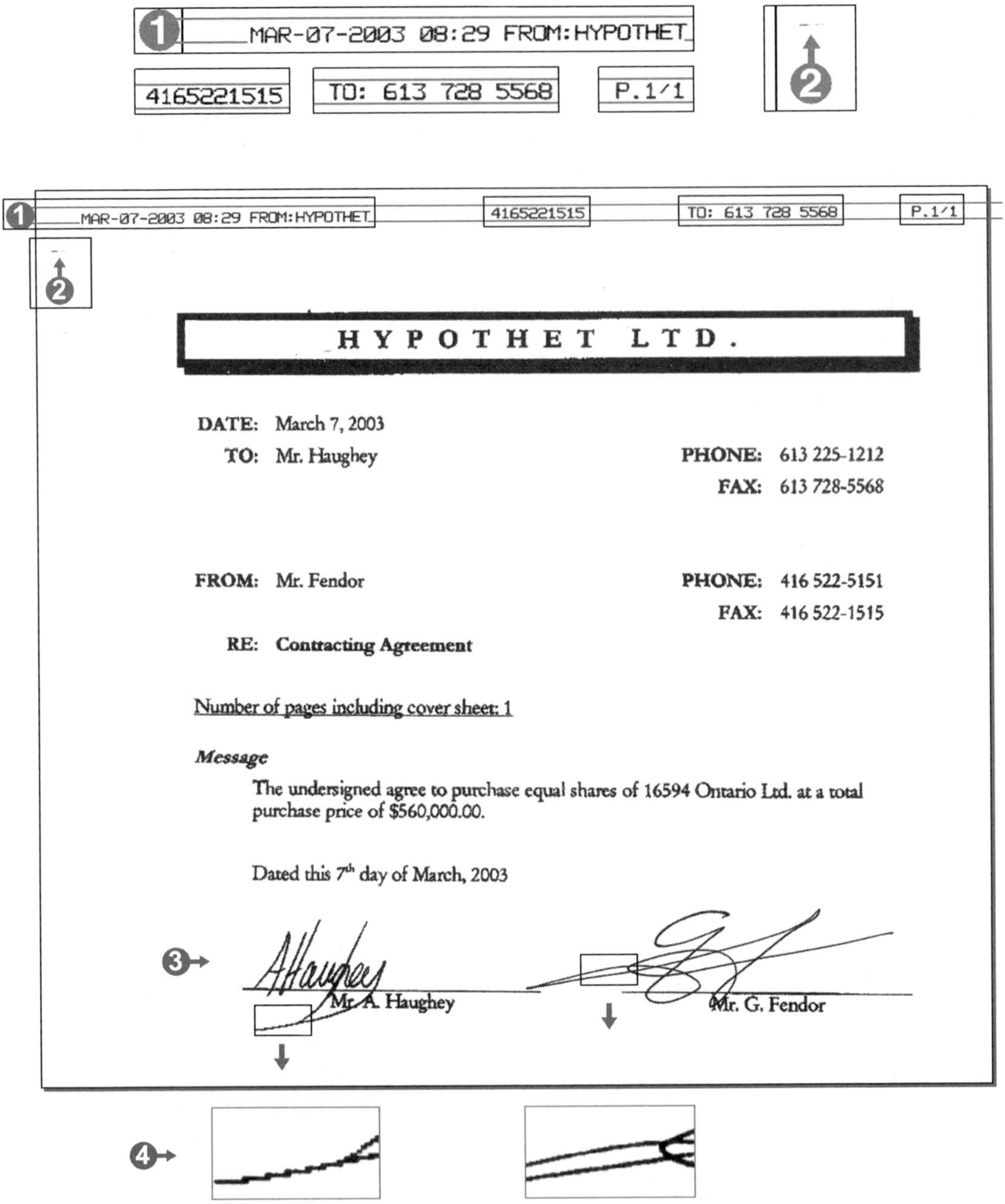
MAR-07-2003 08:29 FROM:HYPOTHET

4165221515 TO: 613 728 5568 P.1/1

MAR-07-2003 08:29 FROM:HYPOTHET 4165221515 TO: 613 728 5568 P.1/1

HYPOTHET LTD.

DATE: March 7, 2003

TO: Mr. Haughey PHONE: 613 225-1212

FAX: 613 728-5568

FROM: Mr. Fendor PHONE: 416 522-5151

FAX: 416 522-1515

RE: Contracting Agreement

Number of pages including cover sheet: 1

Message

The undersigned agree to purchase equal shares of 16594 Ontario Ltd. at a total purchase price of $560,000.00.

Dated this 7th day of March, 2003

Mr. A. Haughey Mr. G. Fendor

Figure 18.4 Presented as an original received facsimile copy, this document shows a number of characteristics consistent with document fabrication. Note the misalignment of the date and sender name (1) relative to the remainder of the information in the TTI. Residual marks (2) appear below this portion of the TTI that may be an indication of cutting and pasting. Although both signatures (3) were allegedly placed on the document before it was faxed, only the Haughey signature shows the pixelated lines that would be expected if this were true. As seen in the enlargements (4), the Fendor signature is more consistent with a photocopy reproduction.

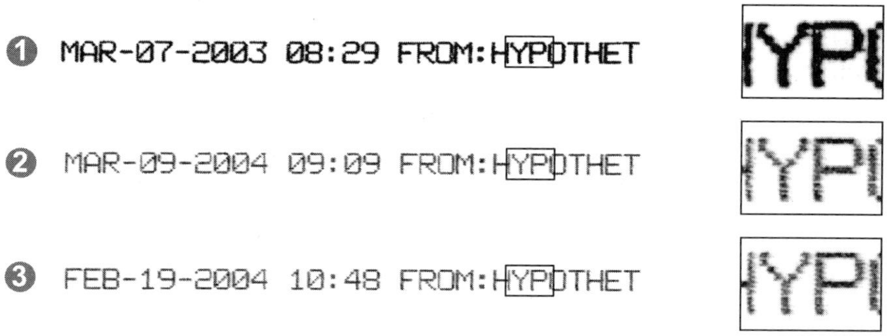

Figure 18.5 Another discrepancy noted in the March 7, 2003, fax (1) is the noticeable lack of definition in the TTI characters when compared with samples of other received fax copies (2 and 3). In both samples the individual pixels are evident, whereas in the suspect fax there is filling in between pixels, demonstrating that the document is a photocopy of a fax rather than the received original.

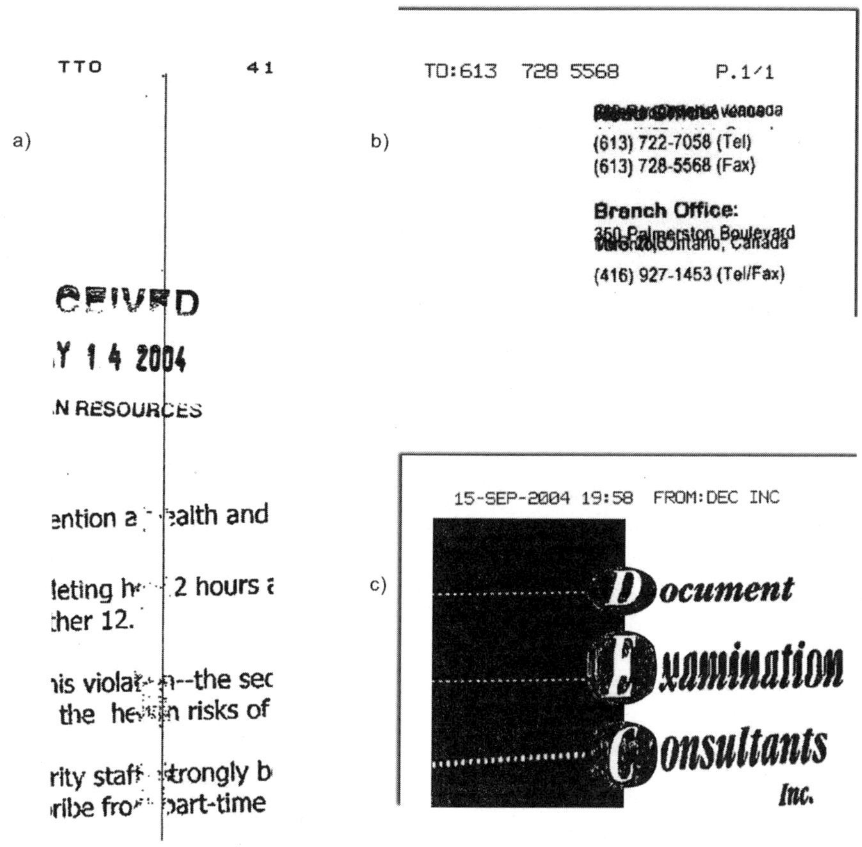

Figure 18.6 Examples of defects that may be found in the product of a facsimile transmission. One of the more common image flaws is a black line or printing void running vertically down the page. (a) An instance in which both are seen in a single transmission. Note that the solid black line does not extend into the TTI header, indicating that the defect lies with the scanning (sending) machine and not the printing (receiving) device. (b and c) The possible effect of paper-feed malfunctions in the receiving and transmitting machines, respectively.

the page, exceeding the image area, it could indicate a malfunction of the receiving machine's printing mechanism.

Other characteristics of identifying value can be found in traces left by the feed mechanism. These marks are similar to those found in photocopies and computer-generated documents. They can be in the form of small black smudges on the reverse side of the page, or appear only as indented impressions detectable through indentation analysis.[7]

18.5 Conclusion

Identifying the sending or receiving fax machine ideally requires submission of both devices as well as course-of-business samples generated or received by them. Sample fax copies dated around the time of the purported transmissions are critical in any comparative analysis. Assistance may be required from industry technicians to identify the origin of a particular defect or to provide guidance in evaluating the machine's function while in diagnostic mode.

References

1. Seaman Kelly, J., Facsimile documents: feasibility for comparison purposes, *J. Forensic Sci.*, 37, 1600, 1992.
2. Lindblom, B., Purdy, D., and Lange, K., Enlargement and reduction characteristics of facsimile transmission copies, *J. Am. Soc. Questioned Doc. Examiners*, 6, 38, 2003.
3. ASQDE, A Collection of Fax Fonts, Part II, paper presented at the American Society of Questioned Document Examiners meeting, Chicago, 1995.
4. Miller, J.L., TTI Database Fax Project III, paper presented at the American Society of Questioned Document Examiners meeting, Ottawa, 2000.
5. Lauterbach, J., A Collection of Fax Fonts, Part III, and Fax Font Database, paper presented at the American Society of Questioned Document Examiners meeting, Ottawa, 2000.
6. Lindblom, B., Facsimile Header (TTI) Examinations, paper presented at the American Society of Questioned Document Examiners meeting, Long Beach, CA, 1994.
7. Laporte, G.M., The use of an electrostatic detection device to identify individual and class characteristics on documents produced by printers and copiers: a preliminary study, *J. Forensic Sci.*, 49, 610, 2004.

Manipulated Photocopies, Faxes, and Computer-Printed Documents

19

BRIAN S. LINDBLOM
ROBERT GERVAIS

Contents

Photocopied, faxed, and computer-generated documents are particularly susceptible to text insertion, page substitution, cut-and-paste manipulation, and the use of multiple genuine documents to form fraudulent new composites. This is due in part to the ease with which such alterations can be made using the above technologies. Equally enticing to the would-be fraudster is the fact that reproductions of this type often conceal many of the common indicia of alterations. It is not at all uncommon to receive documents in which the distorting effect of the copying process and loss of detail have been increased through multi-generation reproduction (i.e., providing a copy of a copy of a copy …). Therefore, it is wise to consider all aspects of a document in order to discover indications of alteration.

19.1 Original Signatures vs. Color Reproductions

One of the first examinations that should be undertaken is an assessment of any signatures or other handwritten information on the document to establish whether the writings are original or reproductions. Many document examiners have been presented with a document that is purported to be an original, but on further inspection shows characteristics consistent with color copying or inkjet printing to simulate, for instance, blue ink. In most cases, the cyan, magenta, yellow, and black dots comprising the process color will be visible under magnification (see color Figure 19.1 following p. 366). In some disputes, simply proving that the signature and handwriting are color copies or prints and not originals resolves the issue.

19.2 Identifying and Comparing the Printing or Reproduction Process

It has, over time, become more and more difficult to distinguish photocopied from computer-generated documents. In fact, as all of the printing and imaging technologies used in one are now found in the other (in many cases even in the same multi-function machine), it is often impossible to make such a distinction. As a result, many aspects of the discussion that follows are equally applicable to photocopied and computer-generated documents.

Any examination of faxed, photocopied, or computer-generated materials should include a general assessment of the printing technology employed. As many printing methods exhibit visually distinctive characteristics, identification can often be achieved through microscopy with the aid of various light sources. Some of the more common printing technologies currently in use include electrostatic, bubble jet/inkjet, thermal wax/pigment, dye sublimation, and dot matrix. The identification of the printing technology used can often shed light on the plausibility of testimony given as well as offer the potential to expose text insertion or page substitution.

In the event that a single printing method is determined, it may be possible to further classify the technology and distinguish differences in printer settings, or even differentiate printers, through the use of more sophisticated examination techniques. In the example shown in Figure 19.2, two lines of text have been generated with an inkjet printer at normal setting, while the final sentence was printed at a higher resolution. Differences in the ink dispersion, along with the location of satellite droplets around the letters, serve as discriminating characteristics between the two outputs.

Toner particle characteristics and dispersion patterns on documents produced with laser printers or photocopiers can be examined and compared using Fourier transform infrared (FTIR) spectroscopy, scanning electron microscopy (SEM), confocal microscopy, or energy dispersive x-ray (EDX). Some marking media, such as bubble jet/inkjet ink and sublimation dyes, can be distinguished from one another through the use of infrared or ultraviolet examination.

With regards to the electrostatic photocopying process, a further subclassification can be made between those machines employing analog technology and the newer digital systems. While it may seem, at first glance, to be a simple matter to identify the pixelation created by digital copiers and contrast it with the non-digitized images created by analog machines, considerable care must be taken to ensure that any pixelation observed is not simply a reproduction of characteristics found in the original or an intermediate generation copy. With modern laser printer output resolutions in the 1200- to 2400-dpi range, the FDE must be cautious when identifying a document as an analog photocopy. At these resolutions the pixel width is so small that toner particle size, toner clumping, and paper fibers may pose significant interference problems in assessing digitization, even at high magnification.

charge of $20 per day for any amount that is more than five (5) days late. Rent will be prorated if the term does not start on the first day of the month or for any other partial month of the term. As a signing bonus, the first 6 months of residency shall be at the reduced rate of $500.00.

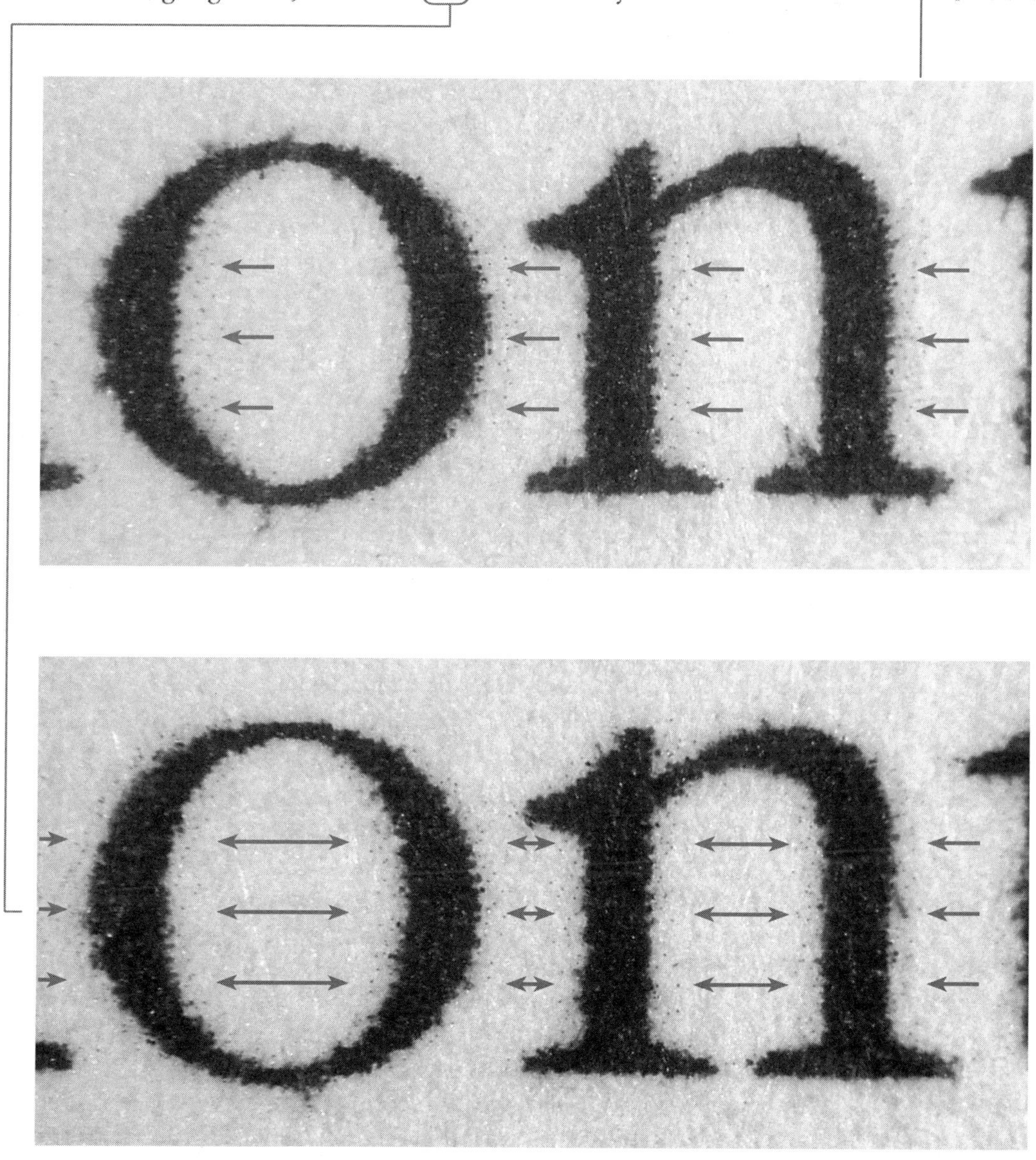

Figure 19.2 The illustration shows an example of text printed using an inkjet printer at normal and fine settings. Note the overspray appears on both sides of the characters in the lower image (fine), while in the upper image the overspray appears primarily to the right (normal).

The ability to determine whether more than one machine has been used to produce text within a multi-page document, or even between one area and another on the same sheet of paper, is sometimes of critical importance. While gross differences may be evident upon microscopic analysis, other differences can only be detected using more sophisticated analytical tools. Leading-edge research is being conducted to measure print quality with the use of specialized machine vision systems. Using these tools, the document examiner can evaluate such parameters as dot position, edge raggedness, overspray, satellites, and text quality, to name a few.[1]

Imaging and printing defects on documents produced by fax machines, copiers, and computer printers have been discussed in preceding chapters. Such defects can have value in identifying material generated on more than one machine and therefore can be used to identify substitution of pages or even interlineations on a single sheet. For example, a printing void may appear on all but one page of a computer-generated last will and testament. This discrepancy, together with other differences (such as in paper, staple holes, and writing indentations), can demonstrate that there has been page substitution.

The true value of these examinations lies in their use as comparative techniques, that is, the comparison of one portion of the document with another to establish whether or not there is consistency throughout. Similarly, the findings may be compared to the output of machines put forth as having been used to create the document — allowing them to be included or excluded as the possible source.

19.3 Typographic Considerations

Once the printing technology used to generate a given document has been fully assessed, the text should be examined to establish whether or not the same typestyle has been used throughout. Given the similarities between many different typefaces, this analysis requires close attention to the fine typographic details. Overlaying repeated letters or word combinations is one method to quickly compare individual characters on a document. The FDE must bear in mind that pages of a different copy generation may be slightly different in size, and this will impact on the comparison. Obviously, the printing of different document sections at different point sizes for formatting purposes must also be considered. Similarly, non-linear reproduction distortion can often be seen within a single page. The later the copy generation, the more difficult it is to determine whether all of the text is of a similar typestyle and point size. This is because of the filling-in effect and loss of fine detail that occurs through sequential copying. A similar problem confronts the FDE when intercomparing text in a facsimile copy, where the resolution is generally between 100 and 200 dpi. Any differences in type fonts, or styles, used must be considered in the context of the document being examined. That is to say, is there a logical reason why more than one typestyle appears?

19.4 Assessing Alignment, Spacing, and Copy Distortion

The next step in the examination process is to assess line spacing throughout the document. Various measurement techniques can be employed, including glass or plastic measuring templates or the use of scanning and graphics software. Regardless of the method employed,

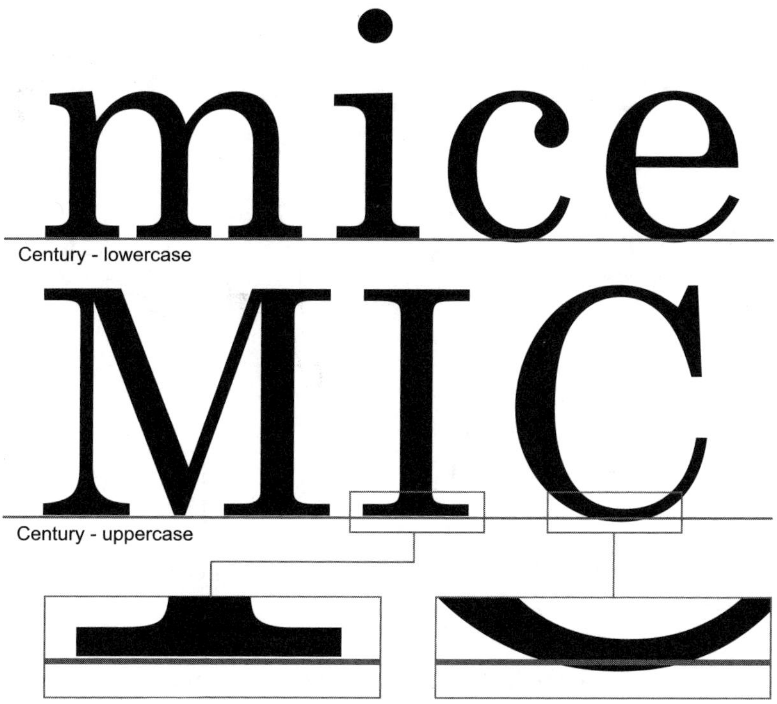

Figure 19.3 An example of optical correction. Note that the letters with curved bases descend slightly below the true baseline, whereas those with a flat bottom do not.

the FDE must be cognizant of any distortion, linear or otherwise, that may be present. This is especially true for multi-generation or fax copies, in which it is not uncommon to see the text baseline undulate across the page. In such circumstances, only very general measurements can be made. To properly assess any distortion present, multiple line measurements should be made. It would be a mistake to focus solely on the entry in question relative to the lines immediately above and below. Differences in line and margin spacing are only relevant if the surrounding text is consistently spaced.

Two elements of typography that come into play when making such measurements are optical correction and special characters. The term *optical correction* refers to the typographic design practice in which characters with rounded bases are positioned so that they actually sit slightly below the true baseline. It is important to recognize, therefore, that different letters and numbers in a given typestyle may vary in their base alignment relative to one another. For this reason, the measurement of baselines should not rely upon characters such as *O*, *e*, or *C* (Figure 19.3). Similarly, special characters may or may not descend below the baseline, depending on the typestyle (Figure 19.4). A further caution that must be considered is that the presence of even a single character at a different point size, or using a different typeface, can shift the baseline of the entire line and, therefore, all those that follow (Figure 19.5).

Whether evaluating a facsimile reproduction, photocopy, or computer-generated text, Adobe Photoshop® or other similar software can be a valuable tool in assessing line orientation. The document is scanned into Photoshop (Figure 19.6a), where the measurement tool is used in concert with the Rotate Arbitrary function to bring the document to a right angle based on a selected line of text (Figure 19.6b). Once this is done, the same set of

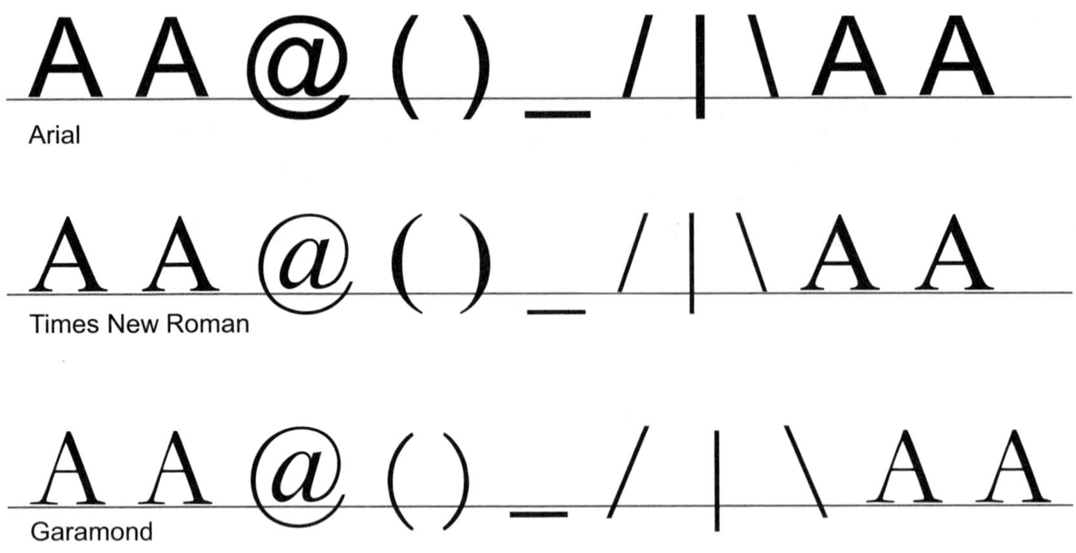

Figure 19.4 Examples in three different typestyles in which certain characters descend below the baseline by design.

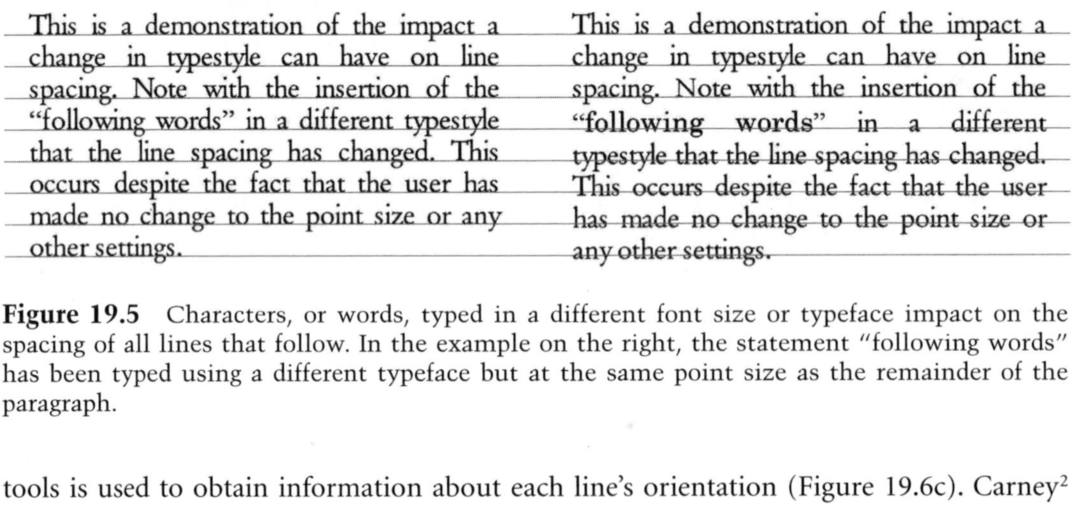

Figure 19.5 Characters, or words, typed in a different font size or typeface impact on the spacing of all lines that follow. In the example on the right, the statement "following words" has been typed using a different typeface but at the same point size as the remainder of the paragraph.

tools is used to obtain information about each line's orientation (Figure 19.6c). Carney[2] recommends performing the measurement of each line multiple times to ensure accuracy.

Graphics software can be effective in the evaluation of line spacing as well. Photoshop, for example, has a built-in grid feature that allows the user to call up a superimposed grid with pixel-level control over its spacing. In other programs, such as Macromedia's Freehand®, the scanned image can be imported and an evenly spaced grid created by marking the baseline of any two lines and then "blending" them with the appropriate number of line increments (Figure 19.7). This method is unaffected by reproduction scaling problems. It not only helps identify any line(s) not conforming to regular spacing, but also provides quick insight into anomalies that may be present resulting from formatting decisions, font or style changes, or distortion. As can be seen, this technique together with the previously cited analytical procedure are key components in the FDE's arsenal.

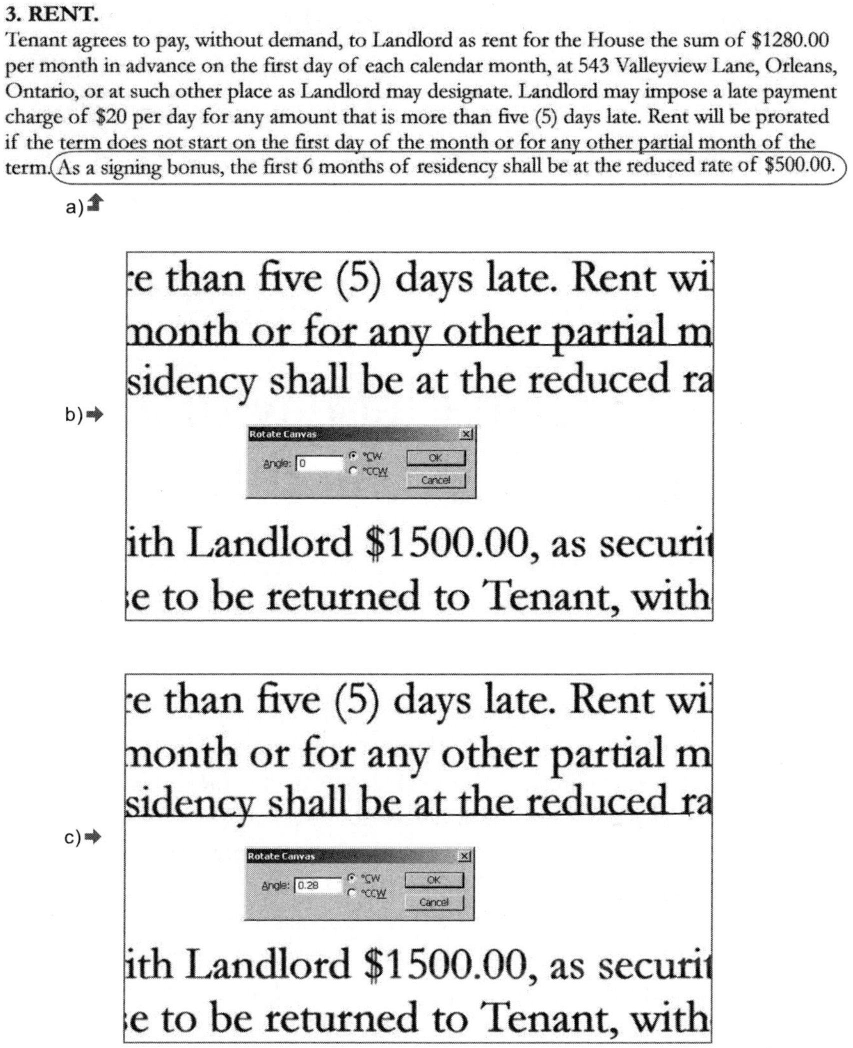

3. RENT.
Tenant agrees to pay, without demand, to Landlord as rent for the House the sum of $1280.00 per month in advance on the first day of each calendar month, at 543 Valleyview Lane, Orleans, Ontario, or at such other place as Landlord may designate. Landlord may impose a late payment charge of $20 per day for any amount that is more than five (5) days late. Rent will be prorated if the term does not start on the first day of the month or for any other partial month of the term. As a signing bonus, the first 6 months of residency shall be at the reduced rate of $500.00.

Figure 19.6 It was suspected that a line of type might have been added to the end of a paragraph. By using the Rotate Arbitrary function in conjunction with the measuring tool of a software program, it is possible to calculate the angle of lines relative to one another, thereby identifying any misalignments that may point to any interlineations.

19.5 Special Considerations for Manipulated Faxes

The cutting and pasting of signatures, paragraphs of text, and portions of the Transmit Terminal Identifier (TTI) header are common alterations to facsimile copies. The low-resolution found in fax transmissions serves, in part, to conceal some of these fabrications. A careful inspection must be made of all of the information in the fax to ensure that there is a degree of consistency throughout. Does the signature(s) show pixelation consistent with having been faxed? Are there any obvious and measurable differences in margins, typestyle, or spacing of the text and TTI (Figure 19.8)? Is all of the TTI detail in the same typestyle? Do the characters in the TTI detail show consistent base alignment? Is the TTI

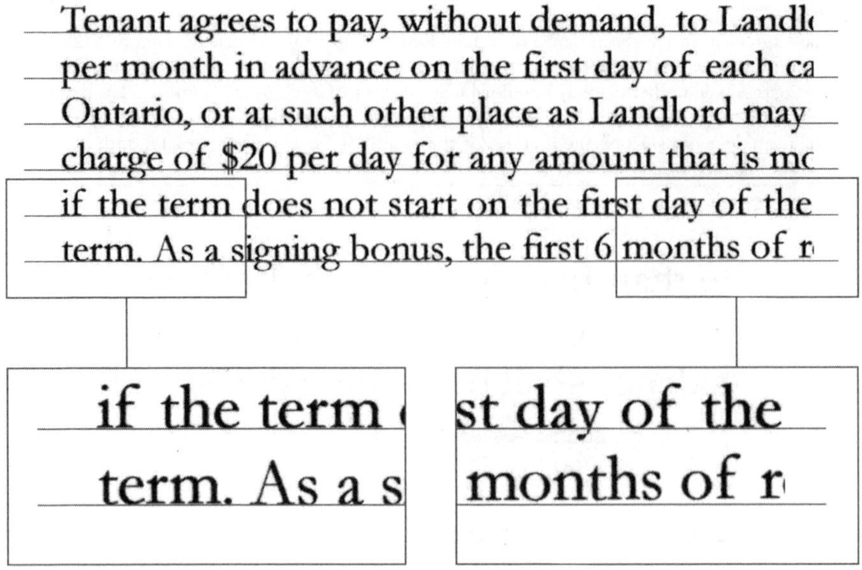

Figure 19.7 Spacing grids can be used to evaluate line spacing and distortion in a computer-generated or typed document. Software versions of measuring templates are now more commonly used than glass or plastic versions due to the fact that they give the user greater flexibility.

a sharp image, or does it show evidence of reproduction (Figure 19.9)? These are basic questions that need to be answered in the examination of a facsimile copy.

Although image resolution is low in a faxed document, this characteristic in itself can have forensic value. For instance, an assessment can be made to ensure consistent pixel orientation throughout the document (Figure 19.10). Side-by-side signature lines, though parallel at a macro-level, may show pixel arrays that are rotated relative to one another, demonstrating that they were not faxed at the same time, but rather are the product of two transmissions. Additionally, assessing the resolution at which the transmission was made may distinguish a substituted page from the original faxed information. The normal low 100 × 200 dpi pixelation will be much more prominent than at the 200 × 200 dpi fine or 200 × 390 dpi ultrafine settings. Intercomparison of the pages within a fax transmission may reveal such inconsistencies.

19.6 Conclusion

A careful, methodical approach must be taken to help ensure that fabrications and manipulations in machine-generated documents are discovered. The information on a single sheet of paper should be intercompared as well as compared against the text on other pages of the same document and, where applicable, with sample outputs from the machines involved. Consideration must be given to the printing technologies involved, alignment and spacing characteristics, typography and formatting, imaging and printing defects, and print quality and resolution.

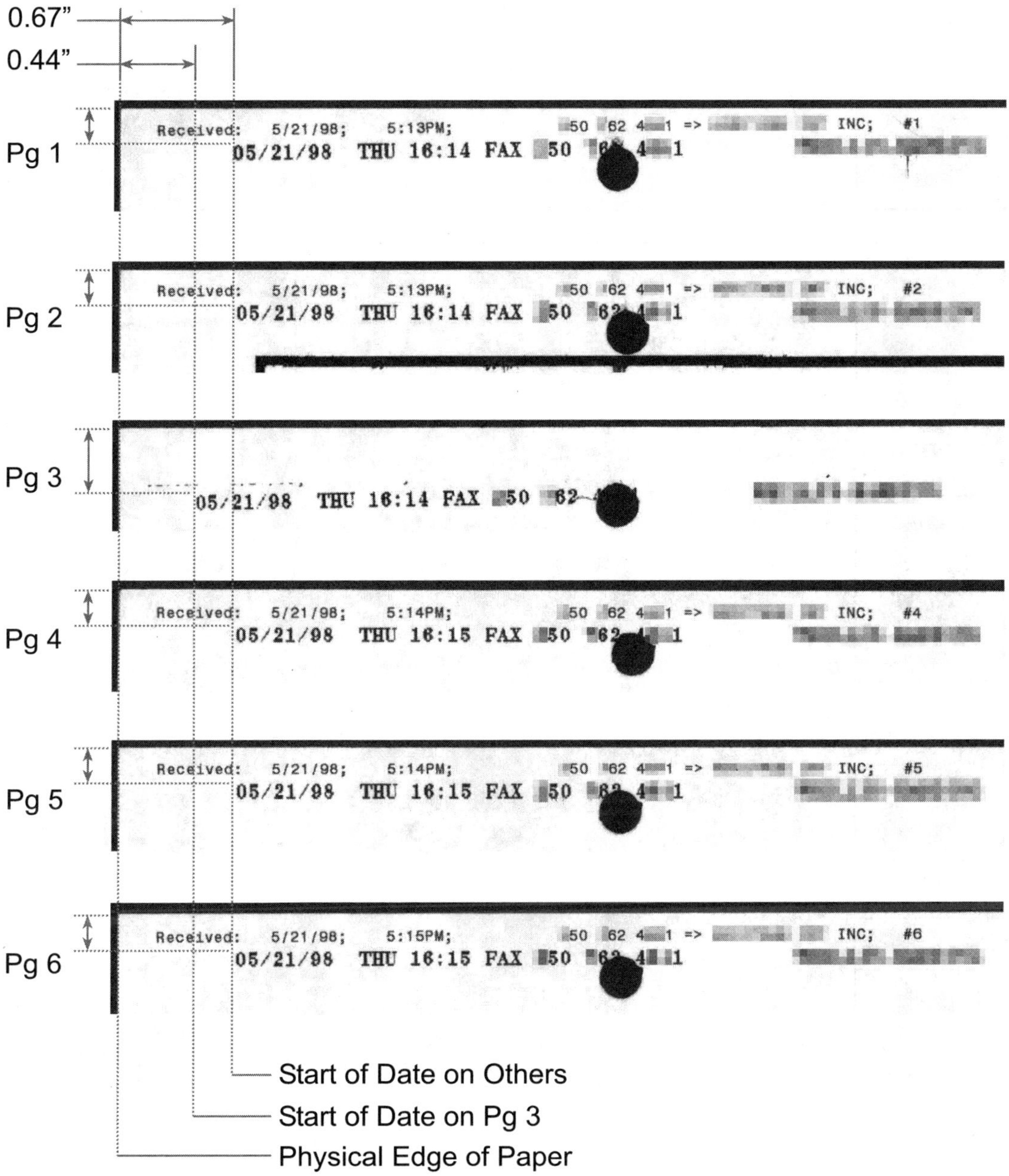

Figure 19.8 An examination of the TTI and Receive Terminal Identifier (RTI) headers reveals several differences between those on Pg. 3 and the other pages in the same fax transmission. These include: both the RTI and TTI are positioned lower on the page; the TTI is aligned closer to the left edge of the page; and the RTI is cropped despite being positioned lower on this page than on any of the others. When considered together, these differences suggest that Pg. 3 may be a second- or later-generation reproduction.

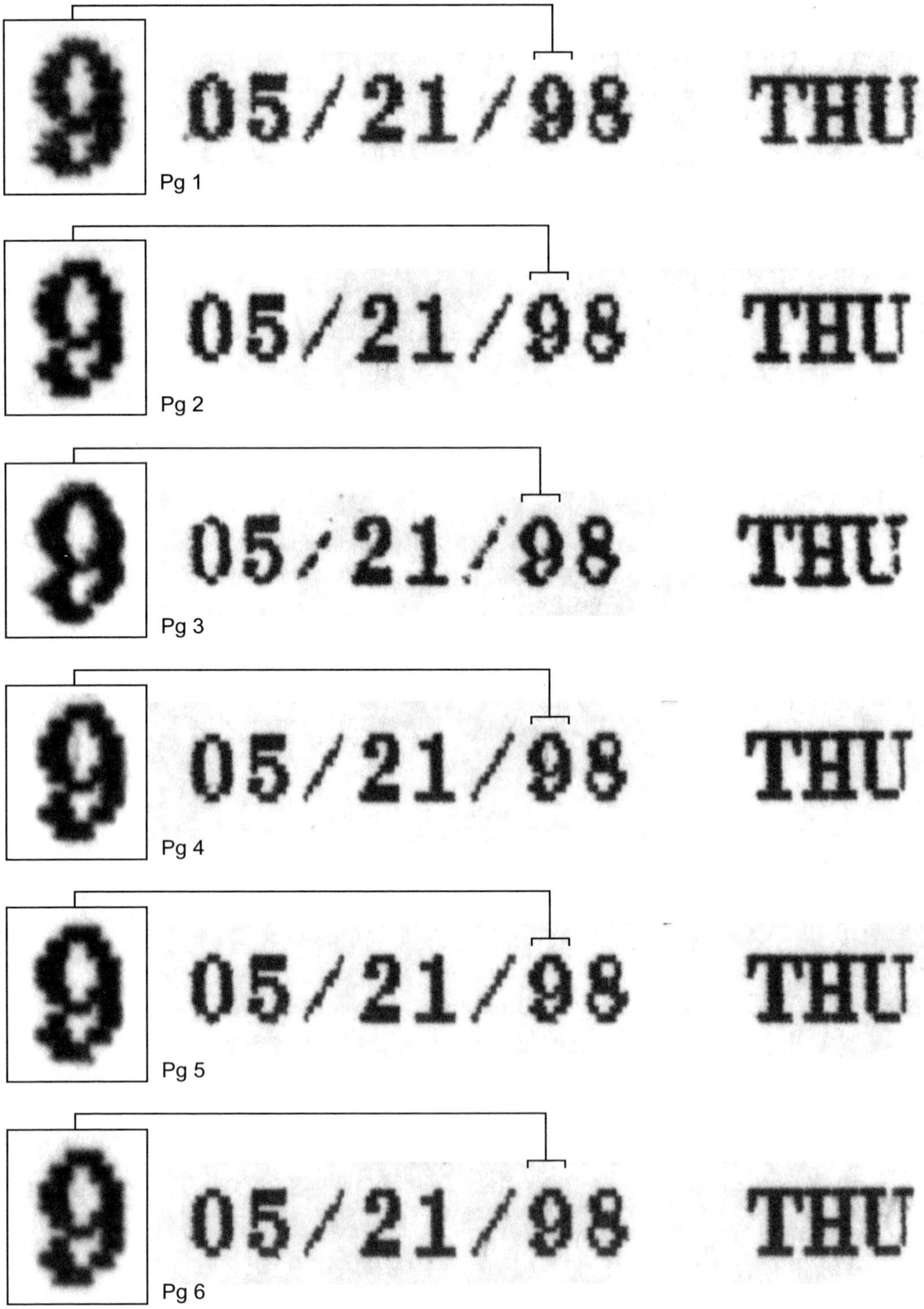

Figure 19.9 A difference can be seen in image quality between Pg. 3 and the balance of the transmission. The TTI shows less defined edges (note the stem), filling in, broken lines, and a less obvious matrix pattern. This evidence supports the position that Pg. 3 may be a second- or later-generation reproduction.

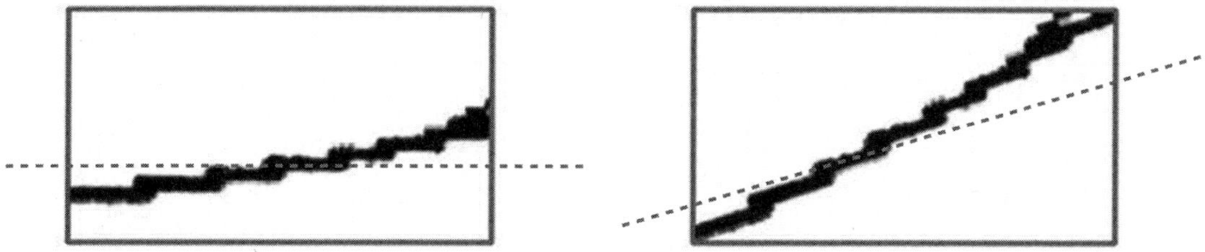

Figure 19.10 Evidence that a document has been fabricated using portions of two different faxes is seen in the different pixel orientation of two signatures that are located side by side. Had both been digitized at the same time and sent in a single transmission, there would not be such a discrepancy.

References

1. Oliver, J. and Chen, J., Use of Signature Analysis to Discriminate Digital Printing Technologies, paper presented at the International Conference on Digital Printing Technologies, San Diego, CA, 2002.
2. Carney, B., Using Adobe Photoshop® to Detect Document Alterations through Interlineations, paper presented at the annual meeting of the American Society of Questioned Document Examiners, Baltimore, 2003.

Rubber Stamps

20

JAN SEAMAN KELLY

Contents

In the 1982 edition of *Scientific Examination of Questioned Documents*, Ordway Hilton wrote:

> As an office laborsaving device, the hand or impression stamp is convenient to use and useful; as a field of investigation for the document examiner, it is rather infrequent but diversified. Much of its diversity lies in the great variety of stamps in use: those of fixed letter design, those made up entirely of loose type, the facsimile signature stamp, those with rotating sections of the time-date stamp variety, the date stamps with movable strips of type, and any combination of these. The diversity is not limited to classifications based upon makeup. The actual material from which the stamp is constructed and its mode of manufacture adds to the possible varieties.[1]

In the above paragraph, Hilton's text refers more to variety in the intended usage of the stamp and not to choice of die materials or manufacturing processes. Variety existed in the mechanical or structural makeup of the rubber stamp in order to meet the applicable needs of business or industrial use. The technological advancements of the last 30 years have brought forth an expansion in manufacturing processes as well as improvements in the materials used by the rubber stamp industry. Prior to 1970, the materials used for conventional rubber stamp dies were either vulcanized rubber for self-inking stamps or pre-mixed gel for pre-inked stamps. The manufacturing process of vulcanization using the Ludlow or the Linotype was used for both materials. The introduction of liquid photopolymer as a die material occurred in the late 1970s and

Figure 20.1 Close-up of a photopolymer die. Photopolymer die can be translucent, clear, or reflect a milk-white, pink, or yellow tint. The photopolymer die is raised. (From Seaman Kelly, J., *Forensic Examination of Rubber Stamps*, Charles C. Thomas Publisher, Ltd., Springfield, IL, 2002. With permission.)

became widely used in the stamp manufacturing process by the mid-1980s. Maureen A. Casey[2] and Jay Levinson and Benjamin Perelman[3] authored the first published articles discussing the ultraviolet process (manufacturing process used to produce the photopolymer die) and the characteristics a document examiner would observe in a stamp impression produced by such a die.

Since 1990, photopolymer stamps make up the lion's share of the rubber stamp market (Figure 20.1). In the U.S., vulcanized rubber dies still exist, but are decreasing in numbers due to the replacement of the Ludlow and Linotype machines by more cost efficient manufacturing processes such as ultraviolet and laser (Figure 20.2). Until the introduction of flat-die stamps, the number of pre-inked stamp manufacturers was small due to the high manufacturing costs associated with this type of process. The introduction of light-burst and thermal manufacturing processes expanded the pre-inked stamp market. The ease of machine cost and operation allows office supply stores and private individuals to make quality pre-inked stamps at minimal cost, minimal training, and minimal space requirements.[4]

20.1 Stamp Classification

Stamp classification is not based on the type of die material, but rather the location of the ink source.[5] Based on the ink source location, there are four main types of conventional stamps encountered by FDEs in their casework: the hand stamp, the self-inking stamp, the pre-inked stamp, and the flat-die stamp. Personal, business, and industrial use rubber stamps are produced by the same manufacturing processes and are classified by the ink

Figure 20.2 Close-up of a rubber die. This raised rubber die is manufactured using the rotary laser. (From Seaman Kelly, J., *Forensic Examination of Rubber Stamps*, Charles C. Thomas Publisher, Ltd., Springfield, IL, 2002. With permission.)

source. However, the die materials and inks differ in the industrial setting due to the marking needs of a particular industry.[6] Even though discussion in this text is limited to personal or business use rubber stamps, the FDE should be aware that other die materials are used in the industrial setting.

Personal and business use hand and self-inking stamps contain dies made of vulcanized rubber or photopolymer. The die of a hand stamp is mounted on a die plate attached to a knob handle mount or a molded mount. Because the die is exposed to its surrounding environment, it is subject to damage and debris contamination. Inking of the die is accomplished by manually pressing the die onto the stamp pad in order to obtain ink coverage. The inking process of the hand stamp is achieved manually, while the inking process of the self-inker is achieved mechanically.[7] The self-inking stamp consists of a container (usually plastic) that houses the stamp die and ink pad (Figure 20.3). Since the die is in a container that usually has a cover, it is protected from excessive debris contamination or cuts and other damage that can occur as a result of colliding with other objects one may find in or on a person's desk. For inking, the die rotates 180° to press into the ink pad when the top of the container is pressed down.

The first two classifications concern dies that primarily use water-based inks. The third classification includes pre-inked stamps characterized by a high relief die and, as a general rule, use oil-based inks (Figure 20.4). Salt-leached rubber, foam and powder, and pre-mixed gel are the materials used to make high relief pre-inked stamps. The flat-die stamp is the fourth classification, and it has a low- or no-relief die, with the ink pad sandwiched between the die and the die holder (Figure 20.5 and Figure 20.6). Depending upon their intended use, water- or oil-based inks are used in flat-die stamps.

Figure 20.3 Hand stamp with a wood handle mount is shown on the right side of the photograph. A self-inking stamp is shown on the left. The hand stamp is dependent upon a separate ink pad for inking. The self-inking stamp has a small pad inside the container. The raised die (vulcanized rubber or photopolymer) rotates 180° to contact the ink pad prior to being pressed onto the paper. (From Seaman Kelly, J., *Forensic Examination of Rubber Stamps*, Charles C. Thomas Publisher, Ltd., Springfield, IL, 2002. With permission.)

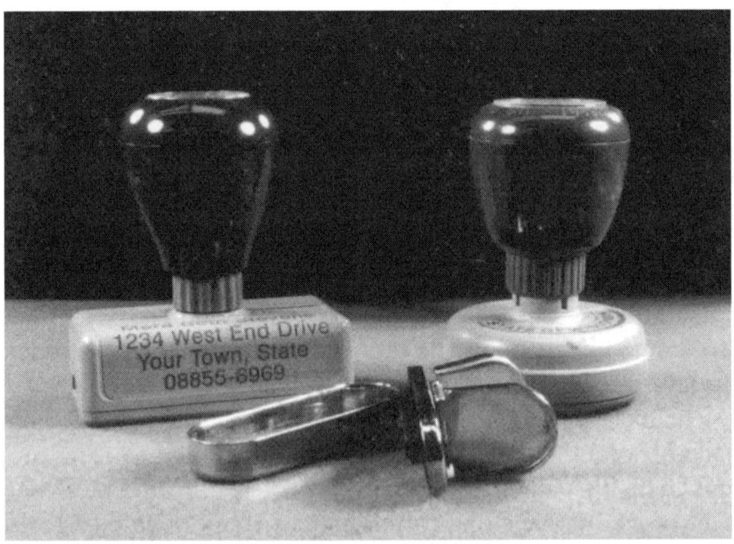

Figure 20.4 Samples of various containers used to house the pre-inked dies. The pre-inked die is made of materials such as pre-mixed gel, salt-leached rubber, and foam and powder, which are raised (type high) dies. (From Seaman Kelly, J., *Forensic Examination of Rubber Stamps*, Charles C. Thomas Publisher, Ltd., Springfield, IL, 2002. With permission.)

20.2 Characteristics of Rubber Stamp Impressions[8]

The characteristics an FDE observes in stamp impressions are influenced by the die material, whether the ink is water based or oil based, the size of the stamp, the type of paper, the interaction of the ink to the paper, and the individual making the impression.

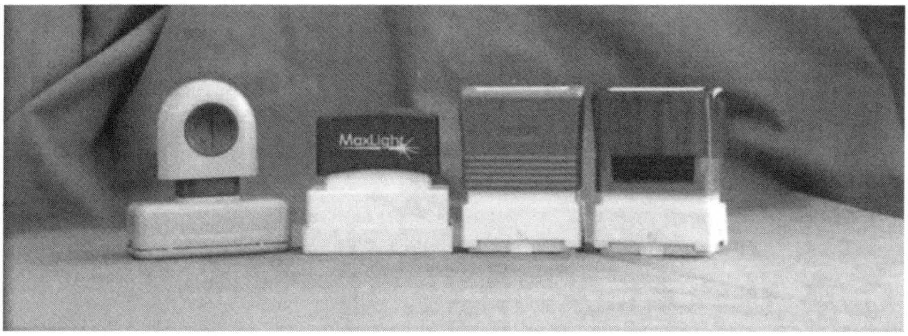

Figure 20.5 Various containers used to house low- or no-relief dies. Low- or no-relief die inks can be water-based or oil-based inks. The containers have a cover to protect the die from damage or environmental debris. (From Seaman Kelly, J., *Forensic Examination of Rubber Stamps*, Charles C. Thomas Publisher, Ltd., Springfield, IL, 2002. With permission.)

Figure 20.6 Close-up of a low-relief die. The thermal printing process was used to manufacture this die. The ink exits through the micropores of the text image. (From Seaman Kelly, J., *Forensic Examination of Rubber Stamps*, Charles C. Thomas Publisher, Ltd., Springfield, IL, 2002. With permission.)

The two most common causes of a less-than-perfect stamp impression involve the mechanics of stamping, i.e., failing to hold the stamp correctly, creating an uneven impression, and using too much force when making the impression, causing the stamp to bounce.

Differences are observed between impressions made by hand or self-inking stamps and pre-inked stamps. As a general rule, it would be difficult to make a definitive statement as to the type of stamp that made the questioned impression. However, characteristics observed in a stamp impression can provide clues as to the die material of the suspected stamp. Vulcanized rubber and photopolymer are the most common materials used for hand and self-inking stamp dies. These two materials are non-porous and do not retain ink. Whether the material is vulcanized rubber or photopolymer, they share the same characteristics in their impressions, as follows (Figure 20.7a and b):

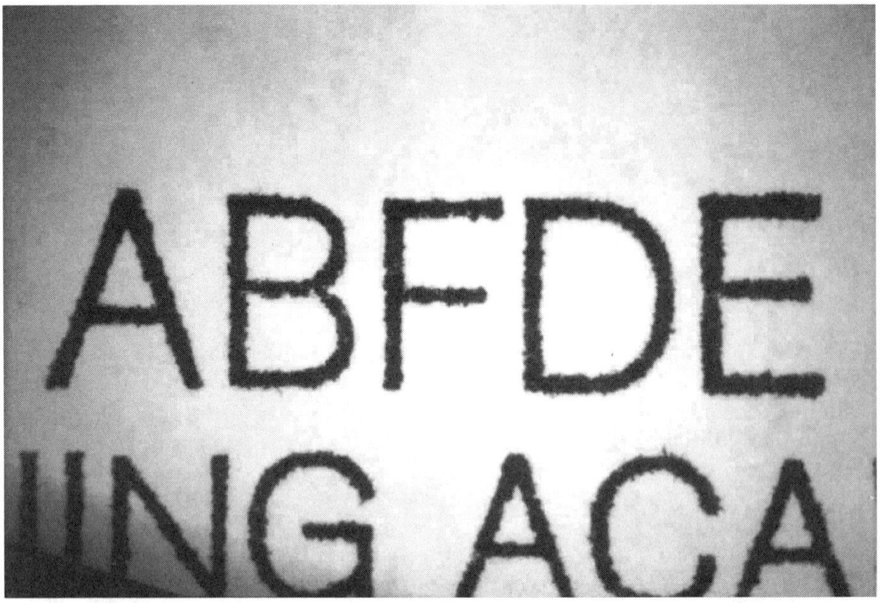

(a)

(b)

Figure 20.7 (a) Impression produced by a vulcanized rubber die. (b) Impression created by a photopolymer die. Even ink coverage with minimal patchy areas, ink filling in the intersection points of two lines, and rounded beginnings and endings of letters are a few of the characteristics that characterize a rubber stamp impression. (From Seaman Kelly, J., *Forensic Examination of Rubber Stamps*, Charles C. Thomas Publisher, Ltd., Springfield, IL, 2002. With permission.)

- Even ink coverage.
- Ring of darker ink outlining the individual letter. Known as the "squeegee effect," this ring occurs as a result of the relief of the printing area squeezing the ink out to the edge of the ink line. This characteristic is difficult to observe if the entire character is heavily inked.
- Absence of an indentation in the ink line.
- Rounded beginning and ending of letters.
- If water-based ink is used, the bleeding of ink through the paper is minimal.
- Ink filling in sharp angles and intersection points of two lines.
- Some patchy areas within the inked impression may be observed.
- Uneven outline of the letter.

Due to its porosity, the die of a pre-inked stamp serves as its own ink reservoir. The materials commonly used for pre-inked dies include pre-mixed gel, salt-leached rubber, and foam and powder. Dies made of foam and powder can contain water- or oil-based ink. The remaining die materials only use oil-based inks for personal or business use. Since the die of a pre-inked stamp is soft and has some flexibility, the impression differs slightly from the characteristics one would observe in an impression produced by a hand or self-inking stamp. The characteristics observed may include (Figure 20.8a to c):

- Clean and concise detail
- Even ink saturation throughout the individual characters
- Absence of a heavier ink line on the inside or outside of the letterform
- Feathering or bleeding of ink on the edges of the letter
- Rounded beginnings and endings of letters
- Absence of an indentation in the ink line
- Blurring or distortion in small typed text
- Oil-based ink bleeding through the paper (Figure 20.9)

Flat-die stamps are divided into two categories: light-burst and thermal printer technology. Light-burst technology uses a Xenon flash to seal the non-print area of the die (Figure 20.10 and Figure 20.11). The only pores left open that allow ink to pass through the die will be the printed areas. Heat is used in the thermal printer technology to seal the micropores in the non-print area of the flat die (Figure 20.12a and b). The characteristics of an impression from either flat-die category closely mirror the characteristics found in the pre-inked stamp impressions. However, a few of the flat-die stamps produce impressions that contain characteristics that are more unique to a particular model. For example, the Brother SC 300 PC makes flat-die stamps using thermal printer technology. This particular model was one of the first flat-die machines marketed to the general public. The impression made from a Brother SC 300 PC flat-die stamp will reflect a dot matrix pattern within the printed text (Figure 20.13). The dot matrix pattern varies in relation to the type size of the text. As the type size becomes smaller, the matrix evolves into a sawtooth pattern (Figure 20.14). Conversely, the larger the stamp, the more dots in the printed text. Aside from the unique pattern, the remaining characteristics of the impression mimic those observed in other pre-inked stamp impressions.

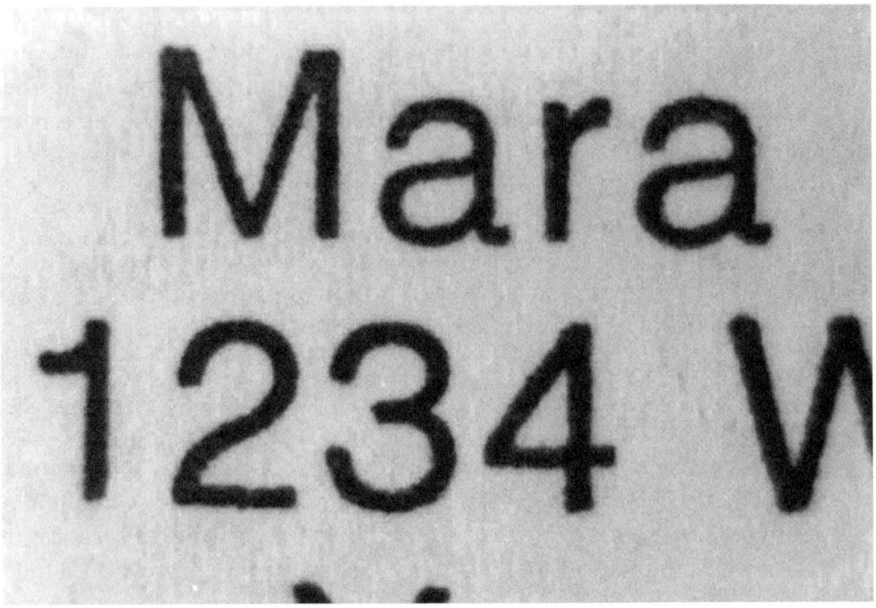

(a)

(b)

Figure 20.8 (a) Impression produced by a pre-mixed gel die stamp. (b) Impression created by a salt-leached rubber die. (C) Impression produced by a flat-die stamp (light-burst technology). Oil-based ink comprises the inked area of the text. Even ink coverage, feathering of ink around the edges of the letter, and rounded beginnings and endings of letters are a few of the characteristics observed in an oil-based ink stamp impression. (From Seaman Kelly, J., *Forensic Examination of Rubber Stamps*, Charles C. Thomas Publisher, Ltd., Springfield, IL, 2002. With permission.)

(c)

Figure 20.8 (continued)

Figure 20.9 Bleed-through of ink observed on the backside of a sheet of paper. It is not uncommon for an oil-based ink, such as the type used in pre-mixed gel, salt-leached rubber, and flat-die stamps, to bleed through the paper hosting the impression. (From Seaman Kelly, J., *Forensic Examination of Rubber Stamps*, Charles C. Thomas Publisher, Ltd., Springfield, IL, 2002. With permission.)

Figure 20.10 Flat-die stamp manufactured using the light-burst technology. A low- or no-relief surface characterizes this type of die. Ink exits through the micropores on the die onto the paper. The ink source is behind the die plate. (From Seaman Kelly, J., *Forensic Examination of Rubber Stamps*, Charles C. Thomas Publisher, Ltd., Springfield, IL, 2002. With permission.)

Figure 20.11 Impression from a light-burst flat-die stamp. The impression contains the characteristics of an oil-based ink stamp. (From Seaman Kelly, J., *Forensic Examination of Rubber Stamps*, Charles C. Thomas Publisher, Ltd., Springfield, IL, 2002. With permission.)

20.3 The Examination Process[9]

A great deal of evidential information is overlooked when a stamp impression on a document is not included in the examination. By taking the time to examine the overlooked evidence, the document examiner may be able to provide additional information to the submitter. For example, a case was submitted requesting handwriting, signature, and

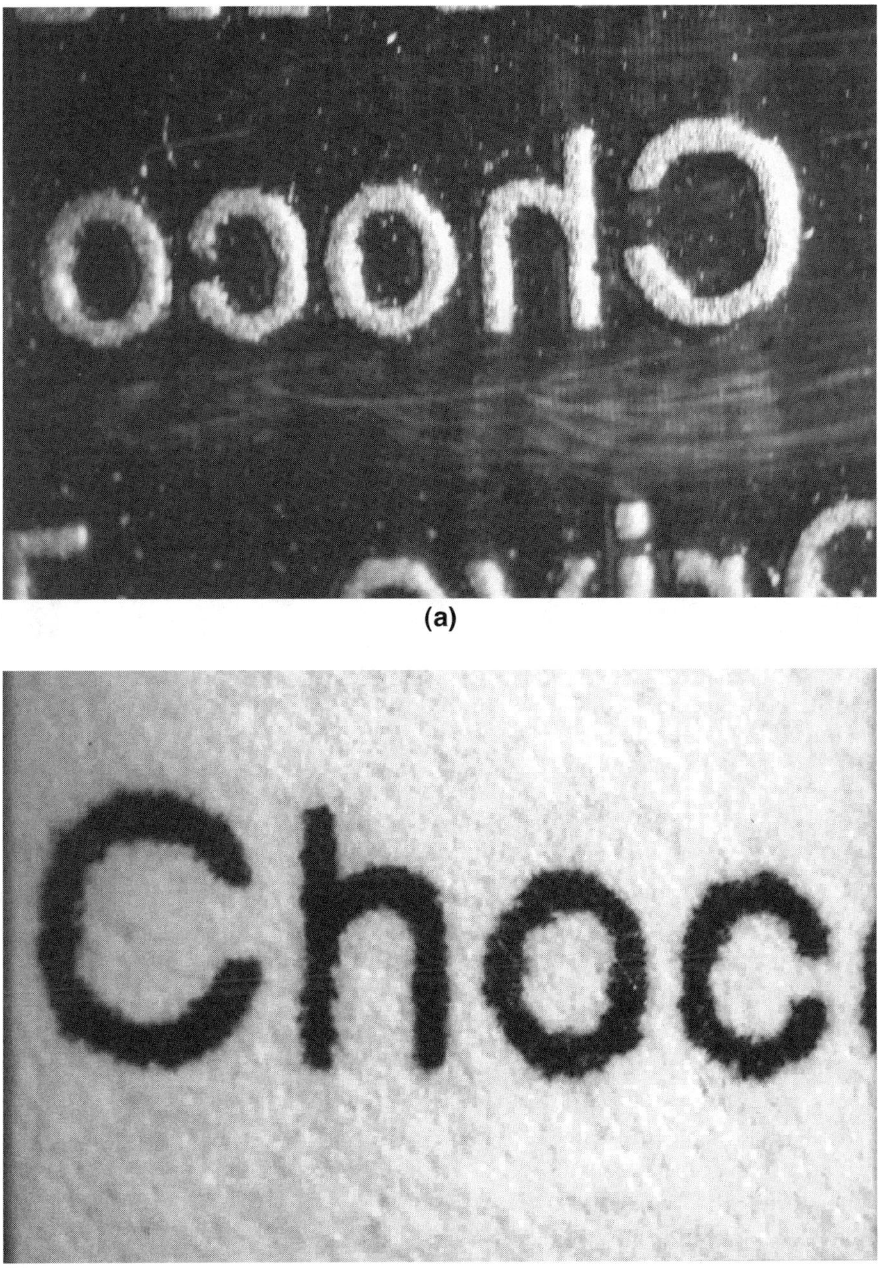

(a)

(b)

Figure 20.12 (a) Low-relief die of a thermal printer stamp. The thermal heating unit on the stamp seals the micropores in the non-print area of the die. Ink exits through the open pores in the text area. (b) Impression created from the thermal printer flat-die stamp. (From Seaman Kelly, J., *Forensic Examination of Rubber Stamps*, Charles C. Thomas Publisher, Ltd., Springfield, IL, 2002. With permission.)

printing process examinations in a group of documents. The investigator did not ask about the notary stamps. Because the stamp impressions were included in the examination, the submitter was given additional investigative information, as the microscopic examination

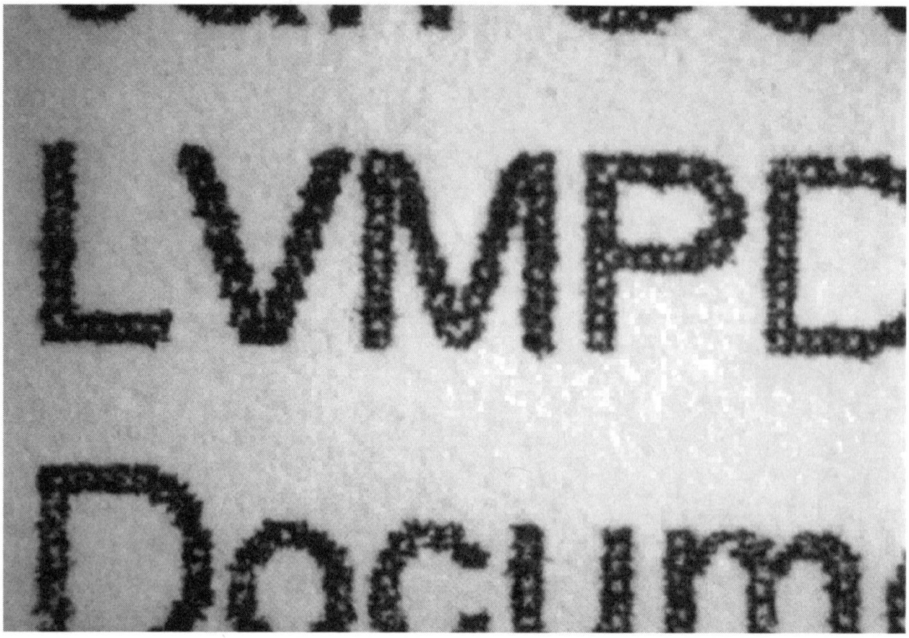

Figure 20.13 Dot matrix pattern within the characters of a printed text is an indication that the Brother SC 300 thermal printer machine may have made the stamp die. (From Seaman Kelly, J., *Forensic Examination of Rubber Stamps*, Charles C. Thomas Publisher, Ltd., Springfield, IL, 2002. With permission.)

of the stamp impressions revealed that they were the product of an inkjet printer and not a rubber stamp.

The *Standard Guide for Examination of Rubber Stamps*, ASTM E 2289-03, published by the American Society for Testing and Materials International®, is a guideline of procedures FDEs should follow in their examination of cases involving rubber stamps and their impressions.[10] The procedure discussed in this section contains more detail than that found in the guide, but is in agreement with the published ASTM guide.

The examination of a rubber stamp or an impression requires a methodological approach. As with any document examination, the FDE must consider the limitations inherent in the evidence. Intrinsic limitations in the rubber stamp impression are determined by a combination of the following factors: the manufacturing process, the material used for the stamp, the type of ink, the type of material hosting the impression, and the stamping style of the individual operating the stamp. Having a knowledge base of the various die materials and the manufacturing processes employed to produce rubber stamps is key to assigning proper weight to any observed anomalies.[2]

There are two main types of defects: those that occur in the manufacturing process and those that occur through use or abuse of the stamp. Sources of manufacturing defects include damage to the original image, defective die material, distortion or misalignment, poor quality control, damage caused by cutting the stamp, bubbles, and impurities such as dirt. In order to classify or determine the source of the defect, the document examiner must have the suspected stamp. Determination or classification of defects cannot be made from the examination of the impression. The defect's significance, i.e., class or individual, is determined by the stage of occurrence within the manufacturing process. For example,

Figure 20.14 The arrangement of the dots is dependent upon the size of the text. For instance, 8-point text will reveal a matrix that has a sawtooth appearance. (From Seaman Kelly, J., *Forensic Examination of Rubber Stamps*, Charles C. Thomas Publisher, Ltd., Springfield, IL, 2002. With permission.)

an air bubble can occur in different stages of manufacturing (Figure 20.15a and b). The air bubble would be a class characteristic if it is in the Bakelite mold used to produce vulcanized rubber stamps. Every stamp made from that particular mold would have the air bubble in the same location. If the air bubble was created by air becoming trapped during placement of the rubber on top of the Bakelite mold prior to vulcanization, the air bubble would be unique to that die and, therefore, would be an individual characteristic.

Individual defects more commonly occur through use or abuse of the stamp. Dirt, paper, fiber, accumulated ink, nicks and cuts, edge wear and breakdown, and stamp distortion are a few of the causes for defects to appear in the impression.

Individual defects can be transitory and care must be given in the handling of the stamp so as not to affect the amount or location of the object causing the defect (Figure 20.16a and b). For example, dirt, hair, fiber, and accumulated ink are transitory and can be removed as a result of handling or cleaning of the die. Nicks, cuts, edge wear and breakdown, and stamp distortion are permanent defects, as they are a part of the die. Both transitory and permanent objects can create individual defects that can assist in the identification or elimination of a stamp making a suspected impression.

The first step in the examination of a stamp impression is to examine it microscopically to make sure you have a stamp impression and not an image created by another printing process, such as an inkjet printer or a copier. If the image was created by another printing process and not by a rubber stamp, the document examiner is limited to a qualified report, as he or she is restricted to examining only the gross characteristics of the text, i.e., the image reflects the same general class characteristics of type style, type size, arrangement, and design of the submitted rubber stamp.

(a)

(b)

Figure 20.15 (a) Air bubbles occur in the manufacturing process of stamps. Determination as to whether they are class or individual defects is dependent upon knowing when the air bubble occurred. If the air bubble is in the matrix board, it is classified as a class characteristic. If, however, the air bubble occurred as a result of air being trapped when the rubber was laid on top of the matrix board, it is an individual characteristic. The air bubble can be observed in the *F* of *OF.* (b) A non-print area was produced in the impression by the air bubble on the stamp die. (From Seaman Kelly, J., *Forensic Examination of Rubber Stamps*, Charles C. Thomas Publisher, Ltd., Springfield, IL, 2002. With permission.)

(a)

(b)

Figure 20.16 (a) A piece of debris is on the stamp die of this salt-leached rubber stamp. The FDE must be careful in the handling of the stamp, as debris or any other transient material could be removed. (b) Debris appears in the stamp impression as a non-print area. The shape, size, and location of the debris can assist in establishing a dating timeline as well as factor into an identifying characteristic if both the questioned and known impressions contain the same void area. (From Seaman Kelly, J., *Forensic Examination of Rubber Stamps*, Charles C. Thomas Publisher, Ltd., Springfield, IL, 2002. With permission.)

If the suspected stamp is submitted, conduct a visual inspection to determine the presence of permanent or transitory defects. Determination of ink source and the condition of stamp die, container, and ink pad also can be made during this visual inspection. The container or mount (if a hand stamp) contains an index (also known as the title) that frequently is made from one of the first impressions of the stamp. The name of the stamp manufacturer or stamp shop that made the stamp may be displayed on the container and handle. This information may assist in locating the business that made the stamp.

A microscopic examination to search for anomalies follows the visual inspection. Notation of both the visual and microscopic inspections should be made. It is a good practice to photograph the stamp and its die in order to record its condition when it was submitted. Once this step has been completed, the next step is to compare the submitted rubber stamp to the questioned impression. Are they in agreement in class characteristics: type size, type style, arrangement, and design? If so, the examination proceeds forward. If there is any disagreement in class characteristics, an elimination is justified, as the submitted rubber stamp could not have produced the suspected impression.

The microscopic examination of the stamp die and the impression should include both direct and oblique lighting. Direct lighting provides even illumination of the area being examined to determine the manufacturing process used to create the die. Oblique lighting coupled with a higher magnification narrows the focus of the examination to detect even the smallest of defects and to determine whether they are permanent or transitory.

Upon completion of a thorough examination of the stamp die, an impression-to-impression examination is conducted. The FDE makes numerous impressions from the submitted stamp on a substrate that is similar to the material hosting the questioned impression. If the submitted stamp die is covered by transient material (dirt, hair, accumulated ink, etc.) and the questioned impression does not reflect defects that may have been sourced to the transient material, it may be necessary to clean the stamp in order to obtain impressions free from transient materials (Figure 20.17a and b). Prior to cleaning, the stamp should be photographed to document the condition in which it was received. It is important to stress that the stamp die should not be cleaned until numerous impressions have been made that reflect the condition of the stamp when submitted.

Ink saturation and the angle at which the stamp die contacts the printing media determine whether defects on the die will be replicated in the impression. A heavily inked impression may not reveal a small defect due to filling of ink of the void area. It is unknown how the stamp was held by the suspect or the angle of the die as it contacted the paper. Because of these unknown factors, the FDE produces numerous known impressions of varying ink saturations and of the die contacting the paper at differing angles. Collecting a series of impressions without re-inking the die of the suspected stamp is the proper method of obtaining impressions with a progressive decrease in ink saturation. For a self-inking or hand stamp, make sure the die is adequately inked prior to producing the first impression. The ink coverage in the impression will be at its optimum in the first impression. By not re-inking between impressions, the ink coverage gradually decreases with each successive impression. From the 10 to 20 consecutive impressions, the document examiner should observe a decrease in the ink saturation with each successive impression (Figure 20.18).

The classification of the stamp will determine the range of ink saturation that can be obtained from a suspected stamp. For example, impressions made from a hand or self-inking stamp will reflect a wide range of ink saturation compared to impressions made from a stamp die that is not inked in between impressions. A pre-inked stamp, such as a

(a)

(b)

Figure 20.17 (a) Cleaning a stamp die prior to making known impressions can remove transient material, including a hair. (b) Impression that reflects non-print areas where the hair touches the die. Known impressions of the die as it was received must be taken prior to cleaning the die. (From Seaman Kelly, J., *Forensic Examination of Rubber Stamps*, Charles C. Thomas Publisher, Ltd., Springfield, IL, 2002. With permission.)

Figure 20.18 Consecutive impressions produced by not re-inking the die of the self-inking stamp. The first impression has the optimum ink coverage. Ink saturation gradually decreases with each impression. (From Seaman Kelly, J., *Forensic Examination of Rubber Stamps*, Charles C. Thomas Publisher, Ltd., Springfield, IL, 2002. With permission.)

pre-mixed gel, has its ink in the die. The document examiner must produce the consecutive impressions quickly in order to observe some decrease in the ink saturation. The range of ink saturation is not as dramatic in pre-inked stamps because the ink-filled die is comprised of impregnated or microencapsulated inked cells. The consecutive impressions must be made quickly to prevent the cells from replenishing ink to the top portion of the die that contacts the paper. Even though flat-die stamps are categorized as pre-inked stamps, obtaining lighter impressions from continuous stamping is difficult. Since ink only exits through the open pores of the text, the ink supply is not depleted. A characteristic of a flat-die impression is that the ink saturation will be the same from the 1st through the 20th successive impression (Figure 20.19).

The next step is to change the angle of the die contacting the paper when making an impression. Obtaining impressions where even pressure is applied to the die allows the examiner to determine how evenly the ink is dispersed throughout the text. Impressions produced by the die contacting the paper unevenly (at differing angles) allow the

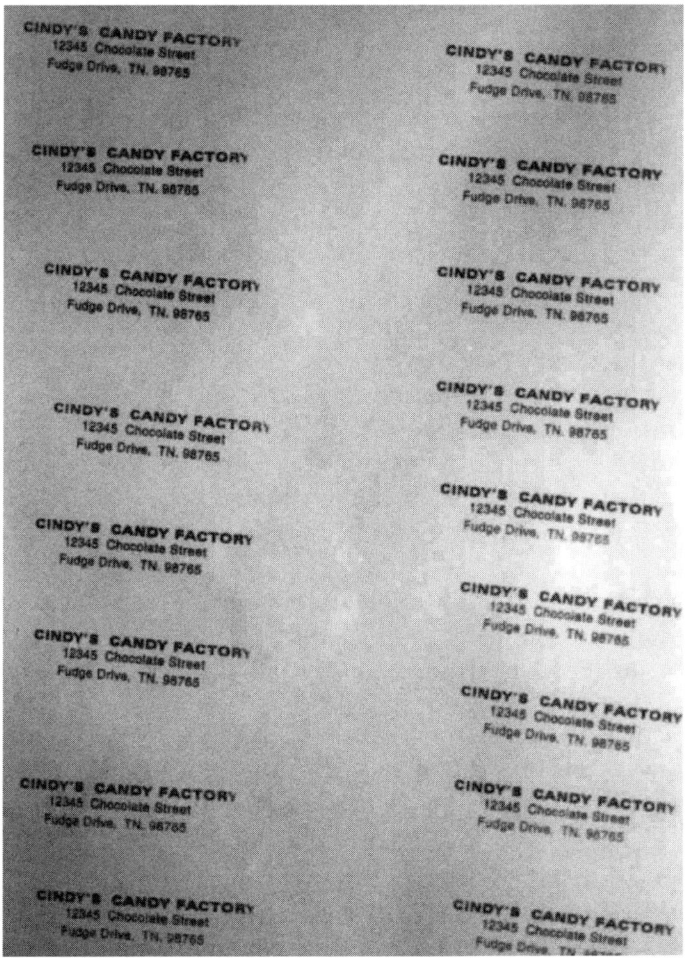

Figure 20.19 Consecutive impressions produced by a thermal printer flat-die stamp. A characteristic of the flat-die stamp is that the 1st impression will have the same amount of ink coverage as the 20th consecutive impression. (From Seaman Kelly, J., *Forensic Examination of Rubber Stamps*, Charles C. Thomas Publisher, Ltd., Springfield, IL, 2002. With permission.)

document examiner to observe how the impression appears when angles and pressure applications change. Rocking the stamp side to side or front to back can cause transient characteristics directly attributable to the improper handling of the stamp during the production of the impression. Once the known impressions have been produced, a side-by-side comparison is conducted with the questioned impression to determine if they came from the same source.

20.4 Criteria for Conclusions[11]

Upon completion of the examination, the document examiner issues a conclusion based on the evidence. The conclusion criteria regarding a rubber stamp impression can be found in the *Standard Guide for Examination of Rubber Stamps*, ASTM E 2289-03, published by the American Society for Testing and Materials International.[10] This particular publication lists the elements needed to support identifications, eliminations, qualified conclusions, and no conclusions.

Differences in only one of the class characteristics — type size, style, spacing, design, or arrangement — are sufficient to justify an elimination. A conclusion of elimination may be issued when there are accountable differences between the questioned impression and the submitted stamp. For example, the questioned impression is evenly inked with no distinguishable individual characteristics. The suspect impression and the submitted rubber stamp are in agreement in class characteristics. Course of business impressions originating from the submitted stamp, and dated before and after the suspect impression, reflect an incomplete printing of the characters along the top edge of the die. The known exemplar impressions reveal the same phenomena as the known course of business impressions. Examination of the stamp reveals the area along the top edge of the die is worn to the point that that portion of the die does not print. Based on the submitted evidence, the wear pattern on the die causing a non-print area along the top of the impression is an individual characteristic that should have been replicated in the suspected impression. Since the questioned impression did not reflect this wear pattern, the submitted stamp is eliminated.

Definitive conclusions of identification are difficult to issue due to the possibility of the existence of a duplicate stamp. According to the ASTM E2289-03 standard, a conclusion of identification is justified when there are no significant differences between the questioned impression(s), the known stamp, and the known impressions, and agreement in all individual characteristics. In accordance with the guideline, the possibility of the existence of a duplicate stamp must be eliminated before a definitive conclusion can be issued.[10]

Qualified conclusions are more commonly issued in the examination of rubber stamps. When the existence of a duplicate stamp cannot be eliminated, the document examiner can issue a report stating a qualified identification. Language similar to "the questioned and known impressions originate from the same source" acknowledges the possibility of a duplicate stamp; reports that the questioned and known are in agreement in class, random, and individual characteristics; and yet properly reflects the possibility of a duplicate.

The examination of questioned to known impressions can, at best, only yield a qualified conclusion. It is difficult to state the two sets of impressions originated from the same source when the FDE did not have the opportunity to analyze the stamp that produced the known impressions. The FDE must keep in mind that defects are not always reproduced in every impression. Even though the investigator may have submitted 20 known impressions, the FDE is unaware of the conditions of the environment when the impressions were made, nor does he or she have the opportunity to examine the die for individual characteristics that may be present in the questioned, but not in any of the 20 known impressions.

A "no conclusion" is issued when the examination reveals that the suspect stamp and the questioned impression both lack individual characteristics or defects. The conclusion level is based on the significance of the individual defects. Class characteristics alone will not support an identification. For example, the questioned stamp impression is evenly inked and the microscopic examination does not reveal any individual characteristics. The submitted stamp produces evenly inked impressions, and there is also an absence of any individual characteristics. In cases where the questioned impression is pristine, the document examiner is limited to reporting that the questioned impressions and the known stamp are consistent in class characteristics. It is possible the known stamp made the questioned impression, but due to the absence of individual characteristics in either the questioned or known, the FDE cannot narrow the field of possibilities to one stamp.

20.5 Summary

In conclusion, the FDE should keep in mind during his or her examination the possibility of the existence of a duplicate stamp. Because of this possibility in the majority of cases, the document examiner should remain conservative in his or her findings. Hilton wrote:

> While inconclusive findings are more often the rule than positive identification, stamp impressions should by no means be passed over without consideration. By their very nature, they are always a potential source of information in both the preliminary field investigation and in the ultimate proof of facts.[12]

References

1. Hilton, O., *Scientific Examination of Questioned Documents*, rev. ed., Elsevier Science Publishing Co., New York, 1982, p. 72.
2. Casey, M.A., The individuality of rubber stamps, *Forensic Sci. Int.*, 12, 134–144, 1978.
3. Levinson, J. and Perelman, B., Examination of cachet impressions, *J. Forensic Sci.*, 28, 235–241, 1983.
4. Seaman Kelly, J., *Forensic Examination of Rubber Stamps: A Practical Guide*, Charles C. Thomas Publisher, Ltd., Springfield, IL, 2002, pp. 50, 66.
5. Seaman Kelly, J., *Forensic Examination of Rubber Stamps: A Practical Guide*, Charles C. Thomas Publisher, Ltd., Springfield, IL, 2002, p.25.
6. Seaman Kelly, J., *Forensic Examination of Rubber Stamps: A Practical Guide*, Charles C. Thomas Publisher, Ltd., Springfield, IL, 2002, pp. 28–30.
7. Seaman Kelly, J., *Forensic Examination of Rubber Stamps: A Practical Guide*, Charles C. Thomas Publisher, Ltd., Springfield, IL, 2002, pp. 26, 28–29.
8. Seaman Kelly, J., *Forensic Examination of Rubber Stamps: A Practical Guide*, Charles C. Thomas Publisher, Ltd., Springfield, IL, 2002, chap. 4.
9. Seaman Kelly, J., *Forensic Examination of Rubber Stamps: A Practical Guide*, Charles C. Thomas Publisher, Ltd., Springfield, IL, 2002, chap. 5.
10. ASTM 2289-03, *Standard Guide for Examination of Rubber Stamp Impressions*, ASTM International, West Conshohocken, PA, 2003.
11. Seaman Kelly, J., *Forensic Examination of Rubber Stamps: A Practical Guide*, Charles C. Thomas Publisher, Ltd., Springfield, IL, pp. 160–171.
12. Hilton, O., *Scientific Examination of Questioned Documents*, rev. ed., Elsevier Science Publishing Co., New York, 1982, pp. 71–72.

Checkwriters*

21

TOM VASTRICK

Contents

Some form of check protection has been used since the early 1800s in order to discourage tampering. Early efforts involved the use of intricate designs in the paper stock, a precursor to the modern safety paper of today. Mechanical devices designed for this purpose began finding their way into the marketplace around 1870. Some of these machines embossed or cut images into the paper, while others cut the entire amount out of the paper, much as a stencil might be used. These early devices ranged in complication and appearance from one device that resembled tongs to one that looked like a large complicated music box. The first major manufacturer of more modern mechanical checkwriters was the Todd Company, with the introduction of the first Protectograph. Todd ushered in the modern mechanical checkwriter industry. Todd was eventually purchased by Burroughs in 1954, which in turn was purchased by Standard Register in 1986.

Mechanical checkwriters ink, emboss, shred, or perforate check information onto the check medium. Toward the end of the century, checkwriter technology joined the computer age. Computerized checkwriters utilize basic computer and printer technology to create printed images. The individual characteristics of computerized checkwriters are applications consistent with other computer-generated documents.

*The content of this chapter and the illustrations are excerpts from Vastrick, T.W., *Classification and Identification of Checkwriters*, American Board of Forenic Document Examiners, 1991. With permission. Update to be published.

21.1 Mechanical Checkwriters: Impression Components

A checkwriter impression is actually a group of impressions that are at least partially independent of each other. A typical impression consists of a prefix, printing element, payee perforator, and the numerical typeface, also called a segment (Figure 21.1).

A prefix is that part of a checkwriter impression that precedes the numeric amount. Major manufacturers have standard prefixes utilizing words and formats that often provide sufficient information, in and of themselves, to determine the manufacturer of an impression. These manufacturers also provide individual prefixes tailored to the needs and requests of the client. Individual prefixes can be any combination of numerals, characters, or images. Many individual impressions are one-of-a-kind products that establish individuality from its uniqueness of design. Some individual impressions will contain a series of numbers that represent a client number or a specific machine number for that client. Manufacturers historically have provided information about the manufacture of any individual prefixes, including the client name and the date of manufacture, to law enforcement.

Most prefixes oscillate to the edge of the left-most numeral through spring action within the checkwriter. An oscillating prefix ensures that there is no extra space between the prefix and the numeric amount. There are machines that do not have oscillating prefixes. These machines can be recognized by an inordinate amount of space between the prefix and the numeric amount. A dollar sign or foreign currency designation located to the left of a numeric amount is routinely a part of the prefix.

As a cautionary note, prefixes can be easily removed from a checkwriter and transferred to other machines of common manufacture. Checkwriters also can operate without a prefix. As such, the absence of a prefix in an impression does not necessarily mean that a prefix does not exist. Because of these precautionary notes, one should never base an identification or elimination of a specific checkwriter as the source of an impression based solely on present or absent characteristics of the prefix. However, the present or absent characteristics may provide the basis for an identification or elimination of the specific prefix.

The printing element consists of commas, periods, and words or symbols within, or subsequent to, the numerals denoting the amount. Common printing element words

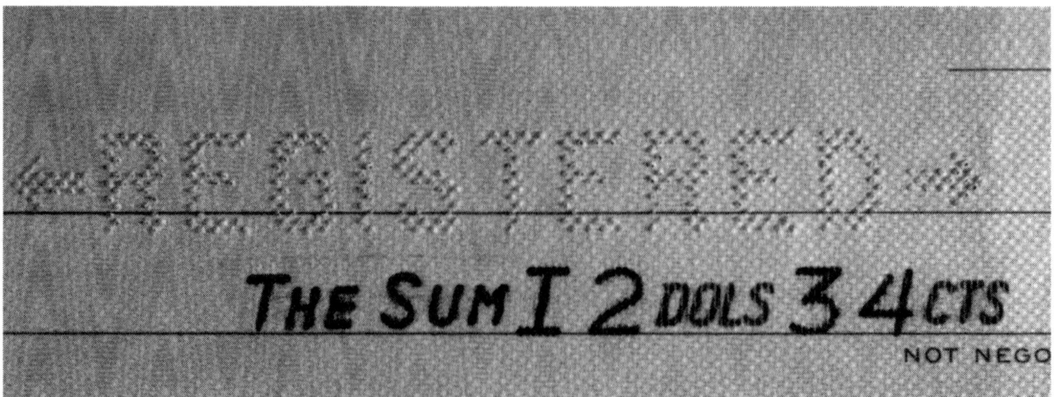

Figure 21.1 A sample checkwriter impression containing a payee perforator (REGISTERED), prefix (THE SUM), numeric typeface (12 and 34), and printing element (DOLS, CTS).

include, but are not limited to, *and*, *dols*, and *cts*. The printing element is stationary and prints on every impression. The designs and words used in the printing element can independently provide sufficient information, in and of themselves, to determine the manufacturer of an impression. Particular attention should be paid to the use of uppercase vs. lowercase lettering and the design of the font.

The numeric typeface consists of numerals (0 to 9) forged onto plates called segments. Each column of numbers has a separate set of numerals. Checkwriters may have 5, 7, 9, 11, or more segments. As such, the 5 representing $50.00 would be a different mechanical entity than the 5 representing $500.00. Certain models of a common manufacturer can be differentiated by size and font design differences utilized in models with different numbers of segments. Often, the design of the font can independently provide sufficient information, alone to determine the manufacturer and, in some instances, the possible model or models of an impression. It is important to note that manufacturers occasionally change fonts, providing some dating possibilities.

The payee perforator is that portion of the checkwriter impression that embosses or shreds the payee area of a check. This portion of the impression is located above the numeric amount. It is the purpose of the payee perforator to make tampering of the payee entry more difficult or impossible. Commonly, the payee perforator will use a different platen from that used on the rest of the impression. Payee perforators will have rectangular patterns or spell out words. The design of the impression can independently provide sufficient information to determine the manufacturer of an impression. Payee perforators can be turned off and on. As such, the absence of a payee perforation on an impression does not necessarily mean that the machine creating the questioned impression does not have a payee perforator.

21.2 Mechanical Checkwriters: Specialized Machines

Mechanical checkwriters come in a variety of forms. The more common forms are keyboard models and lever models, both powered manually or electrically. In addition, there are specialized machines that have unique functions and characteristics.

The most common of the specialized forms is the dialing machine. The dialing machine is the oldest form of mechanical checkwriter and often found among the low-end, inexpensive models, particularly those that are of foreign manufacture. A dialing machine consists of one segment of impressions, to include the currency designation, numerals 0 through 9, and any other designations, such as periods. The number of impressions is limited by design. Dialing machines will not have payee perforators or prefixes. All numbers will be from the same mechanical source. As such, the number 5 in $50.00 will be same impression as the 5 in $500.00, unlike that of keyboard and lever models. A dialing machine functions by impressing one digit at a time. The machine then automatically advances to the next space. This action is repeated until the impression is complete. Dialing machines have poor alignment due their design, and apparent defects of this kind cannot be taken as individualistic in nature. The lack of a payee perforator and a prefix, along with poor alignment, are collectively indicative of a dialing machine.

Heavy-duty machines are checkwriters designed to be used on multi-copy documents. As such, these checkwriters, by design, do not perforate the paper. The reason behind this design feature is to prevent paper from getting clogged up in the machine. Mechanically, this is accomplished through the use of a hard rubber platen. Often the impression pattern will replicate the appearance of the embossment method of the particular manufacturer. Heavy-duty machines will not have payee perforators.

Multi-currency machines are checkwriters designed to be used with numerous different currencies. There are two ways in which this is accomplished. One way is to provide a segment that has casts of numerous currency designs. This method is used on keyboard and lever checkwriters. Dialing machines will have numerous currency designations on the main segment. multi-currency models will not usually have payee perforators, nor will they have prefixes. Printing element features are usually limited to periods, stars, and commas. A star or currency designs of USD or US$ are clear indicators of a multi-currency model. Obviously, an impression with printing element terms, such as dols or cts, will not be a multi-currency machine.

21.3 Mechanical Checkwriters: Impression Formats

Mechanical checkwriters utilize two different inking sources. One is liquid ink and the other is ribbon (Figure 21.2). An inked impression will microscopically display a clear solid inking with well-defined edges. A ribbon impression will clearly display the texture of the ribbon on the impression when viewed under the microscope and will have significantly less defined edges to the impression. While there are standard colors that are used in both liquid ink models and ribbon models, individual coloring schemes can occur. Examinations of checkwriter impressions have found individual coloring schemes to be rare. It is not uncommon to find two-color impressions, and some manufacturers make two-color impressions standard. The change in color should be located between characters, but shifts in the ribbon or inking system will cause the seam to show in an impression. This shift is

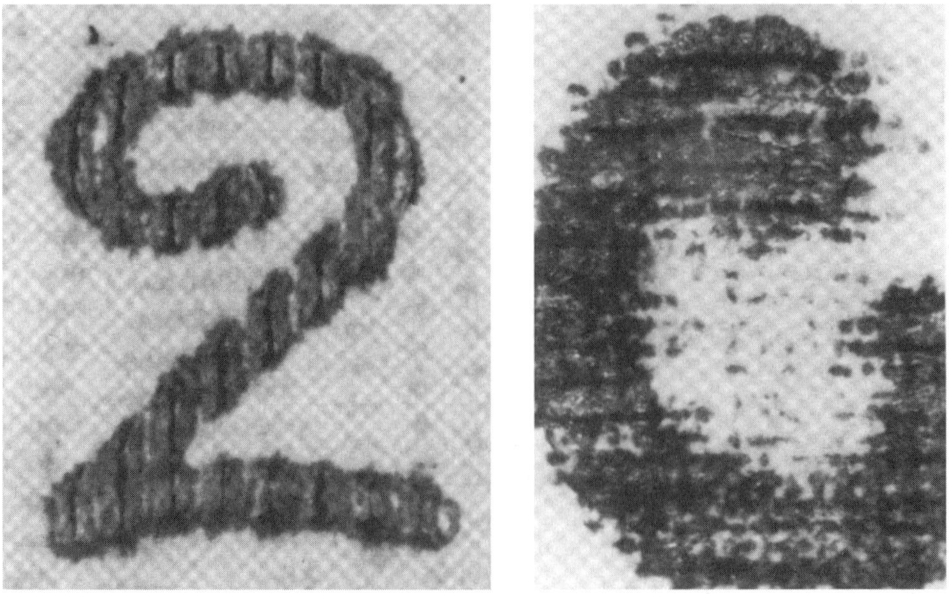

Figure 21.2 Samples of a liquid ink impression (left) and a ribbon impression (right).

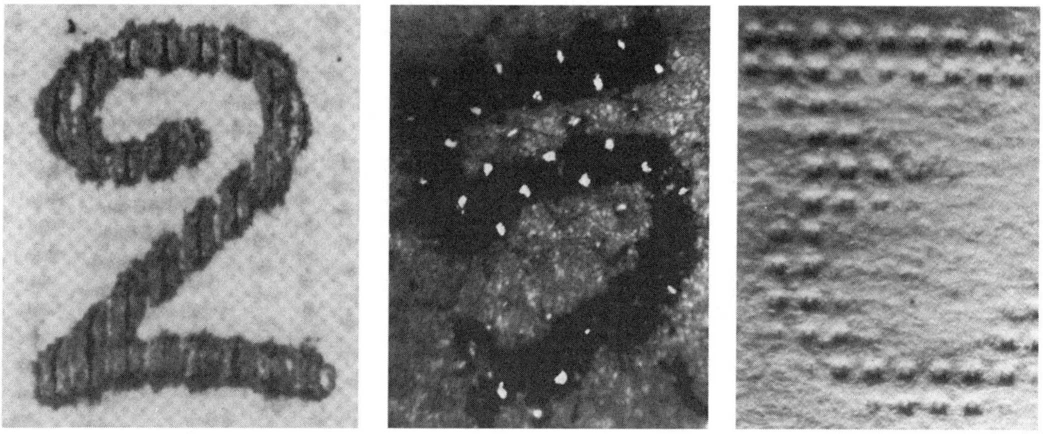

Figure 21.3 Samples of ridge and groove format (left), pinhole format (middle), and dimple design as viewed from the back side (right).

fairly common, but the position of the shift is an individual characteristic and can be treated accordingly in the examination.

Mechanical checkwriters will either perforate paper using a series of pinholes, emboss and shred paper through a ridge and groove format, or emboss the paper with a dimple design. These formats are readably distinguishable even without magnification (Figure 21.3).

21.4 Make Differentiation

Checkwriter manufacturers have created clearly distinguishable designs in their numerals, printing elements, prefixes, and payee perforators. As such, manufacture identification is a simple comparison process of each portion of the impression with proper comparison material. Figure 21.4 provides the comparison material necessary for make differentiation of domestic mechanical checkwriters currently or recently in production. It is important to update this information prior to use in any given examination.

21.5 Model Differentiation

Major manufacturers of checkwriters produce numerous models. Some of these models vary only by the power source or other extraneous feature that does not manifest itself in the impression. Other differences amount to the ability of the checkwriter to "sign" checks in addition to the amount impressions, though the absence of an impressed signature does not necessarily mean that the machine that produced the impression at hand is not capable of signing checks. There are numerous models of which differentiation is possible to the exclusion of all other models.

It would be impossible to provide comprehensive information about model differentiation that would not be outdated before this book is printed. However, general guidelines of model differentiation will provide an invaluable starting point in the examination.

Certex (Figure 21.5) is currently producing one model, the Model 3100. Previously, Certex manufactured the Series 100 and 6000. The Series 100 consisted of Model 100 and Model 101, which do not sign checks. Models 110 and 111 have signing capabilities.

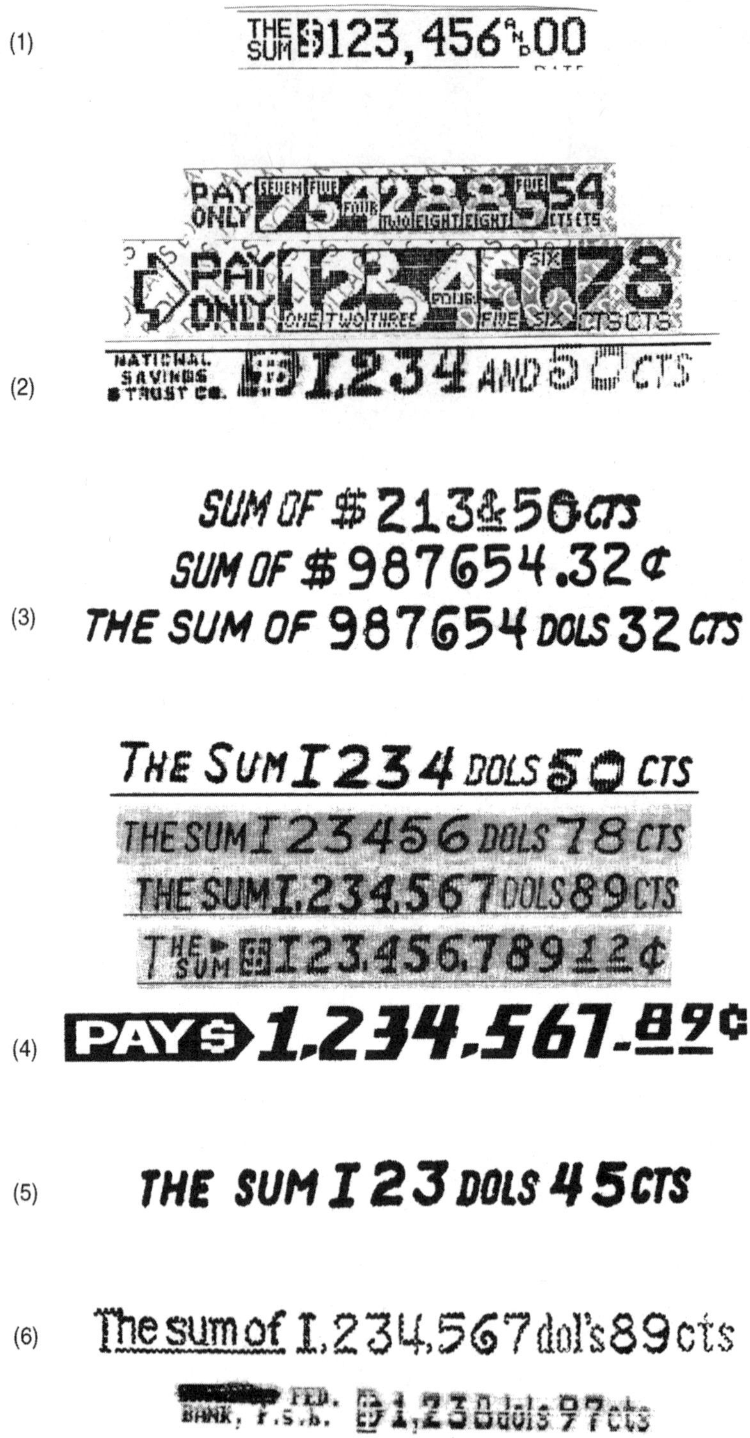

Figure 21.4 A quick reference guide to make differentiation: (1) Certex, (2) Burroughs/Standard Register, (3) Hall-Welter (Speedrite), (4) Paymaster, (5) Speed-O-Print, and (6) F&E Hedman.

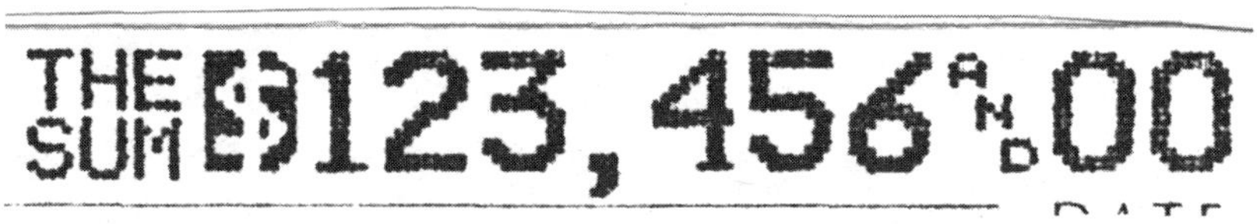

Figure 21.5 Certex Series 100 with the numeric typeface design unique to Certex.

Figure 21.6 Top two samples are from a computerized checkwriter. The third sample is typical of older Burroughs machines. The bottom sample is from the Model TJ510.

Burroughs/Standard Register (Figure 21.6) has historically manufactured so many different models that a model-by-model breakdown is virtually impossible. Many of the computer-generated Standard Register models are noteworthy by the incorporation of a written amount found within each numeral. One unusual feature of some older computer models is their ability to produce two fonts sizes. While the fonts are very similar, they can still be differentiated.

Upon the last inquiry, Standard Register manufactures one mechanical checkwriter and one computer-generated checkwriter. The computer-generated checkwriter, Model TJ510, produces one font, but the size can vary to fit the document on which the impression is to be placed. The distinctive design has not changed.

Hall-Welter (Figure 21.7) has produced two models without change in impression design for over 50 years. These are the Models 900 and 914, which differ only in the material used in the casing. The Model 900 was discontinued in 1996. The Model 914 continues to be manufactured without change. Hall-Welter began marketing a model jointly produced with Standard Register in the early 1990s. The Model 930 is consistent in design features with certain Burroughs/Standard Register models.

Paymaster (Figure 21.8), now Paymaster Technologies, manufactures nine models that are readily distinguishable. This differentiation is based on whether the impression is liquid ink or ribbon, what kind of payee perforator design is present, and the printing element design.

Speed-O-Print (Figure 21.9) manufactured checkwriters from the early 1960s to 1984. During that time, it manufactured two models that never changed during their production.

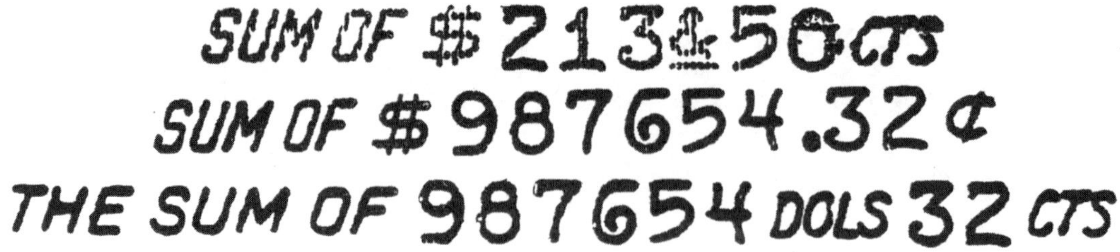

Figure 21.7 Samples of Hall-Welter Speedrite checkwriters, with their unique numeric typeface design.

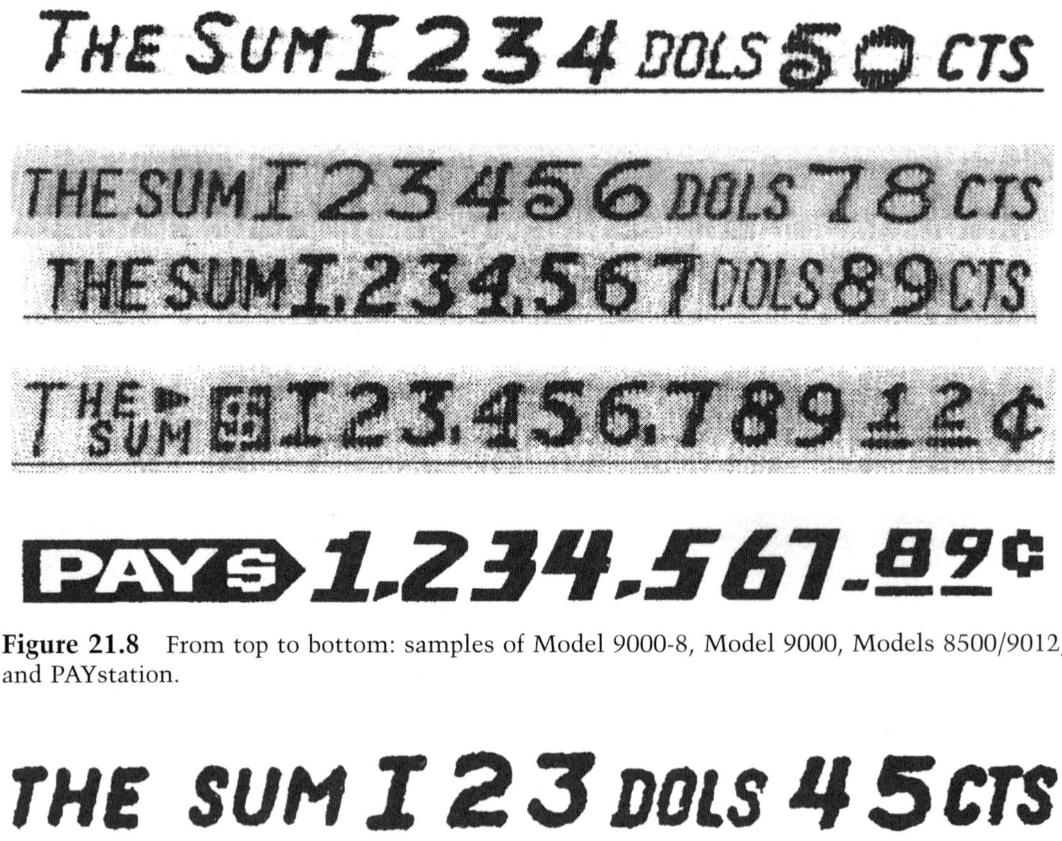

Figure 21.8 From top to bottom: samples of Model 9000-8, Model 9000, Models 8500/9012, and PAYstation.

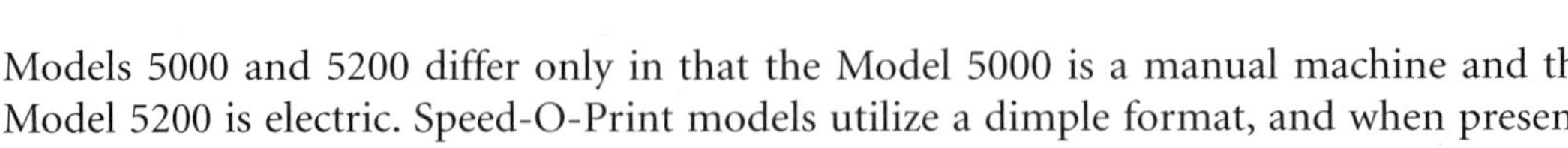

Figure 21.9 Sample of Speed-O-Print checkwriter impression. Noted differences with Paymaster include the even height of the characters of the prefix and the dimple format.

Models 5000 and 5200 differ only in that the Model 5000 is a manual machine and the Model 5200 is electric. Speed-O-Print models utilize a dimple format, and when present, the payee perforator format is 5 × 45 perforations.

F&E Hedman (Figure 21.10) checkwriters have changed dramatically during the 1990s. Earlier models were distinctive by the pinhole format and the typeface design. F&E Hedman currently manufactures 10 computerized checkwriters. Any model differentiation will result from examination of the prefix or of check-signing capabilities.

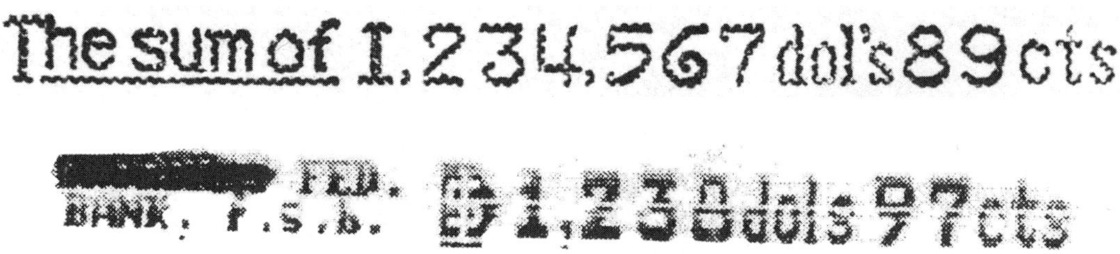

Figure 21.10 Older pinhole model with unique numeric typeface design and pinhole format (top) and the computerized format of more recent and current domestic models with numeric typeface design also unique to F&E Hedman.

21.6 Individual Machine Identification

The final stage of the identification process is the association or disassociation of a specific checkwriter to an impression. Any time two metal objects are brought into contact and pressure, those areas become susceptible to wearing and chipping. Such defects will manifest themselves in the impression as missing perforations, missing portions of a ridge or groove, missing portions of a numeral or character, or drastic change in degree of embossment (Figure 21.11). One should be careful in evaluating a feature as a wear or chip defect, since the presence of foreign objects on the document itself, such as staples or paper clips, can cause similar results.

The perforations of a pinhole or shredding pattern of a ridge-and-groove machine can reveal either a full perforation or shredding or an incomplete pattern, resulting in a reproducible characteristic with strong identifying value (Figure 21.12).

Misalignment can be a strong individual characteristic but requires specific understanding of the mechanics of the checkwriter in order to properly conduct the examination. First, misalignment is not to be expected within the characters of the prefix or within the characters of any of the printing element words. Misalignment will exist in the prefix as a whole or the entire word of a printing element or on each individual segment. Misalignment in checkwriters that utilize a dialing method in which each individual character is embossed one character at a time, as opposed to the one time impression, common to other types of domestically produced checkwriters, is common and not reproducible. If the cause of the misalignment is a loose part, some variation in the misalignment may occur. Misalignment can be vertical in nature and will manifest itself with a distinct difference in ink density and degree of embossment from top to bottom (Figure 21.13).

Extraneous markings have been found on checkwriter impressions that are reproducible. Even though these markings have been found on liquid ink and ribbon machines, they appear much more frequently on ribbon machines. It should be noted that ribbon machines often create numerous extraneous markings that are not reproducible. In rare instances, excessive inking has resulted in excessive ink deposits on portions of the impression. Testing of these machines has found the depositing to be reproducible in both the existence and the location on the impression.

Many ribbon models utilize two-color ribbons. The document examiner should carefully chart the location of any ribbon shift and keep in mind that the phenomenon is common and individual to a machine. Finally, it is important to keep in mind that the shift can change over time.

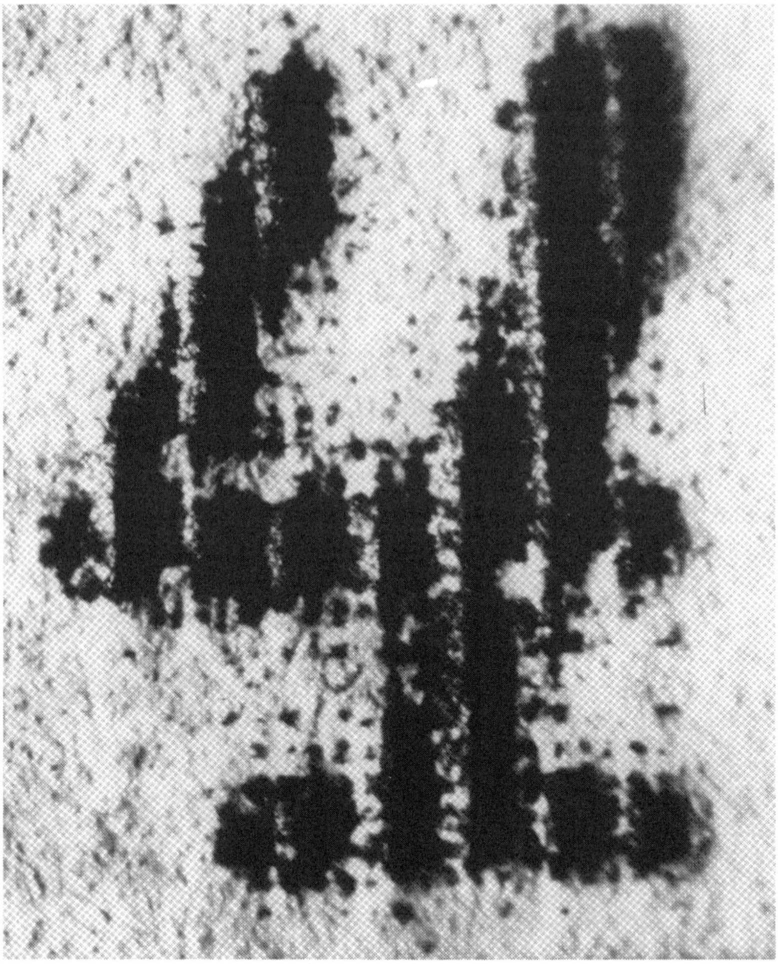

Figure 21.11 Impression with an ink void located near the right intersection of the horizontal stroke of the numeral 4.

If the checkwriter cannot be obtained for personal examination, proper standards are important for comparison. It is recommended that the following guidelines be followed:

1. Specimens should be taken on similar paper stock, as variation in paper can dramatically effect the impression.
2. The first specimen should be the amount that is currently displayed in the checkwriter.
3. Specimens should be taken as soon as possible to limit the time difference between the specimens and the questioned impression.
4. The majority of the specimens should be verbatim.
5. A full strike-up will require specimens with all numerals from all segments.

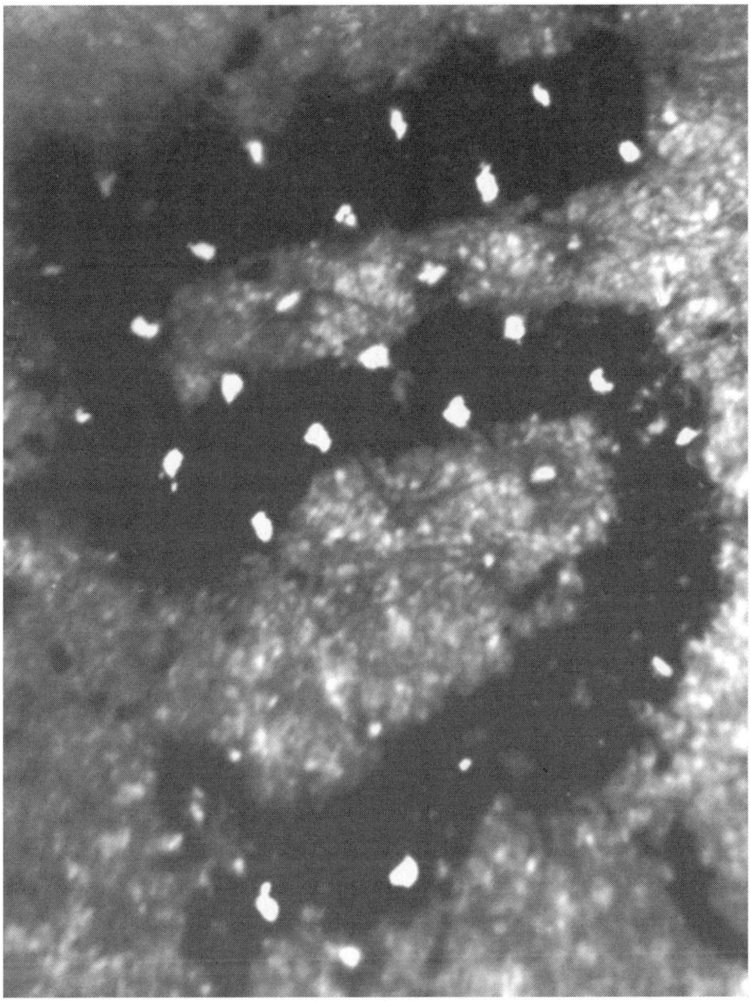

Figure 21.12 Perforation pattern of pinholes through the paper and points that the pinholing did not perforate the paper.

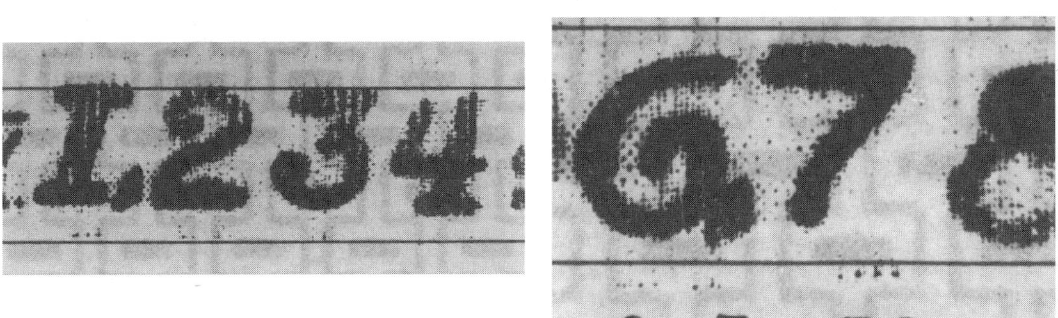

Figure 21.13 Samples of alignment defect of the platen.

21.7 Checkwriters and Alterations

While the purpose of checkwriters is to discourage tampering, this does not mean that impressions cannot, or are not, altered. Impressions without prefixes are susceptible to additions. A careful microscopic examination of the impression will reveal ample distinguishing features. Particular attention should be paid to the embossing, shredding, or perforation. Microscopically, these features are very clear and sharp. Common attempts to imitate the appearance of the registration mark include the use of files to impress the document or hand drawing the amount. Clear differences in the pattern and clarity or the format will manifest themselves under magnification.

A checkwriter can be used not only to discourage tampering, but also as the catalyst to an alteration. Checkwriter impressions can cover up legitimate entries with a raised amount in an attempt to alter the value of a document. The original ink impression can be easily overlooked by the overwhelming impression.

Checkwriter impressions can be flattened out, but the shredding or perforation does not entirely disappear. Testing up to 20,000 psi still left evidence of the original impression. Manufacturers describe their inks as indelible. Laboratory tests suggest that FDEs should hold a healthy skepticism to that claim and carefully examine the impression if such changes are a possibility.

Staples

22

TOM VASTRICK

Contents

A *staple* is a three-sided device used to fasten and secure two or more objects together. They range from large staples used to join pieces of wood together in construction to the more commonly recognized wire-like device used to fasten multiple sheets of paper together. The latter form of staple will be the subject of this section.

A staple is comprised of a crown and two legs. The crown is that portion of the staple that remains across the surface of the top document in the stack (Figure 22.1). The legs are those portions of the staple that penetrate through the pages of the stack (Figure 22.2). Staple guns utilize spring power, or sometimes compressed air, to push the legs of the staple through the sheets. Some staple guns will only come in contact with the surface of the top sheet, and the legs will remain straight or near straight after the application. Others will crimp the legs at a significant angle. An office stapler will utilize the physical force of the user while wedging the multi-page document between the staple tray along the top sheet and the platen along the bottom sheet. Once the legs have penetrated all sheets of paper, a stapler will bend the legs inward. By design, the crimping usually exceeds 90°.

22.1 Staples

Staples will vary in the amount of the material, the length of the crown (as can be seen in Figure 22.1), and the length of the legs. While there are some standard sizes, inquiry of manufacturers indicates that custom sizing is available. Measurements of the crown and leg are partially subjective, as the FDE will have to decide at what point the staple changes from the crown to the leg. The important factor, of course, is consistency in the measurements. Measuring the lengths of the crown and leg from examining the holes may be possible, but only as an estimate. The crown will be approximately the length of the distance

Figure 22.1 Two staples illustrating the difference in length of the crown.

from the center of the two holes. However, these holes are rarely precise and are constantly subject to wear. As such, precision in this measurement is limited. The length of the legs can be estimated in some instances. If the staple has been crimped straight along the back of the last page, leaving a distinct indentation in the paper, a reasonably precise measurement can be made by adding this distance to the distance of the thickness of the document. The measurement of the thickness of the document may prove to be the tougher calculation, as the staple will compress the document pages together, and this must be taken into account. If the back of the final page shows evidence of the end of the staple poking back into the document, the length from the exit hole to the entrance hole can provide a minimum length of the leg when combined with the thickness of the document. The actual length will be slightly longer but often cannot be precisely measured since the extent of the arc of the crimped portion of the leg cannot be accurately determined. If the end of the staple reenters the document through a series of pages, it may be possible to determine the angle of entry and possibly establish, through geometric measurements, a reasonably precise approximation of length of the leg.

Remnants of paper removed from a stapled set of documents may be found wedged in the crown of the staple (see color Figure 22.3 following p. 366). Similarly, examiners

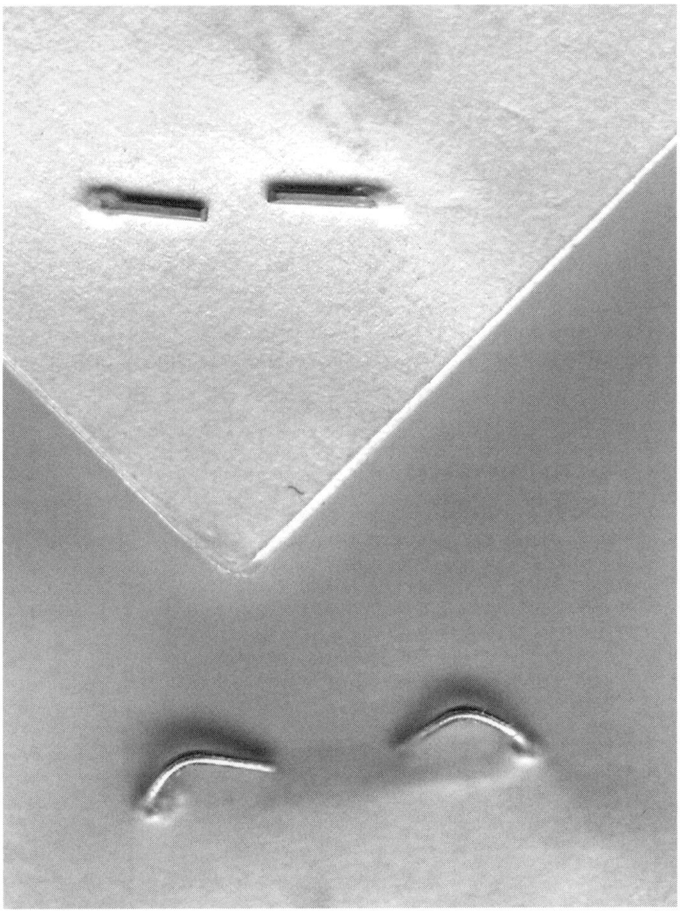

Figure 22.2 Two staple legs illustrating the different ways in which staples are crimped.

should be cautious to protect other forensic evidence that may be wedged or attached to staples, such as blood, skin, or other forms of trace evidence.

22.2 Staplers

Staplers can only utilize certain sizes of staples. Determining whether a stapler can actually operate using the staples in question is an appropriate first step in utilizing basic class characteristic comparisons. Some staplers have limited capacities. For instance, in cases involving numerous pages, experimentation may be warranted as to whether the stapler in question has the capability of binding the number of pages that are in question. Finally, document examiners should keep in mind that manually operated staplers rely on human pressure for their activation, which is subject to variation.

Microscopically, the bottom of the legs of a staple attached to a stack of paper will appear to have ridge and groove patterns on a flattened surface. These patterns are part of the manufacturing process and not a result of the crimping process by the stapler. There is no specific pattern of ridge and grooves characteristic of one brand, one batch, or one

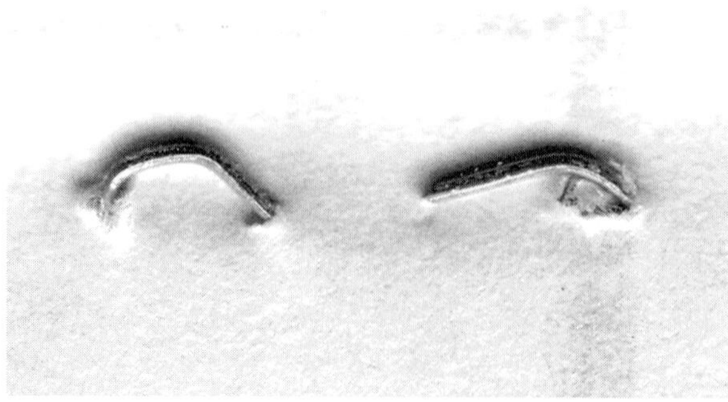

Figure 22.4 The end of one leg is reinserted into the document from the backside.

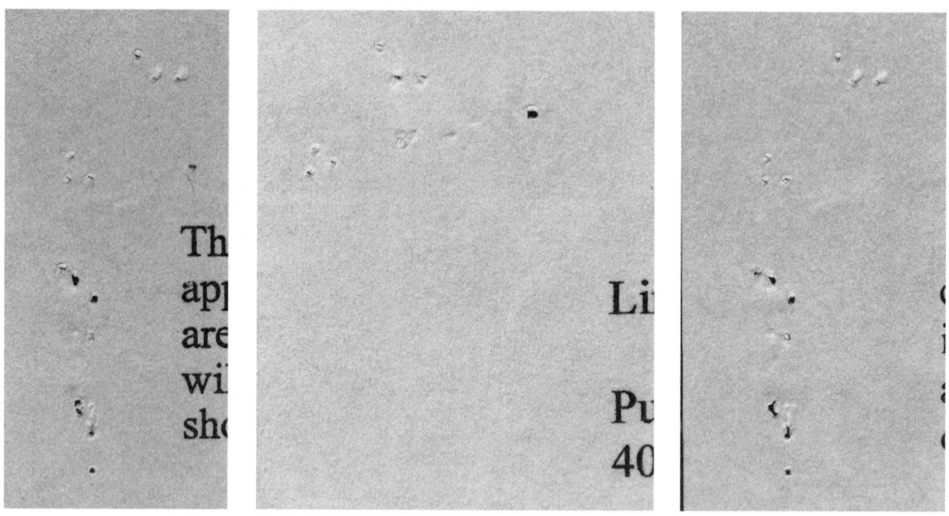

Figure 22.5 Three pages from a multi-page document reveal that the middle document was not always attached to the other documents.

sleeve of staples. FDEs, however, should not overlook the possibility that some significant defect in the crimping bay of the stapler could create unique damage to any staple administered through the device.

22.3 Staple Holes

The most common application of matters involving staples in forensic document examination is to compare the holes of each page of a multi-page document in order to determine if there has been page insertions or page substitutions. If no such changes have taken place, the pattern of holes should be consistent throughout the pages of the questioned document, keeping in mind that the end of the legs can penetrate through the reverse side of the stapled sheets (Figure 22.4). Deviations in the pattern of holes between sheets may be indicative of insertion/substitution or the fastening together of sheets from different origins (Figure 22.5). The first and last sheets of a multi-page document are exposed to outside contact. This must be borne in mind when attempting to associate holes on the outer pages with those on the inside.

Section VI

Conventional Printing and Paper Examinations

Conventional Printing Processes

23

SUSAN L. FORTUNATO

Contents

The document examiner will encounter many different printing processes in the course of analyzing documents. In fact, it is preferable that the FDE be well aware of printing processes and their identification, particularly when documents such as business letters, business checks, credit cards, traveler's checks, travel documents, and social security cards are submitted for handwriting examinations. Prior to the handwriting analysis, an FDE knowledgeable in printing processes may find that the entire questioned document is in fact counterfeit. Most likely, this information would be just as helpful to the investigator as the identification of the author of the handwriting.

In order to make determinations of genuineness or counterfeit, it is important for the FDE to have access to the genuine document, or to have sufficient literature about the genuine document and its features in order to make judgments about questioned material. It is also important to recognize and evaluate not only the printing processes used in the construction of the document, but also the security features that are present including any encoded information.

For the purposes of this chapter, the printing processes discussed will be limited to conventional or commercial processes, which require the use of printing plates. The most common conventional printing processes found on both genuine and counterfeit documents include offset lithography, intaglio, letterpress, and screen printing. References have been included at the end of the chapter for additional reading about these conventional printing processes.[1–10] For information regarding office machine printers and copiers, including inkjet and laser printing, refer to Chapters 16 and 17.

23.1 Offset Lithography

Offset lithography differs from other printing processes in two distinct ways: (1) its success relies on the principle that oil and water do not mix, and (2) the ink is offset from the printing plate to a rubber blanket and finally to the substrate (either a paper or plastic surface).

Offset lithography is a planographic printing process, meaning that the printed and non-printed areas of the plate reside within the same plane. Simply put, a lithographic printing plate appears completely flat to an observer. It is the nature of the surface of the plate that determines which areas will and will not print images. This is where the basic chemical principle that oil and water do not mix comes into play. The hydrophobic image areas of the plate attract oil-based ink and repel water, while the hydrophilic non-image areas of the plate attract water and repel the oil-based ink. The printing plate is mounted on a cylinder called a plate cylinder, which rotates and passes dampening rollers that deliver a water-based solution to the plate. Only the non-image areas are receptive to water and receive this solution. As the plate continues to rotate, it passes a series of ink rollers. The oil-based ink is attracted to the image areas of the plate and transfers only to these areas. The plate is now ready to transfer the image to the rubber blanket, which is mounted on what is called the blanket cylinder, and finally, the image is transferred from the rubber blanket to the substrate. An impression cylinder provides support and counterpressure for the substrate (see color Figure 23.1 following p. 366). Note that in this process the image begins as a right reading image on the printing plate, becomes a wrong reading (mirror) image on the rubber blanket, and is transferred once more to the substrate as a right reading image.

The process of producing a lithographic printing plate generally begins with either a conventional photograph of an original or a digital image. In conventional photography, one or more negatives are produced. With digital image production, there are generally computer data files created. In addition to the printing plates, rubber blankets, blanket cylinders, impression cylinders, inks, paper, and rubbish are sources of forensic information that may be found at a suspected printing operation. Often, partially completed documents or printing mistakes will end up in the garbage and can serve as valuable sources of forensic information. While most of these items are easily collected at the scene, the plate cylinders are extremely heavy and difficult to remove from a press. If images appear on the cylinders of a suspect press, high-quality photographs or digital images can be obtained and submitted for comparison to the questioned documents, printing plates, negatives, and other materials.

The printed images produced by offset lithography appear very uniform and smooth (see color Figure 23.2 following p. 366). The printing has an even, consistent appearance from one side of the printed area to the other, and the edges are very clear and distinct. Offset printing is characterized by uniform, sharp edges and does not indicate any evidence of embossing.

Offset lithographic presses can be single color or multi-color. Single-color presses require a run through the press for each color that is to be printed. A single-color press can have a footprint that equates to the size of a standard floor model photocopier. A press of this size could easily be housed in a print shop, basement, shed, or garage and requires a standard power supply. With multi-color presses, a separate ink unit and a series of three printing cylinders (plate, blanket, and impression) are required for each color. A common multi-color press contains true black plus the three process colors — cyan, yellow, and magenta — to create a four-color process printing press. As a general reference, a commercial press containing four separate ink units with accompanying printing cylinders would be roughly 10 to 12 feet in length. A press of this size and complexity would generally be found in a print shop. When printing using offset lithography, the illusion of tonal qualities can be achieved through the use of halftone screens. These screens create dots of varying sizes to mimic the continuous tone in an image. An example of a halftone screen can be seen by viewing newspaper images with a microscope. Four-color process offset printing typically utilizes halftone screens to produce images (see color Figure 23.3 following p. 366). Note the alignment of each color in a linear dot pattern: the dots are equal distants apart from their centers; however, their relative sizes vary depending on the tone required for the image. This process is most commonly confused with color inkjet technology.

23.2 Thermography

Thermography is an inexpensive form of raised printing that creates special embossed characteristics without the use of engraved plates. Thermography may be used on counterfeit documents to imitate the raised feel of the intaglio printing on currency, social security cards, and birth certificates. In the first step, images are produced on the substrate using either offset lithography or letterpress printing and special non-drying inks. Next, the ink on the surface of the substrate is dusted with a powdered compound or resin. The excess compound is removed by suction. Finally, the tacky ink with the powdered compound adhering to it is passed under a heated element, which melts the powder compound and fuses it to the surface of the substrate, hardening as it cools (see color Figure 23.4 following p. 366). The result is printing that appears raised, but with oblique light, this multi-step process is readily distinguished from intaglio printing (see color Figure 23.5 following p. 366). Sometimes bubbles are apparent in the raised thermographic printing.

23.3 Intaglio

Intaglio printing uses engraved plates to produce images or text. The image areas are recessed or below the surface of the plate. Engraved plates can be hand engraved or chemically etched to render "wrong" reading images on the plate. During the printing process, the viscous ink is applied to the surface of the plate and fills the engraved areas or recesses of the plate. Excess ink is removed from the surface of the plate by a doctor blade. The ink is then transferred to the substrate under tremendous pressure (e.g., 5 to 20 tons) between the plate cylinder and an impression cylinder (see color Figure 23.6 following p. 366). As a result of this pressure, the surface of the substrate may become embossed with an outline of the printed material. This effect can be readily seen with oblique light.

Due to the engraved nature of the plate and the thickness of the ink, the ink rests above the surface of the paper. This characteristic height of the ink can be viewed with oblique lighting (see color Figure 23.7 following p. 366) or felt across its surface. Unlike any other printing process, intaglio printing can produce extremely fine lines and life-like pictures and portraits. In addition, as the ink dries, it begins to settle and flow slightly along the paper fibers into a unique pattern called feathering. This individual feathering pattern can assist an FDE in identifying the original genuine pattern that was used to produce counterfeit documents.

23.4 Letterpress Printing

Letterpress printing produces images in much the same fashion as typographic printing or typewriting. Letterpress printing is known as a relief process, meaning that on the printing plate the images to be printed are raised above the surface, while the non-printed areas are recessed or below the plane of the printed portions (see color Figure 23.8 following p. 366). Only the raised portions of the plate, the designs, numerals, or text characters that are to be printed, receive ink. Since letterpress printing is a direct printing process, the printing plate comes into direct physical contact with the substrate. In letterpress printing, the printing plate contains wrong reading information. Once the inked plate contacts the substrate, the image is transferred. Due to the amount of pressure that is utilized in this printing process, the ink is squeezed out around the edges of the printing plate. This pressure can result in the characters being embossed, or raised, on the reverse side of the printing substrate. The resulting image is not only sharp and crisp, but typically exhibits a darker edge or halo effect around the edges of the printed images or text (see color Figure 23.9 following p. 366). Letterpress printing is frequently used as the printing method of choice for genuine serial numbers on a variety of documents.

23.5 Screen Printing

Screen printing, also commonly known as silk screen printing, requires the use of a stencil, where the image areas are porous and the non-image areas are not porous. The stencil commonly consists of a thin, sturdy fabric or screen made from polyester, silk, or stainless steel. The images are typically produced using a photographic process. A stencil containing a photographic emulsion is exposed to light and rinsed. The emulsion is rinsed away from the printed areas, leaving the hardened emulsion in the non-printed areas. The resulting screen contains a right reading image of the areas to be printed. The screen is then placed in direct contact with the substrate to be printed. Screen printing ink is extremely thick and is spread across the stencil in an even film with the aid of a squeegee (see color Figure 23.10 following p. 366). The ink is forced through the porous areas of the screen, creating images, designs, or text. Screen printing is recognizable by the appearance of a thick layer of ink and the jagged edges, or grid pattern, that are a result of the pattern of the fabric or screen (see color Figure 23.11 and color Figure 23.12 following p. 366). Screen printing is most frequently encountered on plastic substrates, particularly genuine and counterfeit credit cards and driver's licenses.

References

1. International Paper Company, *Pocket Pal: A Graphic Arts Production Handbook,* 17th ed., International Paper Company, Memphis, TN, August 1997.

2. Adams, J.M., Faux, D.D., and Rieber, L.J., *Printing Technology*, 4th ed., Delmar Publishers, Albany, NY, 1996.
3. Stevenson, D.L., *Handbook of Printing Processes*, Graphic Arts Technical Foundation Press, Pittsburgh, PA, 1994.
4. Kasunich, C.L., *Gravure Primer*, Graphic Arts Technical Foundation Press, Pittsburgh, PA, 1998.
5. Crouch, J.P., *Flexography Primer*, 2nd ed., Graphic Arts Technical Foundation Press, Pittsburgh, PA, 1998.
6. Ingram, S.T., *Screen Printing Primer*, 2nd ed., Graphic Arts Technical Foundation Press, Pittsburgh, PA, 1999.
7. Faiola, A., *Typography Primer*, Graphic Arts Technical Foundation Press, Pittsburgh, PA, 2000.
8. Wilson, D.G., *Lithography Primer*, 2nd ed., Graphic Arts Technical Foundation Press, Pittsburgh, PA, 1997.
9. Kipphan, H., *Handbook of Print Media, Technologies and Production Methods*, Springer-Verlag, Berlin, 2001.
10. John, C., The Printworks, Inc., available at www.cjpw.com/printing

24 Paper Examinations

SUSAN L. FORTUNATO

Contents

24.1 History of Paper

Many different materials were used as writing surfaces before paper was invented, such as rock, clay, bones, wood bark, leaves, bamboo, metal, silk, papyrus, and parchment. These substrates were not without problems associated with their use, including varying degrees of surface smoothness, expense, preservation, and storage requirements.

The words *paper*, *papier*, and *papel* are derived from the Greek and Latin words *papuros* and *papyrus*. Although neither papyrus nor parchment is considered paper, they were obvious precursors and an inspiration for the development of true paper. From a historical perspective of paper, the histories of papyrus and parchment deserve a brief overview.

Papyrus was the first pre-paper writing material, dating back to around 3500 B.C. In ancient times the reedy plant, papyrus, grew in abundance in Egypt. Egyptian papyrus was produced from the stalks of the papyrus plant that were cut into pieces about 2 feet long and split down the center. Tissue-thin sections were cut from end to end and laid side by side. The material was then covered with a thin paste of wheat flour and water. Additional strips were laid across in the opposite direction, and the two layers were then pressed or hammered together. This formed a substance that could be written upon, but it is not considered paper, since the process used to produce papyrus is unrelated to the method used to produce paper.[1]

During the second century, Egyptian rulers would not allow papyrus to be exported. Due to its demand, it is believed that others sought to produce a substance that would rival that of the Egyptian papyrus. The king of Pergamum, an ancient city in Asia Minor, is credited with the invention of parchment in 197 to 159 B.C., even though it was probably in use as early as 1500 B.C. Parchment was originally made from the split skin of the sheep. The grain half, or wool side, of the skin was made into a strong leather, while the flesh half, or lining side, was made into parchment. The flesh half was washed and rubbed with lime and any remaining hair was removed by scraping the skin with a curved knife. The skin was then washed and stretched tightly within a frame where the scraping continued in order to remove irregularities in the skin, leaving it with an even thickness throughout the surface. Finally, the skin was dusted with powdered chalk and rubbed with fine pumice. The resulting parchment had a paper-like feel and was extremely durable.[1] Parchment, however, is not considered true paper.

Paper is defined by *Webster's Ninth New Collegiate Dictionary* as "a felted sheet of usually vegetable fibers laid down on a fine screen from a water suspension."[2] Simply put, to be classified as true paper, individual cellulose fibers are collected from a slurry to form a sheet. Cellulose fibers are found in most plants, where they constitute a major component of cellular walls. There are four basic types of cellulose fibers that meet the specifications for papermaking: seed hair fibers (cotton), bast fibers (linen), grass fibers (wheat straw, bagasse, esparto, and kenaf), and wood fibers (hardwoods and softwoods).[3] Each of these fiber sources has its own drawbacks, such as high cost, low strength, dry storage, location and accessibility, or limited crop yields. As a result of these limitations and the sheer abundance of wood fibers, the paper industry uses wood as the primary source of paper fiber. Wood fiber currently constitutes approximately 95% of all the papermaking in the world.[3]

The date 105 A.D. is when true papermaking began, since that is when the invention of paper was officially reported to the Chinese emperor by Ts'ai Lun, who was an official to the Chinese Imperial Court. This early paper was made from bark, hemp waste, old rags, and fish nets.[3] In order to make paper, the source fiber is chemically or physically broken down until the fibers become individual units. The original Chinese papermakers poured the fibrous pulp onto a square of coarsely woven cloth held within a four-sided bamboo frame called a *mould*. It was not long before craftsmen practiced the more expeditious method of dipping the moulds directly into the thin slurry, raising the mould horizontally under the floating fibers. The matted fibers were then lifted from the water as in a sieve and the water drained through the cloth mesh. This dipping method was an important step in the papermaking technique, as it enabled the artisans to not only form more uniform sheets of paper, but also to produce them in greater quantity. The mould, with the thin deposit of matted and felted fibers adhering to its surface, was placed in the sun for drying. During the drying process, the surface of the mould left impressions in the paper. Once the sheet of paper was dry, it was easily stripped from the mould.[1] Since many sheets were made at a time and each sheet of paper took time to dry in these moulds, it was not long before the next advancement in papermaking was established.

The ancient Chinese artisans conceived the idea of a mould that allowed the wet sheet of paper to be removed while it was still moist. This development allowed the papermakers to continue making sheets of paper while reusing the same mould. The mould had to be constructed of a smooth and firm material from which the moist sheet would easily free itself. This type of mould was made in two pieces, a mould frame and mould cover. A mould frame served as a firm support for the mould cover and kept it flat. The mould cover was made by placing thin strips of rounded bamboo side by side and stitching or lacing them together at regular intervals with silk or some type of hair. The bamboo strips, as well as the stitches, left impressions in every sheet of paper made by the mould. These marks or indentations were the origins of the first laid-and-chain type patterns in paper. The entire mould, both frame and cover, was dipped into the vat of pulp and brought to the surface loaded with the wet, fibrous material; alternatively, the fibers could be poured onto the mould. In what is called couching or laying down the wet sheet of paper, the sheet was removed from the mould by a workman who slowly rolled the matting of the mould away from the paper and deposited the wet sheet upon a board. These matted sheets were placed one on top of another and subjected to pressure to expel the surplus water. The fibers, properly treated, allowed the sheets of paper to be readily separated from each other after pressing.[1]

Papermaking was introduced in Spain around the year 1150, and spread to Italy, France, Germany, and England by the 15th century.[4] The first American paper mill was established by William Rittenhouse in 1690, near Philadelphia, PA.[4] Prior to the invention of the paper machine in the 1800s, paper was still made by the tedious hand process, similar to the technique previously described. The Europeans used the same principles as the Chinese in their papermaking method, substituting metal wires for bamboo and hair moulds, since bamboo was not a natural commodity in the West. The paper of the Europeans contained irregular marks, or laid-and-chain marks of the metal wires, just as the papers of the East had upon being impressed with the patterns of the bamboo and stitching from the mould covers.[1]

The introduction of rigid iron wires in the moulds of European papermakers naturally led to the development of the first watermark. In Italy, in 1282, the first use of watermarks occurred.[4] They were simple in design and were incorporated as part of the metal-laid mould.[5] The wirework, in the form of objects or words, was laced to the surface of the mould and created images in the paper. It is believed that the early watermarks were signs or symbols that conveyed meaning among the workers who made them, or among those who used the paper in which the outlines occurred. Much like with advertising, paper manufacturers were proud of their product and watermarking provided a method for identifying their handiwork.

The growing demand for paper soon began to exceed the capacity of hand production and the supply of fibers. In 1798 the first paper machine was invented by Nicholas-Louis Robert in France.[4] It was soon extensively developed by brothers Henry Jr. and Sealy Fourdrinier, who were stationers in England.[1] The modern-day Fourdrinier is a complicated and elaborate machine that performs many of the steps of the papermaking process, including sheet forming or wet-end pressing, drying, and calendering. Rather than creating individual sheets, the Fourdrinier paper machine produces paper in the form of a long, continuous sheet, or web.

The following description of the Fourdrinier papermaking process is a compilation from a number of resources.[1,6,7] Prior to the introduction of the fibers to the Fourdrinier process, the cotton, linen, or wood must be reduced to individual fibers. Modern-day pulping is the process that reduces these raw materials to this form. The cotton, linen, or wood is first cleaned and compacted into felt-like fibrous mattes. These mattes are mechanically cut and shredded into fibers that are again cleaned, bleached, and cut into uniform lengths. Paper pulp is generally produced by chemically cooking the fibers to remove lignin and any non-cellulose materials to leave nearly all pure fiber. As a note, paper mills are notorious for emanating a rotten egg-like odor that is a result of this cooking process. This odor comes from sulfur products (e.g., hydrogen sulfide), the same odor that one would find at a hot spring. Paper mills expend vast amounts of resources in attempting to reduce or eliminate these odors. After cooking, the pulp is washed to separate the fibers from the pulping chemicals (in which the non-fibrous materials are dissolved) and the impurities are washed away. Next, after adding water, the pulp is mechanically beaten to rupture the outer walls of the fibers, enabling the interior to absorb water and swell. The force created by the swelling, combined with the rubbing together of the fibers and force of the beating process, exposes tiny thread-like fibrils that, upon drying, interlock and cohere to form a strong physical and chemical bond. The ultimate effect of this process is to increase the tensile strength, folding strength, density, and smoothness properties of the finished sheet. The paper stock, once it is fully refined, is diluted to a mixture of 99.5% water and 0.5% fiber to ensure uniform fiber dispersion as the paper is formed. This paper pulp is now ready to be introduced into the papermaking machine.

In the Fourdrinier machine, a dilute suspension of the previously macerated fibers is fed from a reservoir, called the head box, through a horizontal slot (the slice) onto an endless moving wire screen. It is at this stage that the combined forward and rapid side-to-side shaking motion of the wire screen results in an even dispersion, coherence, and interlocking of the fibers and fibrils. The water drains through the wire with the influence of gravity. Toward the end of the forming area, additional water is removed by a suction box, which applies suction from below the wire. At this stage, the water content of the sheet is about 85%. Additional water is removed from the wet sheet by a press section. As the sheet reaches the end of the forming wire, the wet web is lifted from the wire to a felt at the couch roll. The wet web of paper then passes through a series of press rolls on the felts. The removal of water is aided by the use of rollers that are perforated, which allows suction to be applied. After the sheet passes through the press section, the water content is reduced to between 65 to 70%. From the press section, the web passes to the dryer, which contains a series of steam-heated, direct-drive cylinders. The paper web is held in contact with each dryer roll by a fabric drier felt, but it passes unsupported between the rolls. After the drying section, the paper contains 5 to 10% moisture. Almost all paper is passed through a calender section as part of the papermaking process. The calender is comprised of two or more cast-iron rollers with hardened surfaces that are arranged vertically at the end of the dryer section of the paper machine. The web passes through sets of these rollers, which are revolving at different peripheral speeds. This process increases the smoothness and gloss of the paper surface. Here, too, external sizing may be added for increased surface strength, folding strength, soil resistance, and printability. The finished paper is wound into a large roll at the end of the Fourdrinier machine.

Watermarking of paper during manufacture on a Fourdrinier paper machine is accomplished by the use of a dandy roll, which can have either a raised or depressed design corresponding to the desired style of the watermark.[4] The dandy roll is rotated above the moving wire and comes into contact with the wet sheet. The design is impressed into the fiber mat while the paper is still about 90% water. If the dandy roll contains a raised design (or wire mark), the watermark will appear lighter than the rest of the sheet. If the dandy roll contains a depressed design (or shaded watermark), the watermark will appear darker than the rest of the sheet of paper. The appearance of the watermark, whether lighter or darker, is not the result of the thickness of the paper, but a result of the concentration or density of the fibers caused by the design of the dandy roll and its effect on specific areas. The raised design disperses the fibers in the web, while the shaded design concentrates the fibers in the area of the design. In contrast to the Fourdrinier process, watermarks can be produced by a cylinder paper machine. This type of machine makes paper by partially immersing rotating cylinders in vats of pulp stock.[4] In this process, the paper forms into a sheet on wire-covered cylinders or moulds as the water drains through, leaving the fibers on the cylinder surface. Cylinder mould watermarks produce varying densities of the paper fibers and result in high-quality images and portraits that are used in security papers for currencies and other security documents throughout the world.

24.2 Analysis of Paper

There are a variety of tests that may be conducted to show similarities and differences in paper. Physical tests, which determine the tensile strength, tear resistance, folding endurance, color, fluorescence, porosity, and opacity, are some of the standard quality control tests that are done by a paper mill and can be used for comparison between papers. However, document examiners frequently find that a set of questioned documents they analyze does not consist simply of numerous sheets of one type of blank paper, but are a single sheet (or multiple sheets) that bear writing, printing, or graphics. As a result, typical questioned documents cannot be subjected to the Technical Association of the Pulp and Paper Industry, Inc. (TAPPI) standards for paper testing that frequently require the destruction of portions of the paper. Nor is there an availability of the number of sheets that are required for many of the industry tests. Consuming all or large portions of the available evidence is a condition that is unacceptable in the forensic community. In taking physical samples and conducting tests, it is preferable to leave at least half, if not over half, of the evidence that is available in order to allow another FDE the opportunity to conduct the same analysis. As a result, testing that compares the overall color, thickness, weave pattern, watermarks, ultraviolet characteristics, basis weight, and fiber analysis, and comparisons using instrumental analyses are the most effective tests used in comparative paper testing. Finally, when rendering conclusions about the aforementioned tests and their results, a document examiner should be aware of the papermaking industry's practices. Although it is less common in higher-quality papers (e.g., business paper, bond paper), the practice of combining multiple rolls (or webs) into finished paper products does occur. As a result, many of the characteristics of paper can vary within a given ream and should be taken into consideration when rendering conclusions. In general, one should expect the paper's weight, thickness, and watermark to remain constant, while the remaining characteristics (e.g., color, weave pattern, ultraviolet characteristics, fiber analysis, and instrumental results) may vary within a given ream of paper.

The overall color of paper (or paper rulings) can be determined subjectively through a side-by-side comparison, or it can be compared and measured against a white standard. It is advisable to not only observe the overall color of the paper, but to split or peel both the questioned and the known papers in half from front to back, using a scalpel, in order to evaluate and compare the internal color of the sheet. Peeling the front and back apart for a distance of approximately 3 to 5 mm is sufficient to evaluate the internal color. This is particularly helpful in evaluating whether the overall questioned document may have been tinted on the surface through a process of printing or staining.

The thickness of the paper is often measured in thousandths of a millimeter. Paper calipers or micrometers are available in many different styles, from manual to electronic, and vary widely in price. When analyzing the paper, it is advisable to measure the thickness of the unprinted areas and to take a number of representative measurements from a variety of areas before calculating the mean thickness for a given sheet. Similar measurements should be taken from all known standards of paper.

Analyzing the weave pattern of paper includes noting whether the paper contains a surface weave, which is intentional on the part of the paper manufacturer or an internal weave. Surface weaves, though not commonly encountered in forensic document analysis, are very distinct and can be easily documented using side lighting. Internal weaves or patterns often exist within the paper and can be viewed and documented with transmitted light. Internal weaves are the result of the felt that carries the wet web of paper as it is lifted from the forming wire of the Fourdrinier machine and into the initial portion of the press section. It should be noted that if changes are made to the Fourdrinier machine or in the papermaking process (i.e., the felt is replaced), the appearance of this pattern may change or disappear.

The value of watermarks as a means of determining the date of paper, books, and documents is worth noting. Watermarks should generally be examined and documented using transmitted light. In the event that a document contains images or text that severely interfere with the visualization of the watermark, a soft x-ray may be taken of the document. A soft x-ray will produce an image of the density of the paper fibers, revealing a clear picture of the watermark that is distinct from the printed text or images.

Watermarks are commonly imitated by printing an image in an opaque ink on the surface, or a light color, such as gray, on the reverse of a document. It should be noted that chemical watermarks also may be used to mimic genuine watermarks by introducing a transparentizing agent onto the finished paper product to simulate a conventional watermark. Using a combination of oblique, transmitted, and ultraviolet light sources should allow an FDE to distinguish between true watermarks and their simulations.

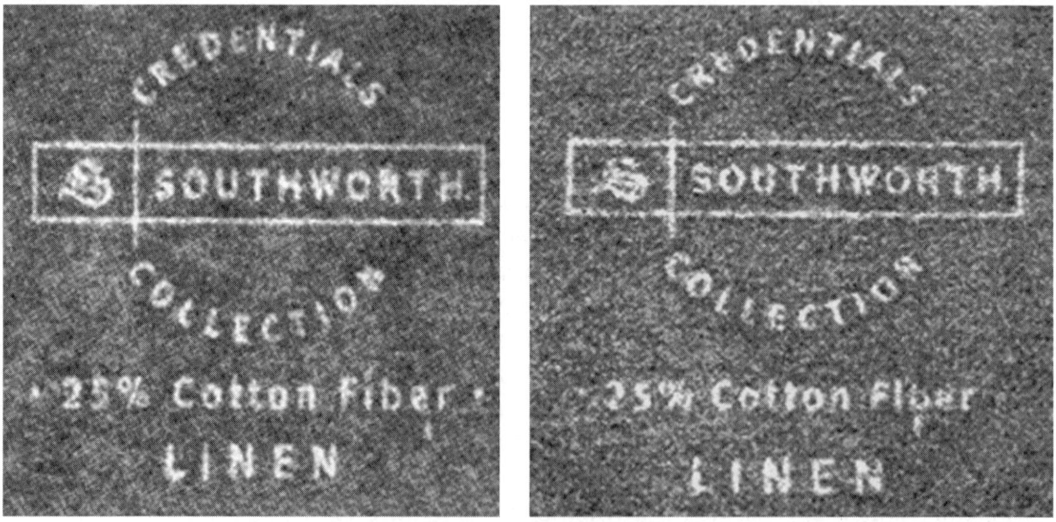

Figure 24.1 A Southworth Credentials Collection Linen with 25% cotton fiber watermark. The image on the left has a vertical mark under the *e* in *Fiber,* while on the right the vertical mark is under the *b*. The placement of the vertical mark can be used for dating purposes. (Courtesy of Jan Seaman Kelly.)

The origins and manufacturers of true watermarks can be identified and can assist in the dating or identification of the source of paper. The specific features of a watermark, such as the design, letter style, arrangement of text and wording, size, terminology, and arbitrary marks, can be used to specifically date paper (Figure 24.1). Information about a specific design may be obtained, including the first date of production, the date of consumer availability, the extent of its use, its production history, and the distributors of that paper in a given area. Sources for this information include paper manufacturers and industry publications such as *Walden's Paper Catalog*,[8] *Lockwood-Post's Directory*,[9] and the *Phillips Paper Directory*.[10]

The ultraviolet fluorescence properties of papers can be subjectively compared using long-wave (366 nm) and short-wave (254 nm) ultraviolet light. Images or photographs of the spectral features may serve to document the relative comparisons. The color of the fluorescence, as well as its relative intensity, should be compared among all of the pages of both the questioned documents and the known documents. It should be noted that even with respect to high-quality bond paper, the fluorescence of the pages within a ream may vary. In addition, the intensity of the fluorescence must be interpreted with caution, because the intensity usually decreases with the age of the paper, especially if the paper has been exposed to light.[11]

Basis weight is the weight in pounds of a ream (500 sheets) of paper in its standard size. When referring to writing paper, the basis weight is based on a 17 × 22-inch sheet and commonly weighs between 20 and 24 pounds (Figure 24.2). By measuring the overall weight and size of a given questioned document, a rough calculation of its basis weight may be obtained. This figure can be compared to known paper standards to evaluate whether the questioned document may have been produced with the same type of paper as the available standards. The analysis of the basis weight is difficult to accomplish in many situations. The ability to accomplish this extrapolation depends on the nature and condition of the document, the amount and type of printing that it bears, and its shape (cut edges vs. torn fragments). Obviously standard paper sizes and pieces of sheets that are evenly cut lend themselves to this comparison, while torn paper fragments, water-soaked paper, and burned or charred documents do not.

Paper fiber analysis is a destructive type of examination that is used to determine exactly which species of fibers have been used to make the paper pulp, the pulping process, and to quantify the amounts of specific fibers. The TAPPI standard T 401 om-93[12] is suggested as an aid to assist an FDE interested in conducting a paper fiber analysis. A small sample of most business papers can be broken down or disintegrated in water to reduce the paper to individual fibers. Generally, papers produced using wood fibers will easily

Figure 24.2 The size, weight, and brightness characteristics of paper are listed on the outside packaging. In the upper example, the weight is 20 lb, brightness 84, and there are 500 sheets in the package. The lower example contains manufacturing information printed on the side of a package of paper. (Courtesy of Jan Seaman Kelly.)

break down in an aqueous or water solution using what is referred to as method A in the TAPPI standard. Papers containing cotton fiber may require more encouragement to break down, and may even require the use of weak alkaline and weak acidic solutions to prepare the fibers, as outlined in method B. Once the paper has been reduced to a suspension of fibers, drops are placed on a glass slide. The water is allowed to evaporate, leaving the fibers evenly distributed across the surface of the slide. The fibers are then stained in order to visualize the specific types and species present. C stain and Herzberg stain are commonly used to differentiate between specific hardwoods, softwoods, cotton, linen, and non-woody or grass fibers. The *Practical Identification of Wood Pulp Fibers*[13] is an invaluable resource when attempting to identify the species of wood fibers in paper, as is the Paper Fiber Identification course offered by Integrated Paper Services in Appleton, WI.

Chemical and instrumental tests can be conducted in order to identify a large number of additional components, such as fillers, binders, sizings, starch, clay, whiteners, and brighteners, used in the making of paper. A variety of methods and instrumentation may be used to compare these components, including chromatography (e.g., gas chromatography [GC], gas chromatography/mass spectroscopy [GC/MS]), spectrometry (e.g., Fourier transform infrared [FTIR]), and electron microscopy (e.g., scanning electron microscopy [SEM], transmission electron microscopy [TEM], and x-ray fluorescence [XRF]). The analyses of various components are treated in more detail in Browning's *Analysis of Paper.*[14] Given the variety of ingredients and the sheer number of combinations of those ingredients, one would not expect two manufacturers to produce paper with the same components in the same relative ratios.[15] However, caution should be used when drawing conclusions as to whether two or more papers originate from the same source. As previously mentioned and with respect to many of the examinations, document examiners should be well aware of the papermaking process and the occasional interleafing of rolls that occurs at the manufacturer.[16] Document examiners should also be aware that business papers, including high-end bond paper that is sold as a unit, may contain physical or chemical variations within the unit. While conducting any forensic examination to determine the possibility of two papers having a common origin, it is important to determine whether the paper was once part of a pad, a tablet, or loose leaf. A visual examination may reveal printing defects, cuts, tears, perforations, staple holes, indented writing, paper clip indentations, residual adhesives, or other characteristics that are common within various sheets, which may lead the FDE to determine that papers share a common origin. Standards for the analysis of some of these conditions can be found in the American Society for Testing and Materials International (ASTM) guides.[17,18] Comparisons of this type also may be useful in determining the source of paper.

References

1. Hunter, D., *Papermaking: The History and Technique of an Ancient Craft*, 2nd ed., Alfred A. Knopf, New York, 1967.
2. *Webster's Ninth New Collegiate Dictionary*, Merriam-Webster, Inc., Springfield, MA, 1989.
3. MeadWestvaco Corporation, *Paper Knowledge*, MeadWestvaco Corporation, Dayton, OH, 1999.
4. Walden-Mott Corporation, *Walden's Handbook for Paper Salespeople and Buyers of Printing Paper*, Walden-Mott Corporation, Ramsey, NJ.

5. Caywood, D., Watermarks and the questioned document examiner, *Int. J. Forensic Doc. Examiners*, 1, 299, 1995.
6. Hunter, D., *Papermaking: Through Eighteen Centuries,* Lenox Hill, New York, 1971.
7. Dawe, E.A., *Paper and Its Uses*, 2nd ed., Crosby Lockwood and Son, London, 1919.
8. Walden-Mott Corporation, *Walden's Paper Catalog*, Walden-Mott Corporation, Ramsey, NJ, published annually.
9. Paperloop Publications, *Lockwood-Post's Directory*, Paperloop Publications, San Francisco, published annually.
10. CMP Information Ltd., *Phillips Paper Directory*, CMP Information Ltd., Kent, U.K., published annually.
11. Brunelle, R.L. and Reed, R.R., *Forensic Examination of Ink and Paper*, Charles C. Thomas, Springfield, IL, 1984.
12. TAPPI Standard T 401 om-93, *Fiber Analysis of Paper and Paperboard*, TAPPI, Atlanta, GA, 1993.
13. Parham, R.A. and Gray, R.L., *The Practical Identification of Wood Pulp Fibers*, TAPPI Press, Atlanta, GA, 1982.
14. Browning, B.L., *Analysis of Paper*, 2nd ed., Marcel Dekker, Inc., New York, 1977.
15. Brunelle, R., Washington, W., and Pro, M., Use of neutron activation analysis for the characterization of paper, *J. Assoc. Off. Anal. Chem.*, 54, 920, 1971.
16. Polk, D., Attard, A., and Giessen, B., Forensic characterization of papers. II. Determination of batch differences by SEM elemental analysis of the inorganic components, *J. Forens. Sci.*, 22, 524, 1977.
17. ASTM International Standard Guide E2288-03, *Standard Guide for Physical Match of Paper Cuts, Tears, and Perforations in Forensic Document Examinations*, ASTM, West Conshohocken, PA, 2003.
18. ASTM International Standard Guide E2291-03, *Standard Guide for Indentation Examinations*, ASTM, West Conshohocken, PA, 2003.

Carbonless Paper

25

TOM VASTRICK

Contents

25.1 What Is Carbonless Paper?

Carbonless paper is a paper stock that is coated with a pressure-activated inking system used as a method of recording an entry on multiple copies of a document at one time. The basis for the carbonless paper systems is microencapsulation of dye components. The process of microencapsulation was invented by Barry Green in 1950. Green, a research scientist for National Cash Register (NCR) of Dayton, OH, worked closely with fellow NCR scientist Lowell Schleicher to turn his concept into the basis of the systems of carbonless paper still in use to this date. The patent for their microencapsulation system was filed on June 30, 1953, and the first commercial sale of their new product was made on March 26, 1954. The term *NCR Paper* is commonly used to describe carbonless paper, but NCR Paper is actually a brand name.

Carbonless paper is a technological bridge between the past and the future. Carbon paper, used since 1803 between sheets of paper, has taken its place in history and is rarely found in today's business world due in large part to the success of carbonless paper (see color Figure 25.1 following p. 366). With the technological advances of the microencapsulation systems, some carbonless papers can be used in printing processes and computer printers. However, it is admitted even by paper manufacturers that non-impact printing technology and other advances in computer printing are quickly making carbonless paper a dinosaur that may have already seen its best years. In spite of this likely demise, it is vital that document examiners understand the carbonless paper system and some of its forensic applications, as matters involving this technology will be potential examination material for decades to come.

25.2 Classes of Carbonless Papers Systems

There are three microencapsulation systems used in the manufacture of carbonless paper. By far the most common is the chemical-mated system, in which colorless dye microcapsules coated on the back of an upper sheet come in contact with a colorless coreactant (also called a receiver material) coated on a lower sheet to create a colored (usually black or blue) image (see color Figure 25.2 following p. 366). Another system is called the chemical self-contained carbonless system, in which the colorless dye microcapsules and the coreactant are both interspersed on a coating on each sheet of paper (Figure 25.3). The final system is called the mechanical self-contained carbonless system, in which a coating containing pigmented dye is physically transferred from the back sheet of an upper sheet to the uncoated front side of a lower sheet. Each manufacturer has proprietary formulations that can be verified through a cooperative arrangement with the manufacturer should product manufacturer become an issue in an examination.

The chemical-mated system utilizes a top sheet of paper, on which a writer physically places the writing instrument, or a printer places its product, which will not have any carbonless system coating on the front side. The backside will be coated with a material that includes the colorless dye microcapsules. This sheet is called CB, which stands for coated back. All of the subsequent pages with the exception of the bottom sheet will contain the coreactant coating on the front of the sheet and the colorless dye coating on the back. These sheets are called CFB, which stands for coated front and back. The bottom sheet has only the coreactant coating on the front of the sheet. This sheet is called CF, which stands for coated front. A three-ply form will have one CB sheet as the top sheet, one CFB sheet in the middle, and one CF sheet as the bottom sheet. A four-ply form will differ from the former example by having two CFB sheets in the middle.

25.3 Forensic Application

Regardless of the type of examination conducted, it is important, and may be vital, to obtain all copies of any forms for examination. While the presence of all copies may not be absolutely necessary to conduct certain examinations, their availability will provide the FDE with the opportunity to look for information of forensic value that may not otherwise be present.

Issues often arise regarding whether certain entries were added or deleted from individual sheets of a carbonless paper form. Certain features common to questions of added entries are appropriate for an analysis of carbonless paper. These include examinations for

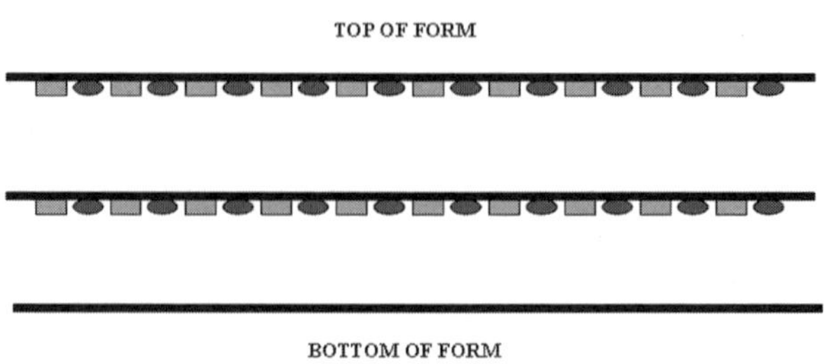

Figure 25.3 The chemical self-contained system of carbonless paper consisting of intermingled colorless dye microcapsules and a colorless coreactant on the same surface.

different inks, writing pressure differences, differences in speed, slant, or similar feature, relative ink dating, and the possibility of different writers (all conducted on the top CB sheet). In addition, the relative position of the handwritten entries are compared to the alignment with the pre-printed material on the form on all sheets except the top sheet. If a form is separated and then reassembled, an alignment shift of the pages, however large or slight, should be present. Any entries added subsequent to the reassembly should show a different alignment pattern throughout the various lower pages (see color Figure 25.4 following p. 366). Document examiners are cautioned that some shifting of forms can infrequently occur naturally, and consideration should be given to this possible explanation to any alignment idiosyncrasies noted. The possibility exists that relative dating technology could be applied to lower sheets (CF, CFB) based on the time differential between the combining of the various chemicals in the pressure activation act. Added entries may be attempted using foreign carbonless paper sheets if not all of the forms are available to the perpetrator. Should certain sheets of the form in question, particularly the top CB sheet, also not be available to the document examiner, an additional examination involving non-destructive spectral analysis or destructive tests, such as thin-layer chromatography, can establish different chemical compositions indicative of the use of a foreign carbonless paper form. If the foreign carbonless paper sheet is not compatible with the sheets at issue, indented writings will exist that can be established through standard indented writing test procedures.

Deleting entries on carbonless paper is an inherently difficult task, as the act, if thoroughly completed, would require separate erasure processes for each sheet of paper. The ink produced by the carbonless paper system historically does not easily disappear. In fact, the ink has a characteristic of smearing easily upon abrasive action. Standard examinations involving spectral analyses would be appropriate methodology to reveal erased entries and evidence of chemical erasure attempts. Microscopic examinations, side lighting, and the use of a light box are appropriate methodologies for the examination of attempts at physical erasure. Document examiners should keep in mind that it is possible for entries to not appear on lower-level sheets of multiple carbonless paper forms if an obstacle is placed between two of the sheets. However, this scenario has its own unique characteristics. Since the obstacle would have to prevent the localized pressure necessary to break the microencapsulations, the barrier would also have to obstruct indentations. In addition, the entries in question would not appear on any other sheets of the form below the location of the obstacle. An obstacle in the form could not account for an entry not appearing in a middle sheet, but appearing on the bottom sheet (see color Figure 25.5 following p. 366).

The effects of carbonless paper cannot be limited only to carbonless paper. Research has found that indented writings with carbonless paper on top of ordinary bond or tablet paper can create not only indented entries, but also entries that can be rendered legible through infrared luminescence. While all such scenarios would be impossible to cover in this book, standard methodologies for examinations involving carbonless paper routinely include spectral analyses and indented writing examinations of each carbonless paper sheet and other papers physically affiliated with the carbonless paper.

Section VII

Indented Writing Impressions

Accidental Markings and Indented Writing on a Document

26

FARRELL C. SHIVER

Contents

The preparation of a document represents a series of planned acts, but at times important elements become part of the document purely by chance and not through the premeditated design of either those who prepared the document or those who subsequently handled it. In this way, latent prints, writing offsets from some other document, impressions from writing strokes that are not part of the page in question, and traces of foreign matter with which the document has been in contact find their way onto it. More than likely their presence is entirely unknown to the document's author, but under favorable circumstances, these chance markings and additions can play a significant role.

26.1 Indented Writing Impressions

Whenever two or more sheets of paper are stacked, traces of the writing executed on the top page tend to become indented on the sheets below. These writing impressions may be useful in connecting a person to a document. The signer of a contract may insist that a particular page of a contract was not present when the document was signed. If the impression of the original contract signature is found on the questioned page, this impression provides clear evidence that the page in question was present at the time the contract was signed. Impressions found on an anonymous note may provide clues to the identity of the writer, such as the impressions of letters previously written. The development of writing impressions may prove that a document has been altered and may not have been made on the date purported. For instance, if the examination of a medical record discloses indentations of an entry from June 2004 on a document created in July 2004, this proves that the June entry could not have been made at the purported time.

A number of factors affect the development of indented writing. The depth of the indented impressions and, thus, the clarity of the indented writing depends upon the pressure of the writing strokes, the number of stacked sheets, the nature of the backing below the papers, the thickness and kind of paper, and the sharpness and firmness of the writing instrument. Depending on these factors, the amount of impressed writing may vary from virtually a whole document to a few weak fragments of letters or words, and its legibility from almost that of the original to a mere indecipherable presence. The sheet immediately below the page of writing often bears the clearest imprint (Figure 26.1), but indentations can often be developed on the second and lower sheets. Indented writings include not only visible indentations caused by the force of writing on a sheet above, but also minute fiber disturbances that can only be visualized through special processing. These impressions are not confined to handwriting alone, but can just as well be indented typewritten characters.

Not all indentations can be deciphered. Clarity of the basic indented impression is a contributing factor, but it is not the only one to complicate the problem. Overlapping indentations from writing on several pages, all of which have lain over the sheet under investigation, can greatly restrict findings (Figure 26.2). The visible writing on the document itself, which appears as a negative or white image on the lift (transparency film), may also interfere. Wrinkles and folds in the document may interfere or preclude the decipherment of the indentation's writing. Since these interfering folds and creases can be introduced with excessive handling, every precaution must be taken to limit handling and to prevent wrinkling or folding of the document.

At times the impressions can be clearly read by allowing light to strike obliquely from one side. The document can be studied while moving the light at various angles and directions. The results of these examinations may be preserved through photography.

Over the years, a number of other techniques have been suggested for the examination of indented writing.[1] Today, the two most common techniques are the use of side lighting and electrostatic processing of the document with an electrostatic detection device (EDD).

Figure 26.1 An indented writing impression developed using electrostatic processing. (Courtesy of Brian S. Lindblom.)

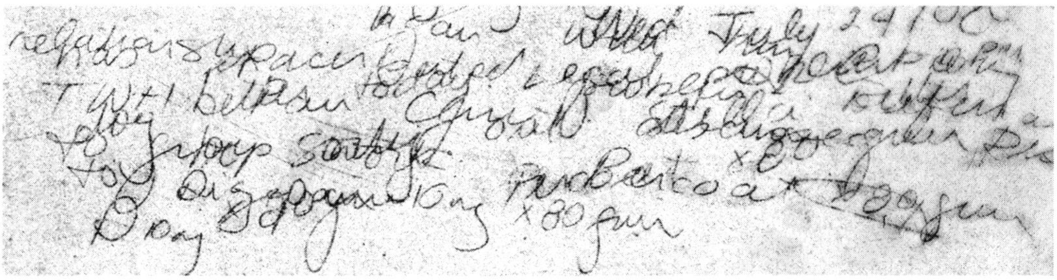

Figure 26.2 The development of overlapping indented impressions can complicate, and sometimes preclude, the decipherment of indented writing. (Courtesy of Brian S. Lindblom.)

A recently published procedure for conducting indentation examinations requires the use of both techniques.[2]

The original and most common EDD is the electrostatic detection apparatus (ESDA), but others also have been manufactured.* In electrostatic processing, the document in question is placed on a vacuum bed and covered with a polymer film. An electrical charge is placed on the surface of the document by passing a wand, which contains a corona wire, over the surface of the film. A black toner is then applied to the film to develop images of any indentations that might be present. There are several techniques for applying the toner, but perhaps the most common is the cascade method. In this method, tiny glass beads coated with the toner are cascaded across the surface of the film. Other techniques include spraying the toner, brushing it on, or, more recently, dabbing the toner onto the document with a small pad that has a self-contained toner compartment. The developed image is fixed by covering the film with a lamination sheet. On some occasions, it may be advisable to photograph the developed indentations prior to attempting to fix the image. Some document examiners preserve the results of EDD examinations even when no images are developed.

Electrostatic processing is considered non-destructive because the images of the indentations are developed on the mylar film, not the document itself. However, it has been found that EDD processing may remove some lead from pencil writing and also may damage documents typed with single-use typewriter ribbons.[2] When documents of these types are processed with an EDD, it is advisable to record images (photographically, by scanning or copying) of the document prior to processing.

Low-humidity conditions may interfere with EDD processing. Several studies have been conducted in an effort to determine optimal humidity conditions.** Humidity chambers are commonly used to humidify documents prior to electrostatic processing. Some laboratories are equipped with humidity-controlled rooms for EDD processing. One technique requires the humidification of the toner rather than the document itself. Moore[3]

*The Indentation Materializer Electrostatic Document Device (IMEDD), the VacuBox, and the DocuStat are commerical alternatives to the ESDA. Some document examiners have custom-made machines that are known under a variety of names.

**Several authors have studied the effects of humidity and temperture. These include Beal, B.L., Effects of water temperature vs. time in humidyfing documents for electrostatic detection apparatus examination, *J. Am. Soc. Questioned Doc. Examiners*, 5, 78–87, 2002; Noblett, M.G. and James, E.L., Optimum conditions for examination of documents using an electrostatic detection apparatus (ESDA) device to visualize indented writings, *J. Forensic Sci.*, 28, 697–712, 1983; Riebeling, I.J. and Kobus, H.J., Some parameters affecting the quality of ESDA results, *J. Forensic Sci.*, 39, 15–20, 1994.

cautioned that excessive humidification of documents may have a deleterious effect on the development of latent prints.

Regardless of the efforts made to produce clear, decipherable results, there are many situations where the resulting lifts require enhancement. Dodge and burn, auto level adjustments, and contrast changes are three functions in graphic software programs that can aid the document examiner in producing legible results for further analysis. These topics are covered in detail in Chapters 29, 31, and 34.

Electrostatic processing of documents can result in the development of fresh fingerprints. The developed fingerprints can interfere with the decipherment of the developed indentations. Document examiners will sometimes wear gloves when handling a document that is to be electrostatically processed. Even when wearing gloves, excessive handling may damage or destroy writing indentations.[4] Depending upon the type of glove used — cloth, rubber, or latex — obliteration of the indentations may occur. Therefore, consideration must be given to the type of glove worn and the handling of the documents. Since latent print processing will also destroy writing indentations, examinations for indented writing should take place prior to latent print processing.

Electrostatic processing can sometimes be used to determine the sequence of ink strokes and indented writing images.* For further discussion on this dating method, see Chapter 29. One study successfully used electrostatic processing to determine the sequence of inkjet images and writing impressions, but failed to do so with laser printing.[5]

26.2 Writing Offsets

When two papers come in contact with each other while the fluid ink on one is still damp, a writing offset may be produced on the other.[6] Storage in damp conditions also can produce offsets from water-soluble inks. Similar offsetting can occur when the graphite from pencil writing is offset to another page.[7] To decipher and interpret correctly what the offset represents stands as a challenge to those interested in determining the facts. The full force of this challenge can be best appreciated when it is realized that these traces are generally very slight fragments of writing. In some cases, the offset proves after accurate decipherment to have only minor significance because of the particular circumstance of the case, but there can be instances in which the interpretation becomes forceful evidence linking two documents to a common source.

With faint or indistinct images, which are the general rule, writing offsets must be intensified by various means. The image may need to be reversed, for an offset is sometimes a mirror image. In the past, this has been done through filter photography, but it is now more common to use computer techniques (see Chapter 31). Each problem requires protracted study and may require repetition of the same method with slight modification to achieve the best decipherment.

Writing offsets have been used to determine that documents purportedly prepared over a long period of time were actually prepared over a much shorter period.[8] When

*Robert W. Radley has conducted research in this area. Two of his articles on the subject are: Determination of sequence of writing impressions and ballpoint inkstrokes using the ESDA technique, *J. Forensic Sci. Soc.*, 33, 69–72, 1993; Determination of sequence of intersecting ESDA impressions and porous tip, fiber tip and rollerball pen inks, *Sci. Justice*, 35, 267–272, 1995.

writing occurs on both sides of a page, there may be a strong transfer of ink to an underlying page if the ink is fresh.

26.3 Foreign Traces

Small deposits or traces of many different substances may be found on documents and may aid in reconstructing their history. Many of these are placed purely through chance contacts with foreign objects during the preparation and subsequent handling or storing of the document. Thus, stains of a similar pattern running across several pages can help to show they have been a unit for some time. Mold on paper is an indication of age. Carbon deposits in and around the outline of a signature cast strong suspicion on its genuineness. With an anonymous letter, small traces of lipstick or the presence of saliva under the sealed flap or stamp may be justification for DNA testing. Paint, rust, grease, and many other stains are encountered from time to time in document examination. By learning what kind of stain is present, it may then be possible to arrive at its cause and significance.

26.4 Latent Prints

Latent prints are commonly referred to as fingerprints, but are actually impressions left on an object when it is touched by any area of friction skin. Friction skin is found on the undersides of the hands and on the bottoms of the feet. In document cases, the most common latent prints are made by the fingers and the "writer's palm."[9]

Latent prints are placed upon papers simply by handling and remain invisible until chemically developed. Paper is one of the best substrates for the deposition and retention of latent prints. The ease with which latent prints are added to a paper and their invisibility are definite assets, especially in criminal investigations. If it is possible to develop an identifiable print to be compared with those of a suspect, a very positive form of evidence may be forthcoming.

Typical processing of documents for latent prints involves the use of chemical processes such as ninhydrin, physical developer, and 8-diazaflouren-9-one (DFO). Occasionally, visible prints, known as patent prints, may be observed on a document without conducting any specialized examinations. In other cases, latent prints may be observed through the use of a laser, or alternate light source, and appropriate filters without any chemical processing. Any observable prints are preserved through photography.[10]

The most common method today of developing latent prints on paper is the ninhydrin technique. Ninhydrin reacts with amino acids in the latent print residue. It has a serious drawback when the documents may need to be subjected to other types of examinations, such as the identification of signatures and writing, examinations for indentation evidence, the determination of possible alterations and erasures, and the comparison of inks.[11] Use of ninhydrin will stain the document, and with the usual method of preparing the ninhydrin solution, inks can be seriously weakened or obliterated. The ninhydrin solution can be prepared in a manner that will minimize the damage to a document, but writing indentations and some trace evidence will probably be lost.

Physical developer is sometimes used after ninhydrin processing. This method will develop latent prints that ninhydrin will not. The reason for this is that ninhydrin reacts with the amino acids in the latent prints, whereas physical developer reacts with salts, fats, waxes, and lipids found in fingerprint residue. The use of physical developer will damage the document.

Another technique for the development of latent prints on documents is DFO. Like ninhydrin, DFO reacts with amino acids. The developed prints are observable at a particular wavelength of light* using a laser or alternate light source. This technique is less commonly used than ninhydrin or physical developer due to the expense of the chemicals and equipment. DFO will also stain documents, but perhaps less so than ninhydrin.

One of the earliest techniques for detecting latent prints on paper, iodine fuming, is sometimes still used. The technique is frequently not used because of the hazards associated with the inhalation of vapors and the fact that the likelihood of positive results decreases as the latent prints age.** The use of iodine fuming may be indicated when a covert examination is required[12] and also when thermal or carbonless paper has been used.[13] It has the advantage that if no prints are developed, other methods can subsequently be used, since the iodine stains will dissipate. The developed prints will quickly fade and should be imaged promptly.

Investigators sometimes attempt to process documents for latent prints in the field, but this is better done in a laboratory setting. Chemical processing in the field can be dangerous and raises health concerns. Further, most chemically developed prints are subject to fading and time becomes a factor. Some investigators continue to dust documents with powders in a manner similar to that used in investigations at the scene of a crime. Powders are less effective than the chemical techniques already discussed and should not be used. Additionally, the use of powders may have a detrimental effect on the development of indented writing images.[14]

Whenever a document is to be processed for latent prints, a top-quality, accurately scaled image of the document should be made so that the handwriting and text that appear on the document can be studied and identified if questions later arise. All processes for developing latent prints on paper raise the risk of damaging or obscuring writing details, no matter how carefully performed.

As a rule, document examinations should be conducted prior to processing a document for latent prints. The document examiner, therefore, must take care in handling the document so that latent prints are not deposited on the document. Each person who handles the paper can leave his latent impressions. With repeated handling, the numerous prints may obliterate the original ones. Therefore, papers that are to be processed for latent prints should be protected from the addition of latent prints during handling and should be submitted for examination as soon as possible.

26.5 Conclusion

The fact that marks and traces have been deposited on a document by chance does not necessarily lessen their value; rather, circumstances peculiar to the particular problem are the controlling factors. Their recognition and identification through careful study and

*Wavelength of 430 to 580 nm.

**Positive results may be expected if the latent prints have been on the document for 30 days or less.

treatment, however, are essential if the investigator is to reveal facts that, weighed in relation to other aspects of the document, determine the true importance of the evidence.

References

1. Brown, J.L., Survey of techniques used to visualize markings, *J. Am. Soc. Questioned Doc. Examiners*, 1, 107–112, 1998.
2. ASTM E 2291-03, *Standard Guide for Indentation Examinations*, ASTM, West Conshohocken, PA, 2003.
3. Moore, D.S., The electrostatic detection apparatus (ESDA) and its effects on latent prints on paper, *J. Forensic Sci.*, 33, 357–377, 1988.
4. Licht, G. and Murano, E., Seeing Ghosts and Reading Demons: ESDA Effects in Light of Current Discussions, paper presented at a meeting of the American Society of Questioned Document Examiners, San Diego, 2002.
5. Mohammed, L.A., Sequencing writing impressions and laser printing or inkjet printing using the ESDA, *J. Am. Soc. Questioned Doc. Examiners*, 1, 40–42, 1998.
6. Moryan, D., Deciphering set-off writing images, *Int. J. Forensic Doc. Examiners*, 2, 342–346, 1996.
7. Gervais, R., Techniques and Terms Useful in the Examination of Graphite Offsetting, a paper presented at a meeting of the American Society of Questioned Document Examiners, Ottawa, Ontario, Canada, 2000.
8. Strach, S.J., Radley, R.W., and Westwood, P.D., Short term relative time of writing determinations by observations of ball-point pen ink transfers, *Int. J. Forensic Doc. Examiners*, 4, 152–153, 1998.
9. Shimoda, S.C. and Franck, F.E., Writer's palmar impressions, *J. Forensic Sci.*, 34, 468–474, 1989.
10. Federal Bureau of Investigation, *Latent Print Processing Guide*, FBI, Washington, D.C., 2000.
11. Horton, R.A. and Shaver, L.C., The effects of latent print processing on ballpoint pen inks, *J. Am. Soc. Questioned Doc. Examiners*, 3, 70–81, 2000.
12. Iodine fuming, *Minutiae*, Summer 2003, Issue 76.
13. Iodine fuming contest winner!!, *Minutiae*, Fall 2003, Issue 77.
14. Dunkerley, M.Y. and Riley, T.P., The effect of fingerprint processing on ESDA impressions, *Int. J. Forensic Doc. Examiners*, 5, 280–282, 1999.

Section VIII

Altered Documents

Alterations in Documents

27

SUSAN E. MORTON
BONNIE L. BEAL

Contents

On numerous occasions documents are altered during their preparation or after their completion with full knowledge of all parties concerned and without an attempt on the part of anyone to perpetrate fraud. However, there are other instances — more numerous than generally believed — in which an addition or other alteration is made with intent to cheat some party in an otherwise fair and legitimate transaction. When such an act has been committed, or is suspected, a document examiner may be consulted. Since these calls for assistance are not infrequent, effective techniques have been devised through extensive research and experimentation not only to disclose evidence of the alterations or additions, but also to demonstrate what has been changed and what the original contents were.

Documents are changed in several ways, and each method requires a separate technique for detection and for the restoration or decipherment of the original material. The most common procedures include the removal of portions by erasure, obscuring the writing by opaque marks, and the addition of material either by interlineation or by extending portions of the document. Now we shall consider the more common problems in which handwriting, typewriting, or printing has been removed from the document or expunged in some manner. Under these conditions the ultimate objective is always to restore the original writing so that it can be read in its entirety, but there are instances when only portions can be deciphered.

27.1 Erasures

It is common knowledge that writing, typewriting, or printing may be removed by either chemicals, abrasion, or scratching with a sharp instrument like a knife. Not all of these methods are successful in removing each class of marking medium, but all are encountered in a variety of problems.

Virtually all erasures can be detected by a thorough examination. On a rare occasion, an extremely skillful removal of fluid ink writing with a chemical erasing fluid or solvent might remain undetected, but the combined effect of two factors should leave some evidence of the act itself. First, the chemicals react with the paper to form visible or latent stains. These stains, and possibly fragments of original ink, can often be discerned visually or under infrared (IR) illumination. Second, eradication fluids usually remove sizing from the paper, leaving it much more porous and absorbent than untreated paper. If an attempt is then made to write over the erased area with a fiber-tip or roller ball pen, the fluid ink is very apt to spread out and be wicked into the bare paper fibers, leaving an evident blotch. Roughened paper fibers also may be observed in the area.

With increasing frequency, checks are being chemically washed, dried, and sometimes pressed. A study conducted by Licht[1] has shown that ballpoint pens are the easiest to remove with no latent remnants left in the paper. These chemical washes are very difficult to detect (Figure 27.1). Sometimes the previous entry is not completely removed and may

Figure 27.1 This image depicts a portion of a check that has been chemically washed. Small remnants of ink from the previous original entry can be observed upon close inspection (top image). The bottom image is the same area under IR luminescence. Remnants of the original ink are not visible.

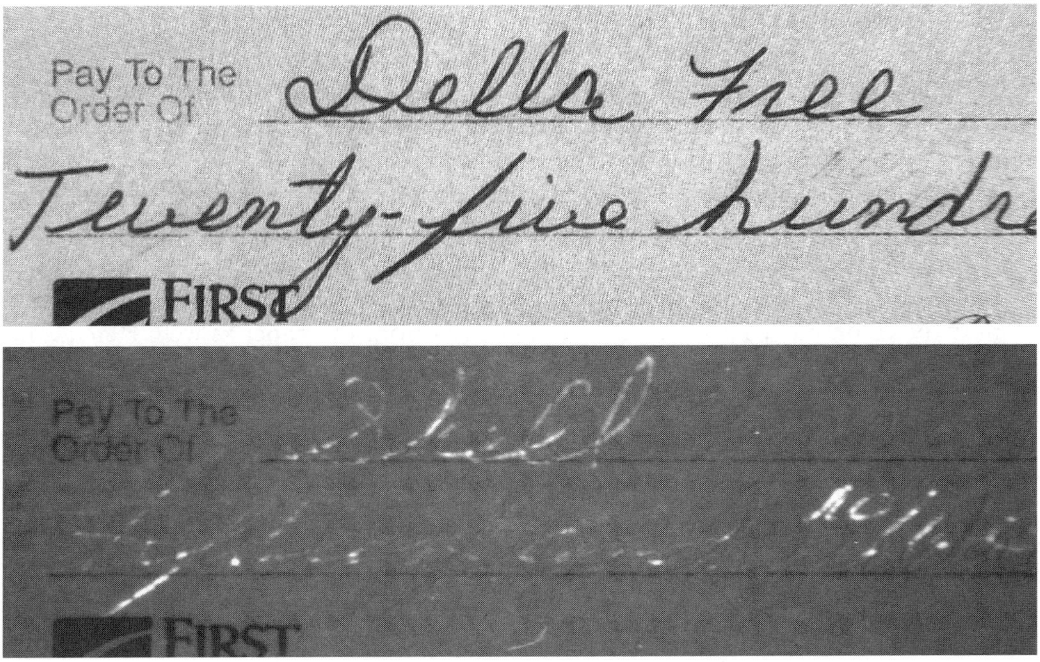

Figure 27.2 In this poor chemical wash portions of the original ink are still visible (upper image). Under IR luminescence the original entries of "Shell" and "Fifteen and no/100" can be seen.

Figure 27.3 The reverse side of the check depicted in Figure 27.2 is shown. Chemical staining is evident.

be viewed microscopically or with IR lighting (Figure 27.2). Staining around the edges of the check may be apparent as well under IR lighting (Figure 27.3).

Overwriting with a ball pen may not appear to be abnormal, but some evidence of erasing probably will remain. Examination with oblique lighting, an electrostatic detection device (EDD), a microscope, photography, ultraviolet (UV) radiation, or IR luminescence should establish in most instances that there has been alteration. In contrast to this, even very skillful erasures by means of abrasion or scratching generally leave much more definite telltale traces consisting of disturbed paper fibers and portions of the original strokes that were not completely removed. These erasures are consequently more easily detected.

Although the first step is always to ascertain that an erasure has been made, in any investigation of this nature the step that is of greatest importance to those who stand to be defrauded by the alteration is the restoration or decipherment of the erased matter. By restoration it is meant that the original writing is revealed on the document either temporarily or permanently, while decipherment includes any method, photographic or otherwise, that permits determination of what has been erased. Whether or not this is to be successfully accomplished depends on both the original writing medium and the manner in which it has been erased.

In the case of synthetic dye inks found in porous-tip or roller pens, bleaching removes not only the color, but virtually all of the chemicals from the paper as well. Here again is an example of the fact that non-chemical erasures, while more easily detected, are less readily deciphered. With these inks, photographic techniques or examination under IR imaging systems are standard methods, although not every erasure can be successfully deciphered.

Carbon ink is very difficult to expunge completely. Chemicals do not bleach it, but an abrasive erasure or a skillfully used knife edge or razor removes the ink effectively. Decipherment of the original ink is usually extremely difficult. If it has been completely erased, the problem may be virtually impossible. However, whenever there are some traces of the original writing, these may be intensified photographically or digitally to bring about at least a partial decipherment.

Ballpoint pen ink does not erase easily. Two papers that examined the erasable ball pen ink were written by Peter Pfefferli and Jacques Mathyer[2] and William J. Flynn.[3] Extensive rubbing is necessary to effect a full erasure because the ink penetrates the paper and is not all on the surface. Common chemicals can completely remove ballpoint pen ink from paper,[1] though this technique may leave the indentation of the ball track. Photographic, UV, and IR luminescence methods, while not foolproof, are the better ways of deciphering the original writing.

The decipherment of erased pencil writing, regardless of whether a black, indelible, or colored pencil was used, presents not only a difficult but also an arduous and lengthy task. The erasure can be made with a soft rubber eraser, which in the hands of a skillful person hardly disturbs the paper surface. Even so, except when the writing was executed with a very light touch, slight indentations of the original strokes may remain after complete obliteration (Figure 27.4), and when incomplete, some carbon or pigment traces are to be found (Figure 27.5). Both the indentations and traces serve as the basis of decipherment. Visual and photographic examination under controlled light, involving both oblique illumination and low-intensity diffused light, are effective methods of reading the erased matter.[4] At times reflected IR illumination is needed and, with colored pencils, IR luminescence.

Utilizing the imaging software with an IR imaging system is a great tool in deciphering erasures and capturing the image for the case notes. EDD examination of both front and back of the paper is often useful in deciphering abraded writing, or at least graphically demonstrating that an erasure has occurred.

With indelible pencil writing some of the unerased dye can be made more visible by moistening with a fine spray of alcohol.[5] The alcohol dissolves the remaining fragments of dye, causing them to spread, thereby reproducing the erased portion.

If during the erasing process the paper surface was badly roughed, the problem becomes more complicated. Under favorable circumstances, a complete decipherment can

Figure 27.4 Impressions left of the initial number that was erased and then written over, as viewed under oblique lighting.

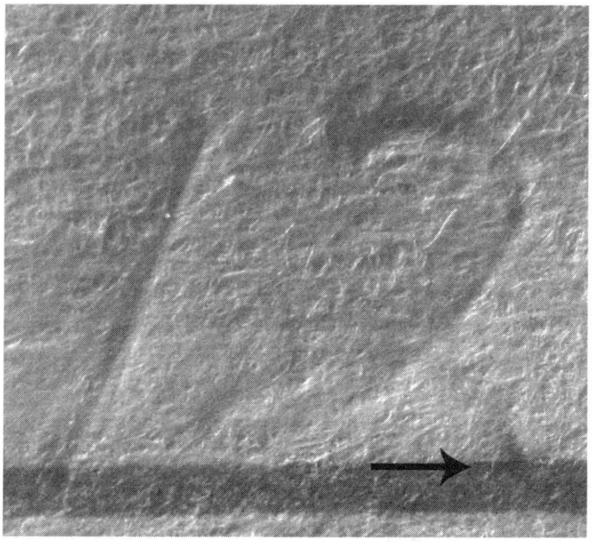

Figure 27.5 Indentations of the original numeric amount are seen using oblique lighting. Traces of the original graphite are seen at the bases of the 1 and 2.

be achieved, but there are many cases where only a part of the original matter can be read. Should traces of the original matter remain, they may be intensified digitally with imaging enhancement software or with an IR imaging system.

Computer-generated text may be easy or difficult to expunge, depending on the technology used to create it. Some inkjet inks are water soluble, though their removal in that way will generally stain the paper. Toner such as laser printing can sometimes be scraped off the page, but again, physical evidence will usually remain. The same techniques used to decipher an erasure may be used here as well, but decipherment is usually very difficult.

When a lift-off ribbon has been used with a correcting typewriter, impressed outlines of the letters usually remain. Virtually all of the pigment is removed, but the impression of the letter may be intensified.[6] Side lighting, an IR imaging system, a plastic-cast technique,[5] or application of the chemical solution used in pencil problems may lead to full or partial decipherment. Nevertheless, cases are encountered in which the erasure has been so thorough that none of the techniques give satisfactory results.

Problems resulting from the removal of a stamp impression or printed matter, or from an erased carbon copy, are less frequent, and the best results are apt to lead only to a partial decipherment. Controlled lighting, and reflected IR, IR luminescence, or UV illumination may provide some results. The latter two techniques are most effective when the impression was bleached out with chemicals, the former when there are some fragments of the original impression. Often with these problems the paper surface has been seriously damaged by abrasive erasing.

Despite excellent results on many erasure problems, there are cases in which the restoration or decipherment of erased writing may not be successful. So much depends on the individual case — the materials erased; the care and handling of the document prior to undertaking the restoration or decipherment; the time between the writing, erasing, and attempted restoration or decipherment; and other factors often individual to the particular problem. Each restoration or decipherment may involve a great deal of experimentation and time. While the previous paragraphs have indicated scientific techniques that may be helpful, much progress is usually made by visual study aided by magnifiers, and sometimes reducing lenses under various controlled-lighting conditions, including oblique or subdued lighting, as well as moderate-intensity daylight illumination. Thus, when an erasure is first suspected, it is well to submit the document for examination immediately. This eliminates unnecessary handling, which may reduce the chances of success, and also allows sufficient time to investigate the problem thoroughly.

27.2 Associated Evidence

While associated evidence may not necessarily involve technical study, it does serve a very useful role in the decipherment and verification of decipherments of erased and altered documents. Associated evidence would be any other documents that relate closely to the altered document. One obvious example would be a photocopy made prior to the alteration. Locating the photocopy would reveal immediately what had been changed. Other types of evidence might involve accounting records in which one such record had been changed but other entries in the bookkeeping system had not been. Thus, anyone familiar with the manner in which the books are kept would be able to locate the associated entries and reconstruct the original entry.

It is always possible that portions of the associated evidence also have been altered. For example, in an accounting record, the erasure and change of a particular entry may require that the totals at the end of a section or a page be changed as well. Although great care may have been taken with the critical entry, it is not uncommon for the changes in the subtotals or totals to have been made carelessly or with only partial erasing. The result is that the original totals can be deciphered more readily than the key erased figure. In such a way, the critical entry could be reconstructed. Many other examples could be pointed out, and often consultation with the document examiner him- or herself may bring to mind where to look for this type of evidence.

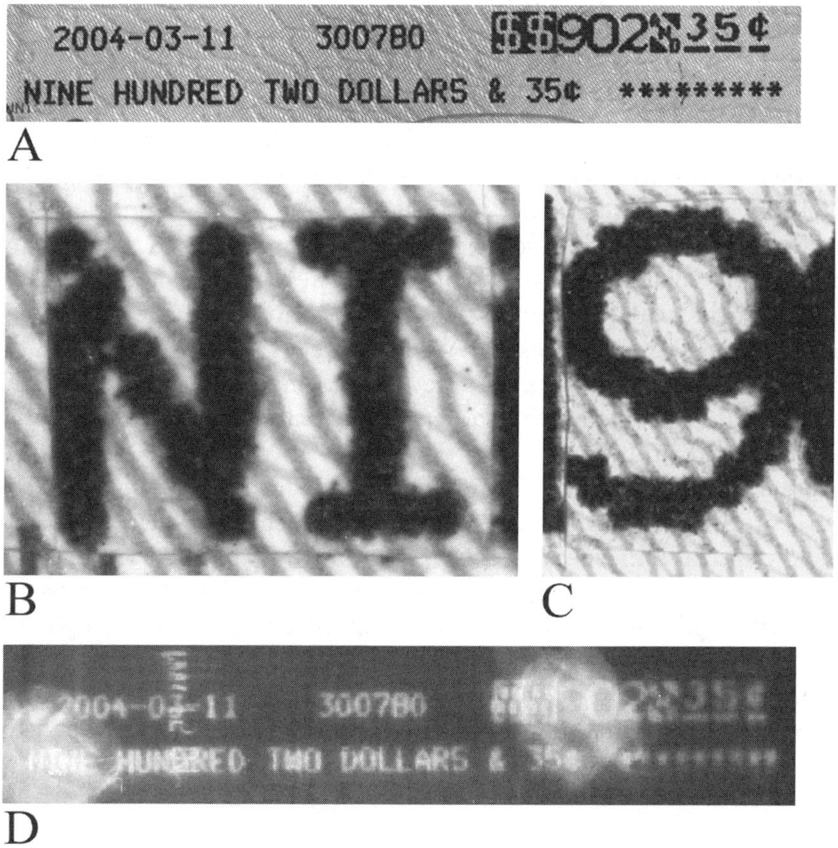

Figure 27.6 An altered money order is illustrated. Image A shows a portion of the document under normal lighting. Images B and C show magnification of the areas altered by cutting the original portion of the document and replacing it with areas from another money order, in this case the letters *NI* and a 9. Image D demonstrates the adhesive on the reverse side illuminated with UV lighting.

27.3 Cutting

A rare, though not unheard of, alteration of a document can be achieved by skillful cutting away of some portions and then inserting new material to fill the gap. Of course, it is necessary to devise some means of securing the inserted material in place. The document may be pasted to another sheet, or stamps or tape may be affixed to the back of the document (Figure 27.6). Checkwriter imprints have been altered by this means.[7] Careful inspection of the paper will reveal the alteration, but casual observers and even those who handle the document in business transactions, businessmen and bank employees, have been fooled by checks altered in this manner.

If the paper is thick, only the top layer need be cut out and the inserted material cut so that it fits accurately into the depression. Contest cards and pari-mutuel tickets are known to have been produced with key numbers modified in this way.[8] Detection of the alteration is possible, of course, but what was there originally normally cannot be ascertained.

27.4 Interlineations and Additions

Fraud can be committed just as effectively by addition as by subtraction. The insertion of a modifying clause or sentence may completely change the meaning of a document in as thorough a manner as the erasure of a key portion. The skill with which these modifications are inserted varies from case to case, but as with erasures, effective methods have been developed by which many fraudulent interlineations or additions are revealed.

Obviously, the crude insert of some important clause between the lines or crowded along a margin immediately arouses suspicion. However, many additions are carefully worked into the form of the document when very convenient space either within it or immediately above the signature was provided by careless preparation. When these insertions are skillfully done, they may pass unnoticed by the casual observer, but still, these inconspicuous manipulations can be revealed by physical faults that are disclosed through proper techniques and study.

To disclose that an insertion or addition has been made may involve an extensive study of the document as a whole. Many of its elements, which have been discussed in earlier sections, assume special importance. The lack of uniformity of ink; the work of more than one pen or output device; crowding, uneven margins, or different spacing algorithms of a modifying section, if printed; evidence of the insertion of pages through study of the paper and fastening devices (Figure 27.7); sharp variation in handwriting; and any of a score of other factors individual to the problem at hand may point out the insertion. There is, however, one other sign that points conclusively to the fact that the document was not

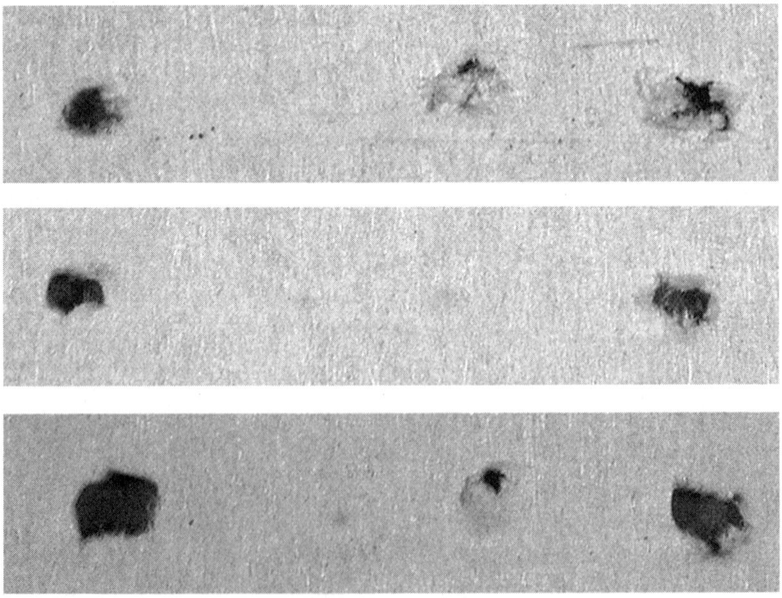

Figure 27.7 The insertion of a sheet of paper is determined by the examination of the staple holes. The top image is from the first page, the middle from the second, and the bottom from the third. Notice the additional perforation between the two main staple holes (top and bottom images). It was created when the end of the staple punched through the back of the sheets. The perforation is not found on the second page, providing evidence that it was not in place at the time the original stapling occurred.

put together in normal order — evidence that the sequence of intersecting writing strokes or strokes across the folds or perforations in the paper is not in the logical sequence that would be consistent with the natural or alleged preparation of the document.

27.4.1 Sequence of Writing

Intersecting writing strokes may have distinctive patterns, depending upon the order of writing, the lapse of time between the two writings, the density of the two strokes, and the kind of inks, writing instruments, and paper used. With the stereoscopic microscope or a hand magnifier aided by skillfully controlled lighting and IR or digital imaging systems, the true order of preparation may be revealed and demonstrated to a lay observer.

What appears to be the obvious solution may not always be the correct answer. For example, the line of deepest color usually appears on top even if it was written first (see color Figure 27.8 following p. 366). Careful study and testing is necessary before reaching a conclusion. A very fluid ink may underflow existing lines so that it lies physically underneath older entries. Some of the more common criteria for determining sequence are considered in the following paragraphs.[9]

If we consider the intersection of two writing strokes or the intersection of writing and printed text, the majority of problems are covered. Substantial, repeated intersections of two writings offer a higher probability of success than a single, indifferent intersection, such as a weak stroke crossing another or two lines barely touching one another, which only very infrequently can produce a clear indication of the order of writing.

Fluid ink flows into the paper, and when such a line strikes another fresh fluid ink or other hydrophilic material, the second line tends to spread out or widen at the intersection. This condition is encountered with fluid ink pen writing and at times with that of porous pens. If the first stroke is completely dry, the phenomenon is less apt to occur. When the first stroke contains a significant groove created by a stiff pen point or a roller pen, a relatively light second stroke may be found to narrow slightly or even to skip at the intersection with the groove. There is also the possibility with highly soluble dye inks and fresh fluid ink that the second ink can dissolve a portion of the first line, spreading its ink beyond the edge of the original stroke (Figure 27.9).

Ballpoint pens with non-aqueous inks are rolled onto the paper. The ink does not flow. When two such pen strokes intersect, the grooved lines caused by pressure on the ball are critical in determining the order of writing. Interruption of the groove edges on one line would indicate that it was written first. If the first groove is deep, the second line may narrow slightly as it crosses, or may show a microscopic skipping. This same phenomenon can occur when the ballpoint pen crosses a significantly grooved pencil stroke.

An additional means of studying the continuity of the edges of two intersecting ballpoint pen strokes involves lifting some of the ink at the point of intersection from the paper. Igoe and Reynolds[10] suggest the following method for achieving such a lift. The glossy side of Kromekote paper, which is a high-gloss backing material used extensively for mounting lifted latent fingerprints, is placed against an intersection in which the edge marks show more prominently, and the back of the paper is rubbed with a blunt point using even pressure. The resulting lift can show more clearly which of the two lines is continuous across the intersection. To improve the results with older ink lines, like those found on documents written several years before, Godown[11] has proposed pretreatment of the Kromekote paper with a dilute solution of thymol in alcohol. The lift is then made

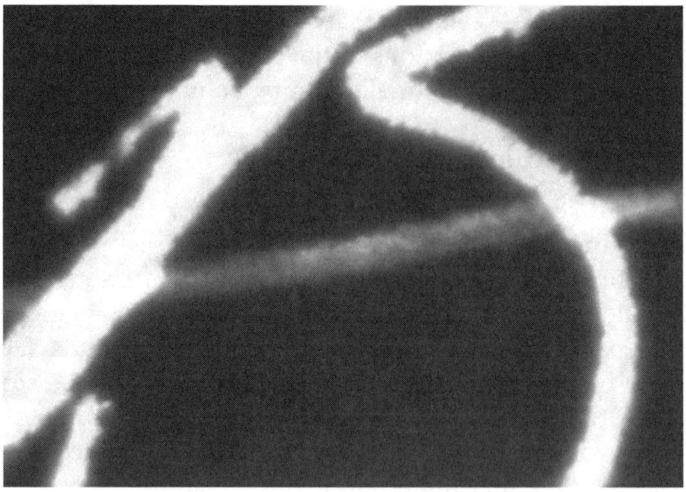

Figure 27.9 Fluid ink that is fresh can be pulled along the line of a pen that passes over the top. This spreading of the ink is seen under IR luminescence. This phenomenon may not be visible if the second ink has a stronger luminescence, thereby masking the spreading of the underlying ink.

by rubbing the back of the sheet with a warm tacking iron for about 5 seconds. At this time there are a number of companies developing instruments and software to determine line-intersecting problems. One paper that deals with one such instrument was written by Veerle Berx and Jan De Kinder.[12]

Pencil-written intersections may contain evidence of continuous and interrupted striations or grooves in the two lines. Occasionally, the second line may drag particles of pigment from the first line, a condition more readily observed if the two strokes are of different colors. In fact, with any kind of intersection, if the second writing instrument drags particles of pigment or dyes from the crossed line, this is significant evidence that it was written last.

Several different considerations come into play when writing and printed text intersect. With fluid ink crossing a relatively fresh typewriting or toner image, there is a tendency for the water-based ink to be repelled by the oils and waxes or resin found in the typewriter ink or toner. As a result, small gaps, a slight narrowing of the fluid ink line, or beading of the ink can be observed. This same condition can occur when a fluid ink crosses a fresh ballpoint pen line. Inkjet images will behave very much like fluid ink writing. If there is a substantial interval of time between the preparation of the typewriting and the fluid ink writing, no repelling may be observed since the oily materials in the typewriting have completely dried out. Because of this complete lack of any repelling of the aqueous ink by the typewriting, it is suggested that the typewriting was prepared first, but it is not a sure indication of this fact. A rare phenomenon of the fluid ink spreading into completely dried typewriting has been reported. This contradictory action is due no doubt to the disturbance of the paper surface by the typewriting. There have been studies that show that when fluid ink is written last across a toner line, the ink may penetrate the toner and absorb into the fibers below.[13]

Ballpoint pen writing across typewriting can be recognized by the spectral reflection revealed by low-angle illumination or somewhat more readily by vertical illumination of the intersection.[9] The lack of such reflection constitutes strong evidence that the writing preceded the typewriting. If both the typewriting and the ballpoint pen writing emboss the paper, the condition of the edges of the embossing may assist in confirming the order of preparation. Again, the lift-off technique can be employed using a relatively weak, tacky, removable pressure adhesive. This material lifts the typewriting except when covered with writing at the intersection. The lifted image shows a continuous outline of the typewriting except when it is covered by ink. The presence of a break in the typewriting is a clear indication that the writing overlays the typewriting, but if there is no break at the intersection, then the opposite conclusion can be drawn.

A further indication that the writing is over the typewriting can be derived when the typewriting embosses the paper and the pen stroke shows a skipping at the center of the typewritten stroke where the pen failed to make contact with the paper in the deepest part of the embossing. This condition is often accompanied by damage to the edges of the typewritten stroke resulting from the pressure of the pen against the edge of the typewriting.

Many of the examinations above can be used when viewing ink strokes intersecting toner. One study conducted by Michelle Novotny[13] demonstrates that some inks will bleed through the toner and soak into the fibers below. Other studies are being undertaken to determine sequence of toner and ink lines that do not intersect.[14,15] Both studies demonstrate the differences in the toner particles when they are below or above inked writing. Color Figure 27.10 (following p. 366) depicts the spectral reflectance, using coaxial lighting, of fluid ink written over the top of toner.

27.4.2 Paper Condition and Sequence

Writing across folds and perforations frequently leaves an accurate picture of the order of events and can well substantiate contentions of fraud or authenticity. Fluid ink strokes leave the clearest picture (Figure 27.11 and Figure 27.12). The flow of such ink across a

Figure 27.11 The smooth, continuous line created by the aqueous ink stroke is seen where it intersects with the fold. This is evidence that the fold was made after the ink was applied.

Figure 27.12 The paper was folded before the aqueous ink stroke was written. Note the skipping and over-inking on the downstroke of the letter *i*.

worn fold in the paper spreads discernibly into the adjacent paper fibers in a manner that immediately establishes that the writing followed the folding. A dried ink stroke that is subsequently bent or broken by a fold is not affected in this way. Instead, it remains unchanged, or with deep folds, the ink film may contain microscopic breaks. Ballpoint pens may skip on the far side of a fold ridge or within the trough of a concave fold. There are times when ink is rubbed off the ball housing at a fold intersection.

The sequence of fluid ink writing and a perforation follows a similar pattern. When the perforations are put in the paper after the writing has been completed, both fluid and ballpoint pen ink strokes are cleanly cut (Figure 27.13), but when the writer tries to write

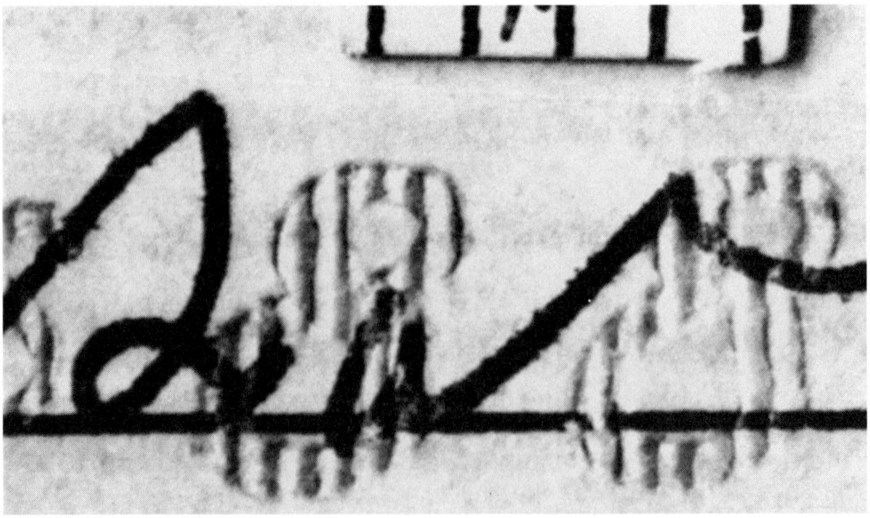

Figure 27.13 The impression of a misaligned checkwriter cut through part of the signature. The ridges of the checkwriter impression break some portions of the ink line, as seen in the 8, indicating that it was applied after the check was signed and raised.

across a perforation, the fluid ink is very apt to flow along the cut edge. With a ballpoint pen, the ink does not flow, but the ball can catch in the perforation, staining the edges. Sometimes the crowding or twisting of the writing reveals the writer's effort to avoid the perforation entirely.

Pencil and printed text strokes across folds and perforations may leave evidence of sequence as well, but the demonstration of the facts is a more complex problem than with ink writing and may be subject to greater limitations. With pencil and toner images, there is no ink that can flow into broken paper fibers. Still, typical differences in the continuity of the writing strokes or the ridge of a fold may allow definite conclusions, although not in every case. With these problems, as with all questions of sequence, experimentation under controlled conditions is often needed to verify that an apparent determination of sequence is the correct conclusion.

Under proper circumstances, sequence can be definitely established through these examinations, and the order is definitely fixed. The weight of this evidence can be great. When an unusual sequence of writing is established from the physical facts, it is not in itself positive proof of fraud. Rather, this physical evidence must be weighed in conjunction with the other testimony and facts in the case. The burden of proof generally shifts to those who oppose this new evidence, for it becomes their task to establish that this situation can exist without conflict with the intent or original purpose of the document.

27.5 Obliterated or Overwriting

Overwriting — that is, the obliteration or blocking out of portions of writing by some opaque material — is seldom used for fraudulent purposes because of its obviousness. Nevertheless, from time to time such obliterations may be encountered in all types of documents. Many times the writer blots out writing simply by striking over it thoroughly with the same writing instrument with which the document is written. The anonymous letter writer may attempt to block out a printed return address on the only available envelope.

These examples serve to illustrate the diversity of problems. Virtually each case has its own ramifications, requiring various techniques and making difficult an accurate, generalized forecast as to the measure of success. The solution at best is a tedious process involving extensive experimentation, and it must be recognized that failures are to be encountered.

Successful decipherment or restoration depends principally upon the medium employed for the original writing and the covering material. Three possible methods of solution present themselves: (1) to penetrate the covering layer so that the original writing lying beneath is revealed, (2) to remove the obliterating material chemically or by some other means while the original writing remains untouched, and (3) to try to decipher the impression from the original writing or typewriting that has not been destroyed by the obliterating action.

Imaging methods can succeed if a difference in color or chemical composition exists between the original writing and the covering material. Separation may be affected with filters of a color similar to the obliterating substances, but this attack breaks down when the covering is a true black. The use of an IR imaging system that is capable of transmitted light, IR luminescence, UV radiation, and oblique lighting may enhance the chance of differentiation success as well. It has been shown that in utilizing scanning and imaging software, some inks may be differentiated. As a general rule, if complete and accurate decipherment is to be achieved, the methods involve repeated attempts under slightly different conditions.

When other methods fail, or are impractical, the obliterating material can sometimes be removed or weakened by chemical or mechanical means. Success depends upon whether the techniques employed affect only the obliterating substance, or at least have a more pronounced effect on it than the writing beneath. Blue-black or aniline inks obliterating carbon inks, pencil strokes covering inks, and many ink or pencil strokes hiding typewriting are combinations that lend themselves to these methods of attack. Ordinary erasers and common solvents that may affect writing inks when skillfully manipulated may achieve the desired results. Thus, under favorable circumstances, this approach to the problem is successful, but as the combination of materials that are not susceptible to this treatment are numerous, success is far from universal.

One form of obliteration has rather widespread use today. This is the use of correction solutions or tapes in lieu of erasing, especially with typewriting. One technique is to type through a correction ribbon or sheet covering the outline of the letters to be retrieved. Usually, new typewriting is placed in the same area. The other technique is to use liquid or tape correction material. These can be either painted on or placed over the area to be removed. Again, new text can be written over the correction material. In most instances a careful inspection will reveal the obliteration. Normally the original material can be read by transmitted light (Figure 27.14) or by photocopying the back of the sheet using color-tinted transparency film, so the image can be turned into a read right position. When the correction material is too thick or there is interference from the reverse side, the document may need to be immersed in petroleum ether,[16] or some of the material may need to be removed for decipherment. A xylene substitute has been found to be effective in removing most solutions with little or no detrimental effect to the underlying ink.[17,18] In some cases the only hope is that the opaquing material may

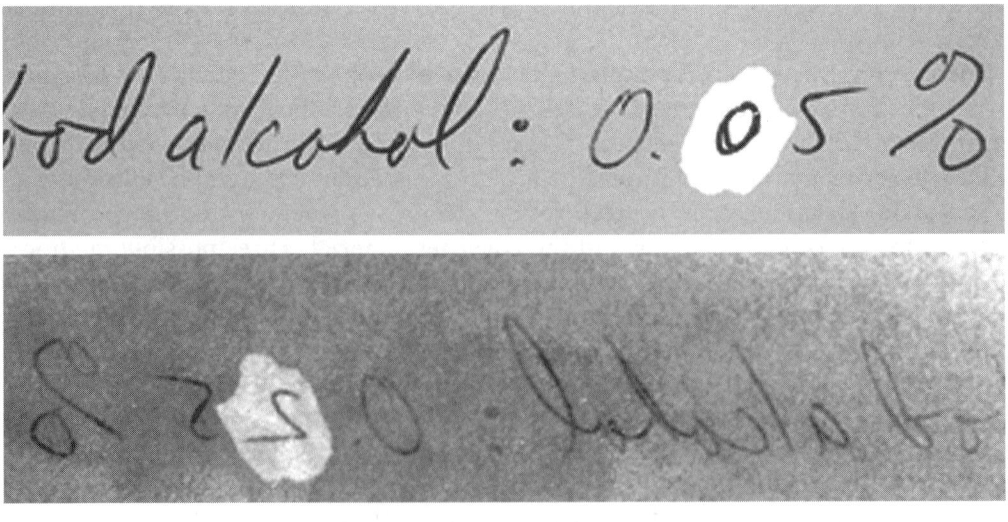

Figure 27.14 An obliteration made with opaquing solution is seen under normal lighting (upper image). The bottom image shows the original number was a 2 when viewed from the reverse side using transmitted light. (Courtesy of Jan Seaman Kelly.)

be faulty and some portions of the original writing may yet be discernible. These partial strokes or weak outlines of semiobliterated letters can be intensified and deciphered. All these obliterations by and large require a diversity of methods combining various techniques, experimentation, perseverance, and often a full measure of luck to ensure ultimate success.

27.5.1 Overwritings and Insertions

Documents may be changed by overwriting words and portions of sentences or by insertion of a character, word, sentence, or more. At times it is necessary to attempt to determine what was originally written. In other instances, it is necessary only to show that the changes were not made at the time of preparation of the document. Insertions in the form of interlineations may be very obvious, but if it can be shown that they were made with another writing instrument, by another writer, or on a different typewriter or printer, it can go a long way toward attacking the value of the present version.

Insertions may be disclosed by differences in the writing material or differences in the handwriting. Crowding of the inserted material compared to surrounding writing suggests an addition. Microscopic study is used to detect differences in ink or writing instruments. Intersecting strokes may disclose the wrong sequence. Using filters, ultraviolet and infrared, is a useful tool. Most of the methods discussed in previous sections may come into play in these problems as well.

Overwriting that is not very obvious may be established by disclosing double strokes. Strokes that are not a part of the letters of the overwritten words assume significance. If there is enough writing, it may be possible to show that there are writing characteristics of someone other than the person who prepared the balance of the document.

Cases of this nature are not common. They are more often found in manipulation of accounting records and check frauds (see color Figure 27.15 and color Figure 27.16 following p. 366). Occasionally, they are incidental issues in document problems of entirely different kinds. They do, however, represent another way that documents can be changed, and despite the obvious appearance, changes of this nature will arise from time to time as evidence in the case of one party to a litigation. They must be accurately evaluated.

Whole pages may be inserted in a multiple-page document. Their detection often depends upon study of binding marks (such as staple holes if the pages are assembled in this way) (see Figure 27.7), the paper for kind and size, the pen and ink, the printed text, or the pencils. Indentations on a following page may be the key.

27.6 Proof of an Unaltered Document

In the previous sections various techniques that may reveal alterations in documents were discussed. The question does arise, however, as to whether it is possible to establish that a document has not been altered and, if so, what procedures are necessary.

Proving that a paper is unaltered is a challenging problem.[19] It is an important one, however, since it is incumbent upon document examiners to be able to prove genuineness as well as fraud. This proof of genuineness is necessary to support the validity of certain disputed documents. Actually, the procedure involves not the application of any single test, but a consideration of all the applicable procedures to determine whether there has been an erasure, a substitution, or any other type of alteration in a document. In each instance, the findings must be that no significant alteration has occurred that in any way would change the intended purpose of contents of the document. It is the cumulative evidence that establishes that the document is unaltered.

Therefore, depending upon how the document was prepared, the FDE must apply those tests that are appropriate to establish that there has been no significant erasure, or if there has been some minor erasure, that it is clear that such an act was merely to correct an error, such as a misspelling, made in the preparation of the document. To accomplish this requires the application of every appropriate test that could disclose the presence of an erasure, and each must show negative results. It is the combination of these tests that supports the conclusion that the document contains no erasures.

By the same token, tests that may reveal additions to the document must be considered, such as those showing the use of more than one writing instrument, the addition of typed or computer-generated text, or the insertion of material by an improper sequence of intersecting lines or lines with folds or perforations. With a handwritten document, was all the writing done with the same writing instrument and by the same writer, and is the document free from evidence of undue crowding of key material? Thus, in dealing with each specific page, the document examiner must be able to say that there is no evidence that a word, sentence, or paragraph had been added.

A further consideration in a multiple-page document is whether any pages may have been removed and others substituted, or new pages added into the document after execution. Such examinations, of course, involve consideration of the writing instrument, printer, paper, manner of binding, and presence of writing indentations that may have resulted from preparation of material on the previous page. There are the problems of determining whether the entire document was prepared at one time in a continuous manner, which involves considering the margins on page after page, the spacing between lines, the manner of handling paragraphs, and, if handwritten, whether there is an abrupt change in the quality of handwriting, which might suggest a different writing episode. In this way the FDE should be able to show that no evidence is present that suggests or establishes that the preparation of any page is inconsistent with any other pages.

Actually, an unaltered document is one that contains no erasures, no additions, and no substituted pages. To establish this situation in a positive and definite manner involves considering a great number of factors. There may be some instances even after considering all the elements in which the FDE is unable to say positively that the document is unaltered, but he or she can certainly point to the preponderance of the evidence that is inconsistent with any change. Thus, the physical facts found within the document itself many times govern just how positively this question can be answered.

27.7 Conclusions

Regardless of how a document is altered — whether it is by erasing, obliteration, or insertion of new matter — it is vital to those who stand to be defrauded that all of the evidence contained within the document itself be brought to light. The extent to which this internal evidence can be extracted has been indicated and the limitations frankly discussed. Despite occasional inadequacies, these techniques are more often potent tools by which fraud can be revealed and, in a number of problems, the facts set forth.

The need to establish that a document has not been altered may involve a complex study. There is no single, simple test. All potential tests for showing that something has been erased, added, or modified in any way must be applied. When the combined results reveal no change, it can be stated that there is no evidence to support that this document was altered.

References

1. Licht, G.A., Common chemicals for common criminals: check washing again, *J. Am. Soc. Questioned Doc. Examiners*, 3, 65, 2000.
2. Pfefferli, P. and Mathyer, J., Eraser Mate un stylo a bille à encre effacable, *Rev. Int. Criminol. Police Tech.*, 4, 407, 1979.
3. Flynn, W.J., Paper Mate's new erasable pen, *J. Police Sci. Admin.*, 7, 346, 1979.
4. Hilton, O., Photographic methods of deciphering erased pencil writing, *Int. Criminal Police Rev.*, 85, 47, 1955.
5. Longhetti, A. and Kirk, P.L., Restoration and decipherment of erasures and obliterated or indented writing, *J. Criminal Law Criminol.*, 41, 518, 1950.
6. Casey, M.A. and Purtell, D.J., IBM correcting Selectric typewriter: an analysis of the use of the correctable film ribbon in altering typewritten documents, *J. Forensic Sci.*, 21, 208, 1976.
7. Harris, J.L., Eyeing the evidence, *South. Calif. Alumni Rev.*, 21, 16, 1940.
8. Casey, M.A., Alteration of pari-mutuel tickets, *J. Criminal Law Criminol. Police Sci.*, 62, 282, 1971.
9. Godown, L., Sequence of writing, *J. Criminal Law Criminol. Police Sci.*, 54, 101, 1963.
10. Igoe, T.J. and Reynolds, B.L., A lifting process for determining the writing sequence of two intersecting ball-point pen strokes, *Forensic Sci. Int.*, 20, 201, 1982.
11. Godown, L., Recent developments in writing sequence determination, *Forensic Sci. Int.*, 20, 227, 1982.
12. Berx, V. and De Kinder, J., The Application of Profilometry in the Analysis of the "Crossing Lines" Problem, paper presented at the ASQDE Annual Meeting, San Diego, CA, August 14–18, 2002.
13. Novotny, M., Determining the Sequence of Original Ink Writing and Toner Printing, paper presented at the ASQDE Annual Meeting, San Diego, CA, August 14–18, 2002.
14. Aginsky, V.N., Determining the sequence of non-intersecting media on documents: ballpoint pen ink and laser toner entries, *J. Am. Soc. Questioned Doc. Examiners*, 5, 1, 2003.
15. Ezcurra, M., Differences between Toner Particles, above and below the Roller Ball and Gel Ink Pen Entries, paper presented at the ASQDE Annual Meeting, Memphis, TN, August 22–26, 2004.
16. Lewis, J.A., Petroleum Ether Immersion: A Technique to Visualize and Photograph Correction Fluid Obliterations, paper presented at the ASQDE Annual Meeting, Ottawa, Canada, August 25–29, 2000.
17. Licht, G.A. and Brown, J.L., Shandon Xylene substitute in document examinations, *J. Am. Soc. Questioned Doc. Examiners*, 2, 94, 1999.
18. Beal, B.L., Removal of Opaquing Solutions from Documents, paper presented at the American Academy of Forensic Sciences Annual Meeting, Dallas, TX, February 16–21, 2004.
19. Hilton, O., Proof of an unaltered document *J. Criminal Law Criminol. Police Sci.*, 49, 601, 1959.

Section IX

The Age of a Document

Multi-Faceted Approach

28

BRIAN S. LINDBLOM

Contents

Questions surrounding the production date of a particular document or entry can vary widely. For instance, the authenticity of a document purportedly written centuries or decades ago may be at issue. In another case, there may be concerns about whether the contents of a medical or legal file have been entirely rewritten some months or a few years after the dates recorded in the file. Other inquiries may pertain to suspicion that a self-serving entry in an appointment or diary note may have been made mere days, or a few weeks, after the recorded date.

A number of backdating methods have been employed in such situations. These range from a relatively minor change to the final digit in a handwritten year to a far more sophisticated endeavor where old paper stock, writing instruments, or pre-printed forms were employed. An entire record representing several pages may have been fabricated using such materials.[1] Several methodologies can be used to address such questions. The analyses are performed by (1) comparing the contested document to other similar documents from the same time frame, (2) intracomparing various entries in a file or multi-page document to determine whether they are the same, (3) referencing manufacturers' and other technical information about the materials used, such as inks, paper, computer printers, fax machines, etc., and (4) examining for accidental markings and indented impressions.

The tests conducted will depend, to a large extent, on the nature of the document itself. By way of example, a multi-page medical record may be comprised of a file folder, index cards, diagnostic forms, handwritten progress notes, correspondence, and facsimiles. Clearly, the document examiner can employ a much larger array of analytical techniques in this case than in one where a single computer-generated page with one signature is contested. Regardless of the number of documents involved, the examination will generally entail a multi-faceted approach. Even where the question is limited to a single page, the document examiner is likely to investigate paper, ink, handwriting, and incidental markings.

Testing should proceed in a systematic fashion, always progressing from non-destructive methods to semidestructive examinations, where warranted and authorized. There are many potential avenues of inquiry. An experienced document examiner may give particular tests priority over others, based on their likelihood of providing important evidence. Some elements to consider include paper composition, watermarks, envelopes, pre-printed forms, inks and pens, photocopiers, computer printers, facsimile machines, typewriters, typestyles, handwriting/signature evolution, stamp impressions, dry seal embossments, staples and staple holes, indented impressions, ink and pencil offsetting, and other incidental markings.

Any part of the document that can be established not to be as old as the date it bears can generally serve as the basis for a conclusive finding. Data from a combination of examination methods are often needed before one can conclude that there has definitely been backdating. It is much more difficult, and often impossible, to establish that a document was prepared on the date it bears. Even when all factors are consistent with the date, the best determination is that it could have been prepared at the time purported.

Many of the earlier chapters in this book explore the examination techniques mentioned above. They should be referred to for detailed information. The chapters that follow consider various components of the document and their relevance to establishing a document's date. The first of these investigates the role of indentation analysis.

References

1. Purdy, D.C., Document analysis, in *Encyclopedia of Forensic Sciences*, Academic Press, London, 2000, p. 571.

The Role of Indentation Analysis, Offsetting, and Transfer

29

BRIAN S. LINDBLOM

Contents

The importance of examining documents for indented handwriting impressions has been addressed in Chapter 26. While this method has the potential to identify the author or creator of a suspect document, it can also produce conclusive evidence relating to the date, or order, of execution. In cases involving several sheets of paper with dated handwritten entries, the presence and location of indented impressions may point clearly to an earliest date of entry. Dating through the use of indentation analyses can be broken down into five scenarios:

1. Indented impressions on a questioned document that can be sourced to handwriting on a document of a known date
2. The presence of indentations on a document of a known date that can be sourced to the questioned document (Figure 29.1)
3. Unsourced impressions sufficiently similar in content to enable identification of the questioned document as a probable substitute page (Figure 29.2)
4. Alignment characteristics of indented impressions resulting from multiple entries
5. Indented impressions, sourced to questioned handwriting, overlapping with known handwriting in a fashion suitable for sequence-of-entry evaluation (see color Figure 29.3 following p. 366).

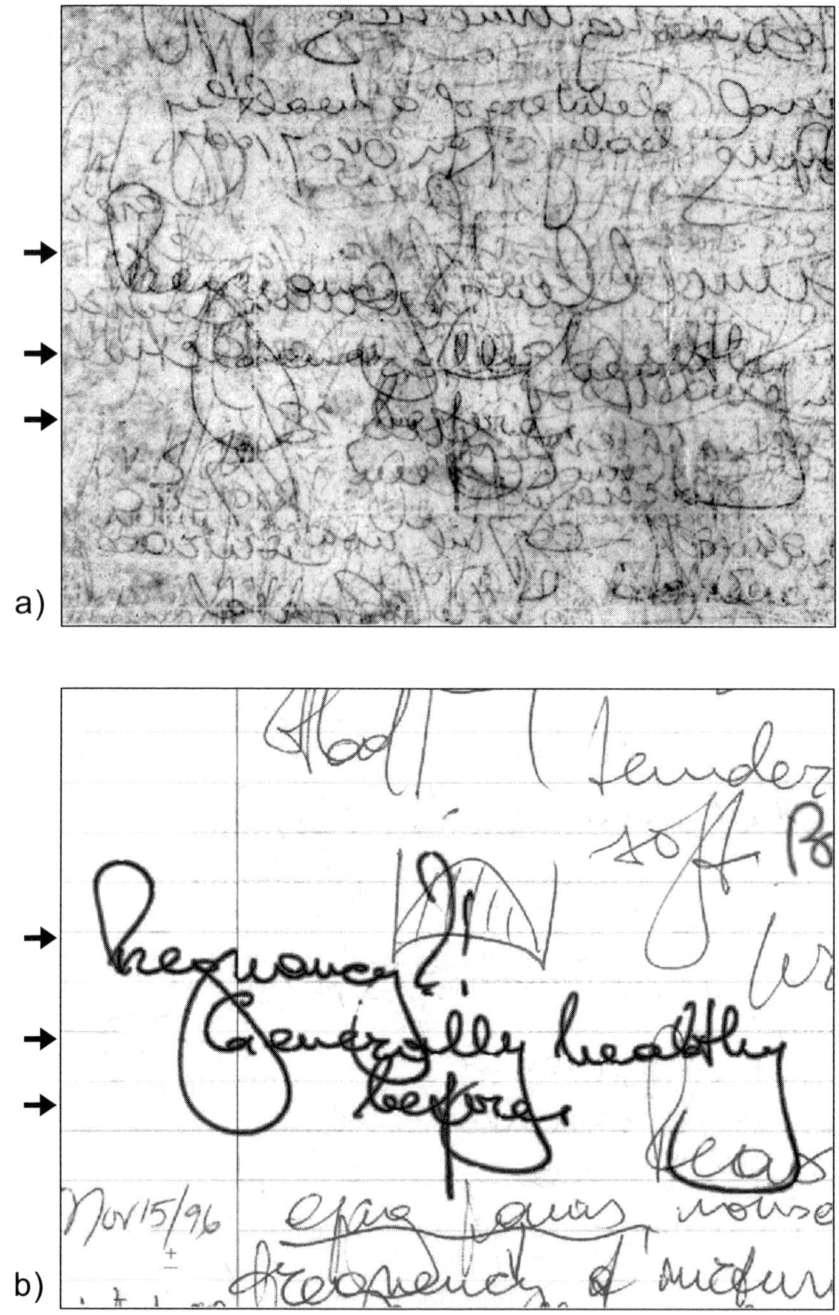

Figure 29.1 The page tested for indentations (a) bears dates from March 7 to May 30, 1997. Among the multi-tude of forward and reverse reading impressions are some sourced to a portion of a progress note dated November 8, 1996 (b). This particular set of impressions is out of date sequence. Such an example provides important evidence that an entry was added after the fact and not before March 7, 1997.

Indented impressions are probably best appreciated by considering some hypothetical case examples.

29.1 Impressions on Questioned Documents

Indented impressions from a handwritten sheet dated June 1, 2003 are discovered on a questioned document that bears a February 15, 2003 date. In this situation the entries are in chronological order, with the older document having been positioned below the more contemporaneous one. While the evidence does not definitively establish that the questioned document was executed on February 15, it does verify that it existed prior to June 1, 2003. If the allegation was, for example, that the disputed page was created in 2004, this assertion would be refuted by the indentation results. However, if the suspicion is that it was written in May 2003, the evidence does not provide an answer, as the questioned handwriting could have been produced any time prior to June 1, 2003.

29.2 Questioned Document Indented onto a Sheet with Known Date(s)

A diagnostic form from a medical laboratory is dated October 1, 2002. It bears indented handwriting impressions that originate from a patient's questioned progress notes dated May 10, 2001. As the form originates from an external source and has an acknowledged, undisputed creation date, it provides a reliable chronological reference. The questioned entry could not have been made on the date associated with it, given the indentations present on the October 1, 2002 form. Here, the earliest creation date for the questioned entry would be October 2002.

29.3 Unsourced Impressions with Content Similar to Questioned Entries

Five pages of a stock broker's handwritten telephone log become the subject of a dispute. The dates span February 11, 1999 through February 20, 2000. The contested entry concerns a telephone consultation dated December 12, 1999, appearing on page 3. Indentation analysis reveals that all sheets except page 3 bear indentations. The impressions are found to be in date sequence. In other words, newer entries are indented onto sheets with older dates. Page 4 contains no indentations from the current page 3, but does have impressions from another sheet, not found in the file, with similar dates and content to those of the current page 3. These anomalies strongly suggest that the questioned entry was produced at a different time and that page 3 is a substitute. However, no estimation of the time difference between writing and rewriting can be made.

29.4 Alignment Groups within Indented Impressions

A physiotherapist's patient chart contains several sheets, each bearing a number of dated appointment treatment notes. The pages are unbound and all handwritten. A claim is

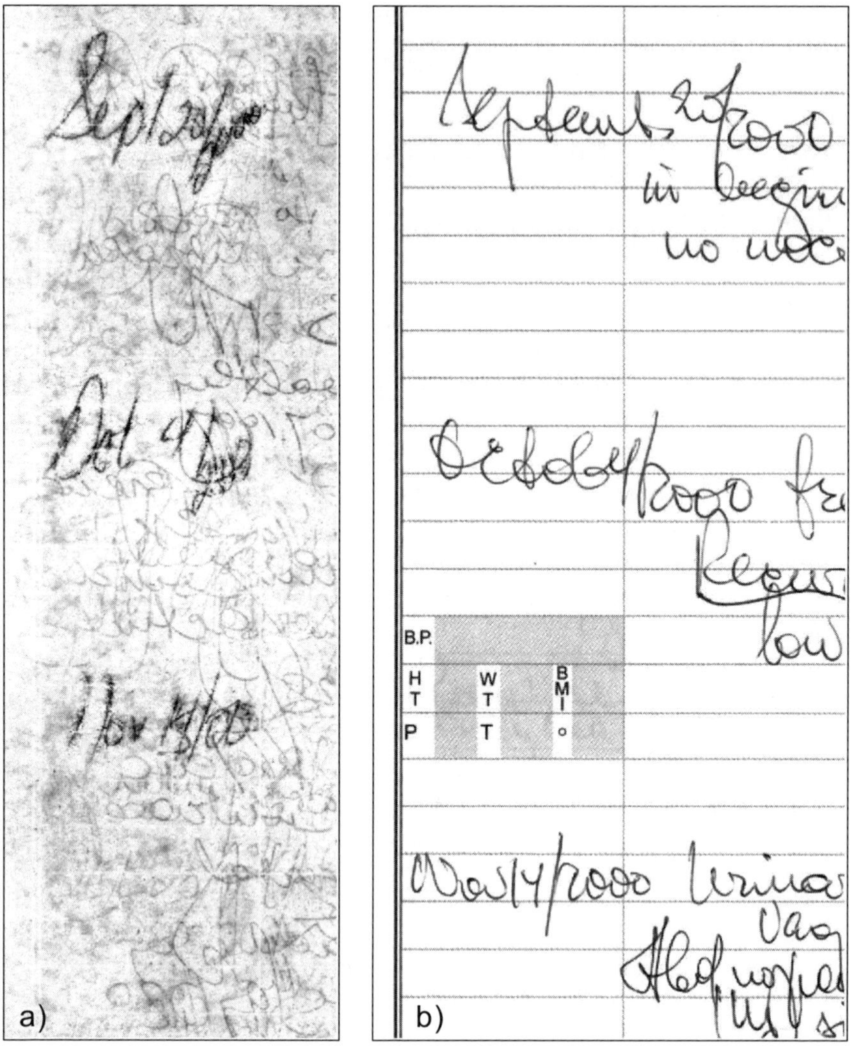

Figure 29.2 The indentation analysis of a patient's progress notes revealed indented impressions (a) that include dates corresponding to, but not originating from, entries found on another page within the chart (b). Page b was found not to bear any indented impressions. Based on these findings, it was determined that a page had been substituted. Precisely when the backdated document was included in the chart could not be established.

made that several notes do not reflect the treatment provided. Many pages bear impressions that originate from other progress notes in the chart. With the exception of one sheet, the indentations are askew and scattered across the page relative to one another as expected with loose-leaf multiple entries made on different dates. In other words, they do not align in the same way that they appear in the original notes. However, one full set of treatments, representing four dates, are found indented such that they fully register, one relative to the other, when superimposed with the handwriting from which they originate. This phenomenon strongly suggests that the notes may not have been made over the period of time indicated, but rather in one sitting.

29.5 Indentation Sequencing

A personal diary becomes an essential piece of evidence in a criminal prosecution. One of the entries makes reference to the victim being attacked by her common-law partner. Her boyfriend denies the charge and argues that the victim made this up and added the notation to her daily diary at a much later time. The specific notation comprises the bottom two lines of a three-paragraph entry. As one would expect, handwriting on the left side of the diary is indented onto underlying pages on the same side. Similarly, handwriting on the right is indented onto underlying pages on the right. The entry in question is dated September 16, 2002, while the date on the underlying page is September 18.

The indentation analysis of the September 18 page reveals a complete set of impressions originating from the questioned September 16 entry that intersect with the visible handwriting on the page tested. In this situation a study of indentation/handwriting sequencing can be conducted.[1,2] Where indentations sourced to the contested statement intersect with the visible handwriting of September 18, there are several instances of the indented lines (in black) being solid and unbroken by the ink strokes. The opposite is noted at intersections involving the non-contested portion of the diary entry. The results show that the questioned statement was added some time after the September 18 diary entry. Precisely when cannot be determined.

29.6 Offsetting and Transfer

Ink offsets typically occur when freshly written entries come in contact with another sheet of paper, resulting in the partial transfer of ink marks to the contacting page. The offsets can have value in establishing whether a series of entries were written at once or over a period of time. For instance, a loose sheet of paper bearing three dated handwritten entries is found to have offsets on another loose page within the file. The offsets register, or align, as a group with the source writing. This would occur because all three dated entries were written at the same time and, therefore, presented fresh ink for transfer to the contacting page.[3]

Ink and graphite transfer occurs when a moving writing instrument intersects with an existing entry on a page below and displaces ink or graphite from that point of intersection onto a page in contact with the earlier writing. The transfer can have importance in determining the order or dating sequence of one entry relative to another. Furthermore, in the case of ink transfer, it can provide evidence that one entry was done soon after another rather than days later, as may be suggested by the entry dates.[4]

References

1. Radley, R.W., Determination of sequence of writing impressions and ball pen inkstrokes using the ESDA technique, *J. Forensic Sci. Soc.*, 33, 69, 1993.
2. Radley, R.W., Determination of sequence of intersecting ESDA impressions and porous tip, fibre tip and rollerball pen inks, *Sci. Justice*, 35, 267, 1995.

3. Radley, R.W., Strach, S., and Westwood, P., Short term relative time of writing determinations by observations of ball-point pen ink transfers: a simple relative dating technique, *Int. J. Forensic Doc. Examiners*, 4, 152–153, 1998.
4. Gervais, R., Techniques and Terms Useful in the Examination of Graphite Offsetting, paper presented at the Annual Meeting of the American Society of Questioned Document Examiners, Ottawa, Canada, 2000.

Dating by Materials: Identification, Comparison, and Examination of Changes

30

BRIAN S. LINDBLOM

Contents

A typical contemporary document is made up of handwriting and computer-generated text on paper. An examination of these components, along with stamp impressions, preprinted forms, letterheads, and other constituents of the document, may provide information that establishes an earliest possible date of production.

30.1 Paper

The materials that make up paper contain information about when the paper was first manufactured. The fiber content and various other materials, such as the substances for coating, loading, or brightening, all have first dates of use.[1–3] Any study of the paper on the basis of formulation requires two factors in order to establish a significant date. First, the manufacturer of the paper must be known. Second, the manufacturer must have records that disclose when certain combinations of materials were first introduced. Occasionally, this information can be developed and is of value, but many times it fails because either the manufacturer cannot be identified or its records are incomplete.

A common method of determining the source of paper and its possible time of manufacture involves the watermark found in better grades of writing, typing, and computer printing papers. Today only a few of these watermarks include a dating code that allows the manufacturer to determine the year in which the paper was made.[4] Records are generally available about when a brand of paper, designated by the watermark, was first placed on the market, although most brands have been sold for a number of years. Documents have been dated, or the dates proven fraudulent, by changes in design of the watermark (Figure 30.1) or by defects developed over time in the watermark dandy roll. Here again, success depends on the accuracy of the plant records. Since the watermark can be displayed in scans and photographs, the dating information can be demonstrated effectively to all concerned.

30.2 Writing Inks

During the 20th century a number of new chemicals were developed and used in the production of writing inks. In recent decades, new classes of inks have been developed for use in a variety of pens now on the market, including ballpoint, roller ball, porous tip and

Figure 30.1 Trojan Bond Paper before (top) and after (bottom) the change in its watermark in 1976.

gel. Each may be identified from examination of the written strokes or chemical composition. Their presence on a document of a particular date may be inconsistent with the earliest availability of the writing ink or pen. See Chapter 13 for a more detailed discussion on the history of these pens and inks.

In the early 1970s rare-earth elements, and later special chemicals, were added to some inks in order to tag them for identification and dating purposes. The rare-earth component was dropped early on due to problems in detecting their presence and other analytical problems. The U.S. Laboratory of Alcohol, Tobacco and Firearms initiated this process, allowing for a relatively precise way of identifying the year of manufacture. Although the program was discontinued in the mid-1990s, it was reinstituted in 2002 with a major American ink manufacturer. The U.S. Secret Service is coordinating the program. The special chemicals can be identified using a combination of thin-layer chromatography (TLC) and ultraviolet excitation. Not only does the presence of the tag establish the ink's production year, but it also allows for the identification of its manufacturer.

Extensive ink reference libraries have been developed by government laboratories in the U.S. and overseas as well as by a small number of private practitioners. The application of TLC allows for the differentiation of the various dye components that make up the ink's color. The results can be compared with the ink reference library and probable matches made, which may allow the document examiner to establish the earliest date of commercial availability. These reference specimens are most helpful in proving that a document could not have been written on a certain date because the ink was not then available.

During the 1980s and 1990s, research was conducted on the extractability of inks from paper as an indicator of age. Ink chemists at the Bureau of Alcohol, Tobacco and Firearms initiated much of this research. The basic theory underlying relative dating of ink is that the longer ink has been on a document, the more difficult it will be to extract the dyes. Alternatively, a newly written entry will extract more rapidly. Therefore, the age of the latter should be distinguishable from an older ink entry. To perform such tests, it is necessary to have a benchmark entry or entries to compare against that are on the same paper, composed of the same ink formula, subject to the same storage conditions, and of a known date(s).[5,6] Three measuring parameters were developed: (1) R-ratio, (2) percent extraction, and (3) dye ratio.

A modification to the technique, known as accelerated aging, has since been advanced to assist in cases where dated comparison entries are not available.[7,8] For example, it would be applied to a document that is computer generated and bears only an ink signature. The accelerated aging technique does not require known entries for comparison, but rather uses a heated sample of the questioned entry as a comparison standard against which the unheated questioned ink is measured. In essence, the analyst is actually comparing a questioned sample against another sample from the same entry. This adaptation has not been widely endorsed in the forensic community given that there are questions about the effect heating dyes has on any effort to establish dating timelines. None of the government laboratories currently involved in ink dating employ this analytical method.

Concerns have been raised about the reliability of the dye ratio method. There are instances where results have varied widely when multiple runs have been conducted. In some cases a reverse aging curve has been documented. The latter suggests a significant problem with this test.

In recent years some practitioners have further modified the method, often without having conducted significant, verifiable research on the reliability of the technique.

Consequently, the method and its variations are now the subject of considerable controversy, in both the forensic document examination community and the courts of more than one country.[9,10] The technique has been rejected by a number of courts due to its lack of scientific reliability and questionable statistical interpretations.[11–13] Two lengthy judgments on the method have been handed down.[14,15]

Promising work is surfacing that focuses on dating inks by study of the volatile components rather than the dyes.[16–18] This technique examines the ink's vehicles, such as phenoxyethanol, the theory being that a difference between a new entry and an older one will be reflected in their solvent concentrations. The method has some limitations in that the volatiles are detectable for only a limited period. For instance, a document dated 2002 but actually written in 2004 and tested in that same year would be suitable for solvent loss testing. Employing this technique in 2004 to a document purportedly dated in 1991 but actually written in 1999 will almost certainly result in an inconclusive finding. This is due to both the rate at which the solvents evaporate and the sensitivity of the instruments.

In summary, ink dating must be approached with caution. Only methods that have been thoroughly researched and subjected to multiple blind tests (for reproducibility) will be accepted in the forensic document community.

30.3 Typewriting

Though the majority of type encountered today is computer generated, typewriting still appears on a variety of documents, such as pre-printed forms, envelopes, receipts, deeds, transfers, and other real estate-related forms. As discussed in Chapter 15, a study of the typestyle or design may lead to identification of the typewriter, typeball, or typewheel manufacturer. Each design has a specific date of introduction. It may be discovered that some aspect of the imprinting on the document is inconsistent with the introduction of that typewriter or element (Figure 30.2). For example, the discovery of a typewritten last will and testament dated in the 1960s and prepared using a daisy wheel would be proof that the document was not prepared at the date indicated, but rather some time after the daisy wheel's introduction in the early 1970s.

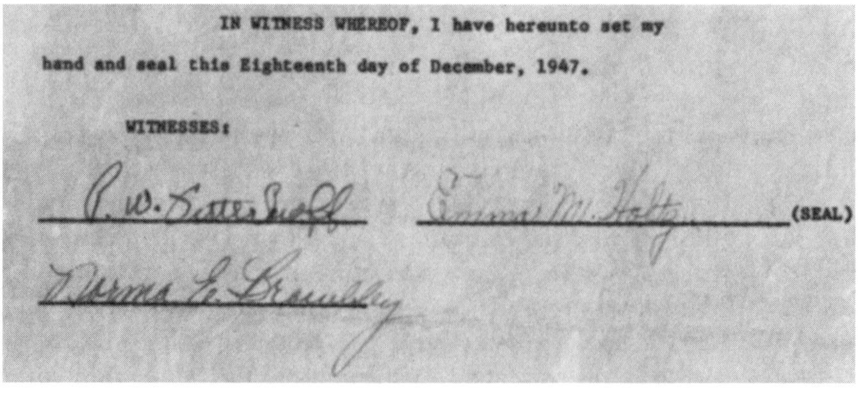
IN WITNESS WHEREOF, I have hereunto set my hand and seal this Eighteenth day of December, 1947.

WITNESSES:

(SEAL)

Figure 30.2 A will dated in 1947 was typewritten with a machine equipped with Underwood Esteem type, first released in March 1958.

Computerized databases of typestyles are available to the document examiner to aid in manufacturer identification and to provide information on dates of introduction.* These records do not establish when the document was typed, but can provide an earliest possible date of creation.

A second method for dating a document through study of the typescript is the evolution of defects developed through progressive wear and tear of the machine or element. These defects can include broken or bent characters, filling in of letters and numerals, misalignments, bead defects, etc. As these tend to develop over time, it may be possible to establish a defect chronology and, in turn, date the questioned document with that chronology. In this circumstance, depending on the number and quality of comparison samples, it may be possible to date the document within a defined time frame.

As with any comparative examination, a critical first step is to establish conclusively that the questioned typewriting is being compared with the actual typewriter or element used to prepare it. Likewise, a series of typewritten samples must have been produced on the same machine or element that was involved in the production of the questioned document.

Not only the typewriter or element has dating value, but the ribbon may provide invaluable evidence as to the timing of a typewritten document. Equipment (such as the RAW) is available to decipher multi-strike typewriter ribbons and, when employed, can be useful not only in identifying the machine involved, but also in providing indisputable proof about a document's date.[19] For example, the text from the typewritten document in question may be found on a typewriter ribbon that bears other text from documents prepared immediately before and after. If these documents bear dates, then the possible execution time is greatly narrowed.

30.4 Photocopiers and Printers

Like pens, inks, and typewriters, photocopiers and printers have evolved over the decades. In fact, many models now incorporate the two, along with faxing capabilities, in a single multi-use machine. The merging of these technologies has made distinguishing between a photocopy and a printed document a considerable challenge. Because some imaging systems are common to both applications, certain defects may be found in both printed and copied output.

Some of the earliest copier models used a liquid toner. In later years dry toner became the standard. Another change saw a move from analog to digital image processing. In the past two decades, we have seen advances in inkjet technology such that it has taken a prominent position alongside electrostatic imaging. Dot matrix technology has, to some extent, fallen by the way, though it is still used in situations such as imprinting on multi-part NCR (no carbon required) forms. As a result, it is still encountered from time to time in questioned document cases.

A study of the document will generally reveal the copying or printing method used. It may be possible to determine that a document could not have been produced on the date suggested based on the availability of the copying or printing process at that time.

* One such example is a computerized version and hard copy edition of the Haas Atlas (which includes a number of research publications authored by Philip Bouffard) distributed by the American Society of Questioned Document Examiners.

For instance, a document put forward as being a first-generation copy prepared in 1975, but showing clear indications of being a digital reproduction, cannot be what it purports to be, given that the technology was not introduced until many years later. Likewise, a color copy may reveal technological advances as well as manufacturer and date coding that can be evaluated in order to determine whether or not the product was on the market around the time of the suspect document.

Like many electronic office machines, photocopiers and printers develop defects such as trash marks, printing voids, and streaks that have dating significance (Figure 30.3 and Figure 30.4). By acquiring a representative sample made from the same machine as the suspect document, and executed over the relevant time frame, it is often possible to establish whether the copy/printed page is consistent with the date it was allegedly made. Such examinations must take into account servicing, replacement of the imaging cartridge, and the presence of transitory defects. As with most forensic comparisons, a large quantity

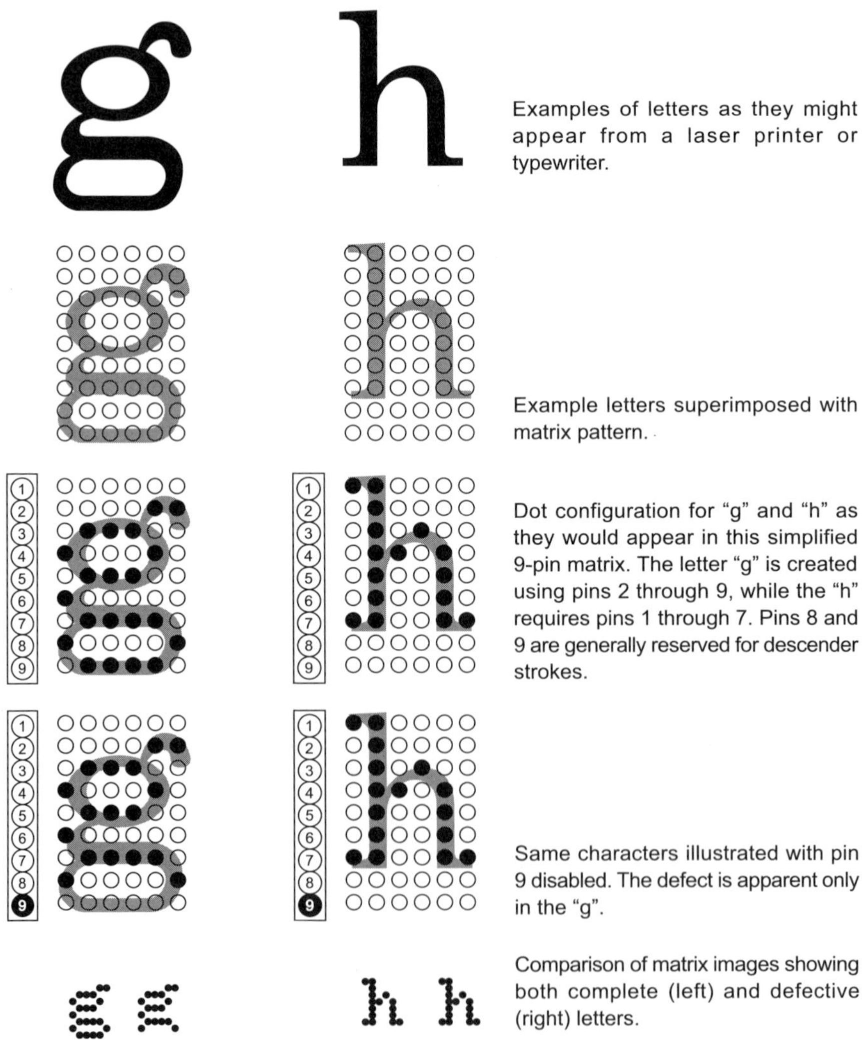

Figure 30.3 Simplified explanation of how dot matrix printers generate characters.

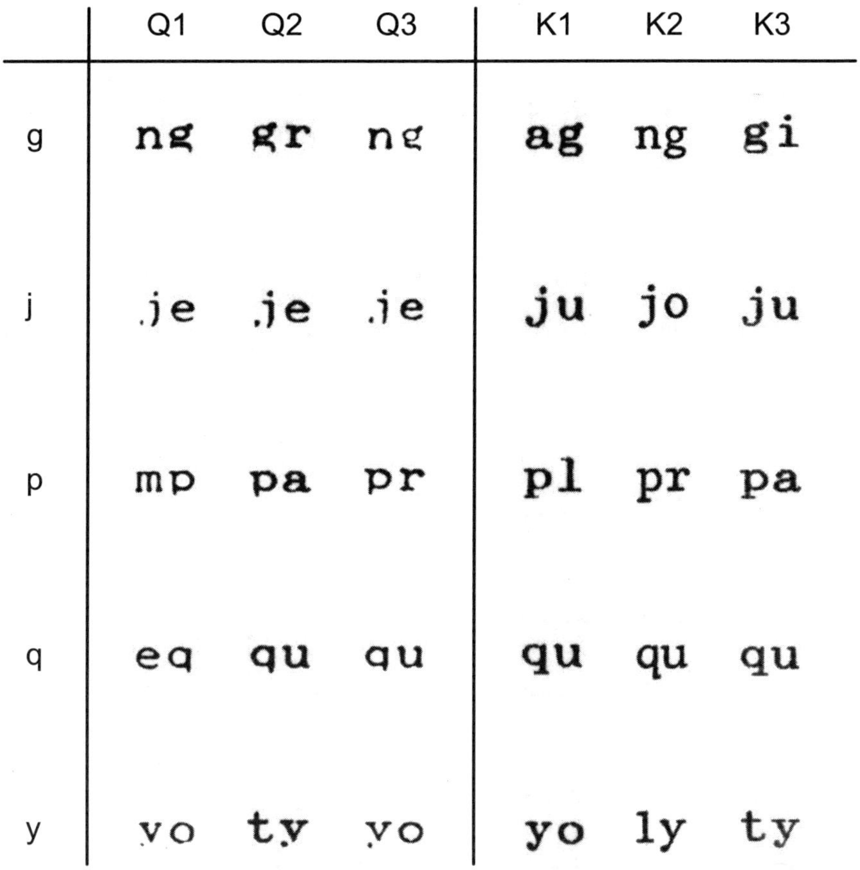

Figure 30.4 In this case, the questioned documents all exhibited a dot matrix printer defect, manifested as incomplete descenders, which was not found in any of the many specimens submitted from the same time period.

of standards, spanning a broad range of time, will allow for a complete assessment of defect evolution and potentially provide the necessary data for a strong opinion regarding the timing of a contested document.

30.5 Facsimile Machines

Most modern offices now have one or more stand-alone, or multi-function, fax machines. The popularity of faxes makes them the subject of scrutiny in many litigation cases where issues arise about when a document was generated or transmitted. Establishing whether a particular imaging system or model was available on the date at issue is an important forensic inquiry. There have been cases where the printing technology postdates the disputed document's date. For example, an inkjet-printed facsimile supposedly transmitted in 1970 could not have been, as at that time this printing method was not found in fax machines.

During the past decade or so, members of the American Society of Questioned Document Examiners (ASQDE) have developed an extensive database of Transmit Terminal

Identifiers (TTIs) and other technical data on facsimile manufacturers and models. While the collection is not all-inclusive, it nevertheless provides a very valuable tool to the document examiner. Searches can be made based on the formatting and typestyle characteristics of the letters and numerals making up the TTI. Once a potential match has been found, further inquiries can be made directly with the manufacturer to identify a probable make, model, and introduction date.

It is not uncommon for TTI headers to be altered, through either the programmable features on the machine or manipulation of the printed image. The use of manufacturer specifications can be of considerable assistance in detecting inconsistencies in the suspect faxed document. The identification of these inconsistencies can potentially result in an opinion that the document was not generated at the time suggested by its date.

Defects in either the sending or receiving machine may help to establish dating parameters for the document in question. A representative sample of faxes, over a reasonable time range, from both machines may allow for the development of a chronology of defects. Such an examination is sometimes challenging because one or both of the machines may be unavailable for inspection or it may be difficult to acquire samples that can be definitively associated with a particular machine. This is especially true when a considerable period of time has passed between the alleged production of the document and the time when suspicions were aroused.

30.6 Printed Matter

Documents prepared on letterhead stationery or on pre-printed forms frequently contain information regarding the earliest date at which they could have been prepared. Pre-printed forms may bear a code indicating the date of design/printing. Changes in design or wording may provide chronological touch points that can be used to aid in establishing possible ranges of creation dates (Figure 30.5). Likewise, changes to telephone numbers, area codes, addresses, etc., on letterhead related to the individual or business establishment concerned will sometimes assist in establishing that a document is backdated. Such information may be inconsistent with the date on a fraudulent document. Those using pre-printed forms many times overlook this information, as they have no direct knowledge of the significance

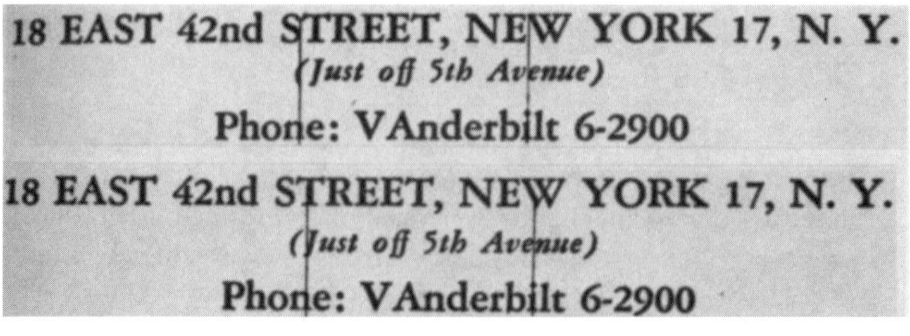

Figure 30.5 The top letterhead bore an undisputed 1947 date; the lower, 1948. There are slight differences in the arrangement of the two printings as indicated by the ruled vertical lines running through the *ST* of *STREET* and the *EW* of *NEW*. In addition, the spacing between the first and second lines differs by 1/16 inch. The information helped to authenticate the date of a disputed document.

of these changes. For example, doctors often prepare patient progress notes on forms that feature advertisements for pharmaceutical and other medical products. Periodically the sponsor updates the forms, a fact that has been known to escape the notice of the doctors.

30.7 Handwriting and Signatures

Changes in handwriting, and more particularly signatures, sometimes occur over time. The evolution may be gradual or quite sudden, depending on the circumstances. An accident, illness, or the influence of alcohol and drugs can result in a profound change in writing ability. This change may be short lived or there may be a gradual improvement, but with residual characteristics remaining. Other traits develop over a much longer period. One example is a writer who finds himself or herself in an occupation that requires the signing of his or her name numerous times throughout the workweek. It is not uncommon to see such a person's signature change from a fully legible style to one that is more abbreviated and scrawled. The pictorial differences need not be so apparent, as subtle changes may be introduced with time, often manifesting themselves in more stylized individual letters or the combining of characters at the end of a signature. All have the potential to assist in dating a document.

When studying the evolution of a signature, it is important to have a generous supply of known standards for comparison that are dated over a considerable period of time. One or two samples from a 3-year period is unlikely to provide a sufficient foundation to date an individual's signature, as the document examiner could not be certain that changes relate to a particular time frame, rather than being a reflection of natural variation.

In some cases, information on the person's age and medical history can be of considerable importance, for instance, in the examination of a last will and testament. The signature may prove to be more consistent with an individual's writing ability at an earlier time than is reflected by the will's date.

Aside from acute changes in writing qualities, noting subtle transformations in extended handwriting is difficult in the absence of a considerable amount of questioned and standard material.

30.8 Miscellaneous Items

Every element of a document has dating value. Thus, stamps, embossments, binding materials (staples, paper clips, etc.), correction fluids, and postage/cancellation stamps on envelopes each have dates of introduction or particular dating parameters. Consequently, a thorough analysis of the entire document is necessary when issues of a document's date arise.

30.9 Summary

There is no universal approach to verifying the date of a disputed document, but the various techniques discussed have, under appropriate circumstances, resolved issues about when a document was created. Not every document can be dated by laboratory techniques. To establish from the document itself that it was prepared on a specific date is a rare accomplishment, possible only under a unique set of circumstances. Far more often, when any dating is possible, the findings must be expressed within a time bracket — not before one date and not after a later one. It is such evidence that the document examiner strives to elicit from the conditions and materials that surround and make up the document at hand.

References

1. Grant, J., *Books and Documents*, Grafton, London, 1937, chap. 2.
2. *Paper Knowledge*, 1st ed., Mead Corporation, 1990.
3. Nickell, J., *Pen, Ink and Evidence*, University Press of Kentucky, Lexington, 1990, p. 71.
4. Purdy, D.C., Document analysis, in *Encyclopaedia of Forensic Sciences*, Academic Press, London, 2000, p. 571.
5. Cantu, A., On the relative aging of ink: the solvent extraction technique, *J. Forensic Sci.*, 32, 1151, 1987.
6. Brunelle, R., Ink dating: the state of the art, *J. Forensic Sci.*, 37, 113, 1992.
7. Brunelle, R., A sequential multiple approach to determining the relative age of writing inks, *Int. J. Forensic Doc. Examiners*, 1, 94, 1995.
8. Brunelle, R. and Speckin, E., Technical report with case studies on the accelerated aging of writing inks, *Int. J. Forensic Doc. Examiners*, 4, 240, 1998.
9. Stewart, L. and Fortunato, S., Distinguishing between relative ink age determinations and the accelerated aging technique, *Int. J. Forensic Doc. Examiners*, 2, 10, 1996.
10. Lindblom, B., Practical Problems in the Relative Dating of Inks by the Study of Dye Components, paper presented at American Society of Questioned Document Examiners meeting, San Diego, 2002.
11. *Aptix Corp. v. Quickturn Design*, U.S. District Court, Northern District of California, San Francisco Division, 2000.
12. *Learning Curve Toys, L.P., v. Playwood Toys, Inc.*, U.S. District Court, Northern District of Illinois.
13. *Multiut Corp. v. Yehuda Draimain et al.*, Circuit Court of Cook County, Illinois, Chancery Division.
14. *Equal Employment Opportunity Commission v. Ethan Allen, Inc.*, U.S. District Court for the Northern District of Ohio, Eastern Division.
15. *Wang Din Shin and Nina Kung (Wang)*, High Court of the Hong Kong, Special Administrative Region: Court of First Instance.
16. Laporte, G., Wilson, J., Cantu, A., and Mancke, A., The identification of 2-phenoxyethanol in ballpoint inks using gas chromatography/mass spectrometry: relevance to ink dating, *J. Forensic Sci.*, 49, 155, 2004.
17. Aginsky, V., Measuring ink extractability as a function of age: why the relative aging approach is unreliable and why it is more correct to measure ink volatile components than dyes, *Int. J. Forensic Doc. Examiners*, 4, 1998.
18. Gaudreau, M. and Brazeau, L., Ink Dating Using a Solvent Loss Ratio Method, paper presented at the American Society of Questioned Document Examiners meeting, San Diego, 2002.
19. Hunton, R.K. and Puckett, J.T., Restoring texts of typewriter ribbons: a reliability study of the Raw-1 ribbon analysis workstation, *J. Forensic Sci.*, 39, 21, 1994.

Section X

Digital Photography and Enhancement

Cameras, Scanners, and Image Enhancement

31

FRANK HICKS
BRIAN S. LINDBLOM
ROBERT GERVAIS

Contents

31.1 Cameras and Scanners

Cameras, whether they be conventional or the newer digital designs, are a necessity for document examiners. The clearest record of evidence examined can be obtained through the use of photography or digital imaging (scanners). Much has been written on conventional photography in textbooks and journals, so the focus of this chapter will not be on that area. Rather, the discussion will be directed toward the new imaging technologies. Where necessary, contrasts will be made between digital and conventional photography.

In conventional photography different film speeds, exposure settings, and lighting conditions all impact on the quality of the photograph. In the preparation of charts for incorporation in a report or for presentation in court, a greater number of steps is required than is the case in digital photography. Conventional photography is widely recognized as having the capability to capture minute details and subtle differences in color and contrast. It was the mainstay of forensic photographers and document examiners for many decades.

Until the introduction of the scanner and digital camera, file copies of documents examined were traditionally made on a photocopier. However, now the resolution of even low-end digital cameras can be better than that of analog photocopy machines. There has been a general shift from analog technologies to digital. Even copiers now employ digital systems equivalent to those found in scanners. Furthermore, the digital copier often functions as a scanner as well.

Both forms of photography can record the fine details in typewriter fonts, the type of printing process used on a computer-generated document, and the presence of torn paper fibers in an erasure. Digital imaging can be used to prepare work charts to aid in examinations and to create demonstrative charts for use in conjunction with reports and for testimony purposes.

Today's digital cameras and scanners are an outgrowth of spy satellites used in the 1960s and 1970s and the U.S. space program, NASA. The needs of these programs led to the creation of the charge-coupled device (CCD). Essentially, the CCD is a computer chip that converts light into electric signals and is the image sensor for most of today's digital cameras. The CCD has an array of light-sensitive diodes known as pixels. The resolution of the camera is a measure of the number of pixels. These pixels detect the quantity of light that hits them, but do not interpret the color of this light. It is only through the use of one of many different filtering methods that the color information can be interpreted. In most digital cameras, the color information is an interpolation of the information gathered from a single CCD array. Only high-end 3 CCD cameras are capable of returning true color information.

Digital cameras offer advantages over conventional film cameras in some applications. Probably the most useful advantage is the immediacy of the image. The photograph can be viewed on, in most cases, a liquid crystal display (LCD) screen on the back of the camera as soon as the exposure is captured. This allows the user to ensure that the proper area of the target has been framed in the image, a problem not typically encountered in conventional film photography. It may be difficult, however, to observe small areas of blurriness on this very small image. Many of the newer models have a viewfinder, which to some extent lessens the problem. To assess the quality of the images, they can be immediately downloaded into a computer and viewed in a much larger format. If an image is deficient in any way, another exposure can be taken. The images are captured directly to a digital format that can then be processed with image-editing software on a computer. Also, various attachments enable a digital camera to be connected to a microscope to allow for instant microphotography. Typically, digital models are somewhat smaller than most 35-mm single-lens reflex (SLR) cameras and certainly smaller than the larger-format cameras. This can be a big advantage when the camera is taken into the field to photograph documents. Small portable scanners are now available as well and can be an alternative to digital and conventional film photography.

There are several companies that are now beginning to offer digital SLR units that look and operate much like 35-mm SLR cameras, even using the interchangeable lenses. In fact, the settings found on many non-digital cameras, such as F-Stop and ISO, are now also present on digital models. Of course, one of the most practical advantages of digital units over film cameras is cost. Digital cameras do not require repeated purchase of film, developing chemicals, enlargers, and darkroom equipment.

The digital SLR models now address one of the biggest limitations of past digital versions compared with their film counterparts — resolution. The newest digital models are capable of 10-megapixel or greater resolution. Though there is some disagreement as to the specific value, resolutions in the 6- to 10-megapixel range are generally considered equivalent to that offered by 35-mm film. Until the advent of these multi-megapixel cameras, digital images could not be enlarged to much more than an 8 × 10 size before they started to show pixelation — a circumstance in which the individual pixel elements that comprise the image begin to show as noticeable squares. As the resolution capabilities have continued to rise, it is now possible to enlarge digital photographs with clarity equal to that of conventional film images.

One photographic consideration regarding the use of digital cameras for forensic purposes is the need for macro-capabilities. Most mainstream point-and-shoot models are not capable of this type of close-up focusing. If the camera is capable of macro-focusing, it may have as an option a macro-flash unit, since the built-in flash will be useless at macro-distances. The FDE must recognize the difference between optical zoom and interpolated, or digital, zoom. Optical zoom is a function of the lens on the camera, whereas interpolated zoom is a function of software and results in deteriorated image quality. Another consideration is the storage formats available with the camera. Most cameras can store digital images in JPEG format, but this compression method discards varying amounts of image information, depending on the quality setting. A much more useful format for maintaining the image quality needed in forensic sciences is TIFF (Tagged Image File Format). While this convention also offers compression options resulting in data loss, its application in most digital cameras results in all of the information in each pixel of the image being recorded. The primary drawback of TIFF images is their size. A high-quality TIFF image can easily exceed 10 megabytes in size. This becomes important when considering the storage medium of the camera. A new format that is becoming popular for those seeking truer image reproductions is RAW. This format records the information as it appears directly from the CCD. Consequently, it is not subject to the color interpolation performed in creating the other file types. Surprisingly, the RAW format often produces smaller files sizes.

While some cameras have small internal storage capabilities, most storage is done with removable memory devices referred to generically as flash memory. These devices come in different formats, such as compact flash (CF) and secure digital (SD) and memory sticks, to name just a few. Many contemporary computers and printers now offer direct input from the camera, its podium, or the memory medium.

Another consideration in the purchase and use of a digital camera is the settings that can be adjusted prior to taking the photograph. Some cameras offer a variety of features, such as white balance, sharpening, resolution, image quality, exposure, and flash level. Similar options and adjustments are present in some scanner applications. Some of these options may alleviate the need to later employ enhancement techniques on the image with a computer imaging program.

31.2 Image Enhancement

Since images captured with digital cameras and scanners are already in digital format, they can be downloaded into a computer and enhanced using a variety of image-editing programs. This process can be advantageous in many ways. With a digital camera or scanner and an image-editing program, a computer becomes a powerful tool in the document lab.

There are a number of procedures and techniques that can be applied to a digital image. These techniques can be applied for two reasons. The first is to aid the FDE in the forensic examination. For instance, by separating an image into its color components, it may be possible to differentiate two inks that appear to be the same visible color to the naked eye. Alternatively, enhancements may be applied to assist in presenting findings. For example, the enhancement technique described by Sperry and Tolliver[1] can be used to improve faint images from electrostatic indentation impressions.

As described in Chapter 34 on chart preparation, any enhancements to an image must be logged. This can be done manually through the use of features built in to the graphics application or with the aid of third-party software.

31.3 Image Enhancement as an Examination Tool

Various image enhancement and digital handling techniques can play an important role in the examination process. The creation of virtual grids over a typewritten or computer-printed image can aid in the detection of misalignments. By digitally overlaying and superimposing images using layers, the shapes of printed characters, stamps, and seals can be compared. The layering tool is also valuable when associating indented impressions with their source handwriting. Specialized "stitching" software allows multiple images, such as those that might result from infrared examinations, physical matching, or macrophotography, to be easily aligned and seamlessly stitched into a single image.* In some instances, computer scanners together with image-editing software can be used to differentiate similarly colored inks, such as black ballpoint.[2] Selective supersaturation of individual color channels in the digital image can accentuate subtle color differences. By taking advantage of the various color models used by graphics software, selective color filtering can be used to isolate overlapping elements of a questioned document. These are just a few of the techniques currently being applied by FDEs in the examination process, with new and innovative processes constantly being developed.

31.4 Image Enhancement as a Presentation Aid

Often an FDE's findings can be difficult to convey to the intended audience of attorneys, jury members, or a judge. While the preparation of court charts is explored in detail in Chapter 34, some specific image enhancement techniques well suited for chart preparation are described below.

It is not uncommon for an FDE to find faint indented impressions on a questioned document. In some cases, these impressions can even be sourced to the handwriting that created them, or deciphered as important relevant information. The untrained eye of the layperson, however, requires that this type of impression be enhanced for viewing. The Tolliver and Sperry technique, described earlier in this chapter, is ideal for such situations. Similarly, an overlay and superimposition of indented and source writing can allow the viewer to readily appreciate the relationship.

The overlay-and-superimposition technique is useful in many situations. It can be used to show the relationship between residual marks left behind after an obliteration relative to the original document, to compare typefaces, and to demonstrate a match between signatures in a cut-and-paste manipulation. In order to make such overlays easy

*Though an image-stitching function has now been incorporated into the current version of Adobe® Photoshop®, specialized software, such as PanaVue Image Assember by PanaVue software in Quebec, can yield excellent results with minimal effort.

to comprehend, the electronic images are often colorized so that one can easily discern one from the other. Adobe® Photoshop® allows the user to change the "blending mode" of each layer so that it affects the layers beneath in various ways. For FDEs, one of the most useful settings is "multiply." Using this mode, areas of a layer that are white have no impact on layers below. Colored portions of the image, however, blend with those of the underlying layer. This allows the relationship between the two to be better appreciated.

Distortions, such as those resulting from page curl when copying a book or general reproduction distortion, can be compensated for using graphics software. This can allow the layperson to more readily compare different reproduction generations of the same original. The techniques used to accomplish this can include simple transformations or more complex displacement mapping. These are often done solely for the benefit of presenting the evidence, as the FDE may have little or no difficulty in appreciating the relationship, even on a distorted reproduction.

31.5 Summary

Digital photography and scanning capabilities have added to the arsenal of equipment in modern forensic laboratories. Continued development and improvements of digital image capturing and the increasing versatility of enhancement software will increase their application to a greater variety of document cases.

Conventional photography still has its place in document examination. For instance, capturing images of digitally generated or reproduced documents (created on a fax machine, computer printer, or digital copier) is sometimes best done using conventional photography. This approach avoids any interference between pixelation introduced by a digital image-capturing device and that of the document itself.

For archival applications, conventional photographic prints are longer lasting than digital images printed on computer printers, which can fade or degrade much more rapidly.

Some level of familiarity with conventional photography will aid in producing good-quality digital images. Many of the settings now present on digital models, such as ISO, are simulated representations of adjustments important in conventional photography. Without a basic understanding of film cameras, the adjustments and options available on digital models may have little or no meaning to the FDE.

References

1. Sperry, G. and Tolliver, D., Applying specific digital enhancement techniques to ESDA-developed impressions, *J. Am. Soc. Questioned Doc. Examiners*, 4, 12, 2001.
2. Bodziak, W., Using a Flatbed Scanner and Adobe Photoshop Software to Distinguish Black Inks, paper presented at the American Society of Questioned Document Examiners meeting, Ottawa, Canada, 2000.

Section XI

ASTM Guidelines for Forensic Document Examination

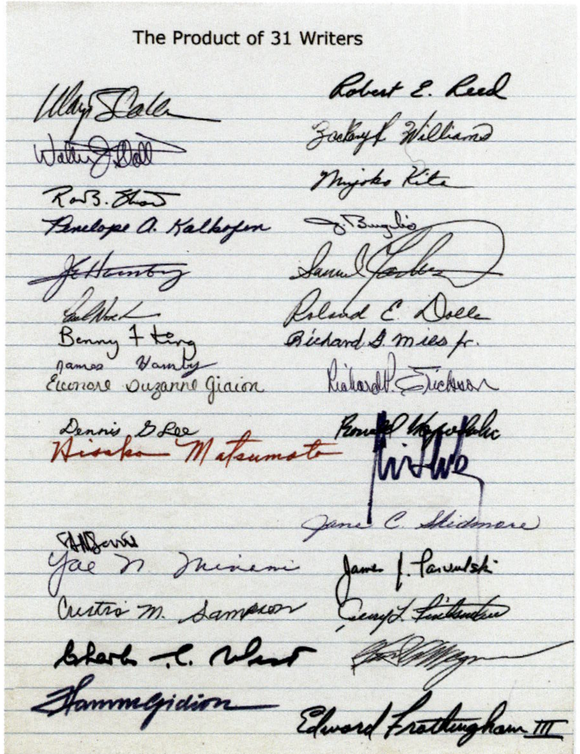

Figure 9.12

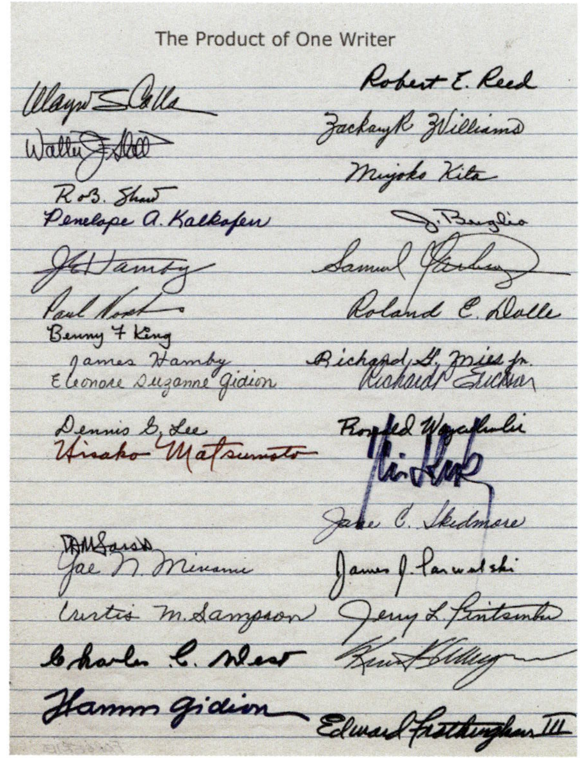

Figure 9.13

Figure 9.12 and Figure 9.13 The above figures demonstrate the skill of one individual in replicating the signatures of 31 other individuals. The "forger" has very good hand-and-eye coordination and an attention to detail. He prepared Figure 9.13 without a great deal of study or practice. His simulations undoubtedly would have improved with practice. This exhibit demonstrates points to consider, not the least of which are the need for multiple examples of known signatures, the importance of line quality, and the effect that particular writing instruments can have on the appearance of a signature, as well as the difficulties that they can create in their evaluation.

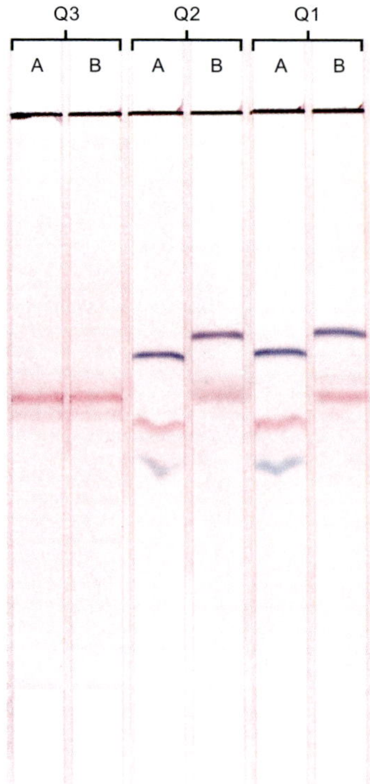

Figure 13.8 Ink samples from various signatures, authored by two writers (A and B), could not be differentiated visually, microscopically, or through the use of optical filtering. However, with the aid of thin-layer chromatography (TLC), at least three inks could be distinguished at this level of testing. The signatures on Q1 and Q2 were signed using two inks, while those on Q3 were written with a different third ink.

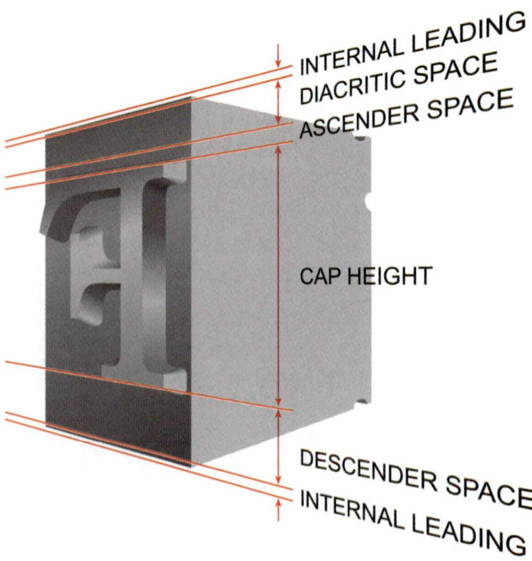

Figure 14.8 A 100-point font can actually measure more than 100 points of vertical height if the character has diacritic marks.

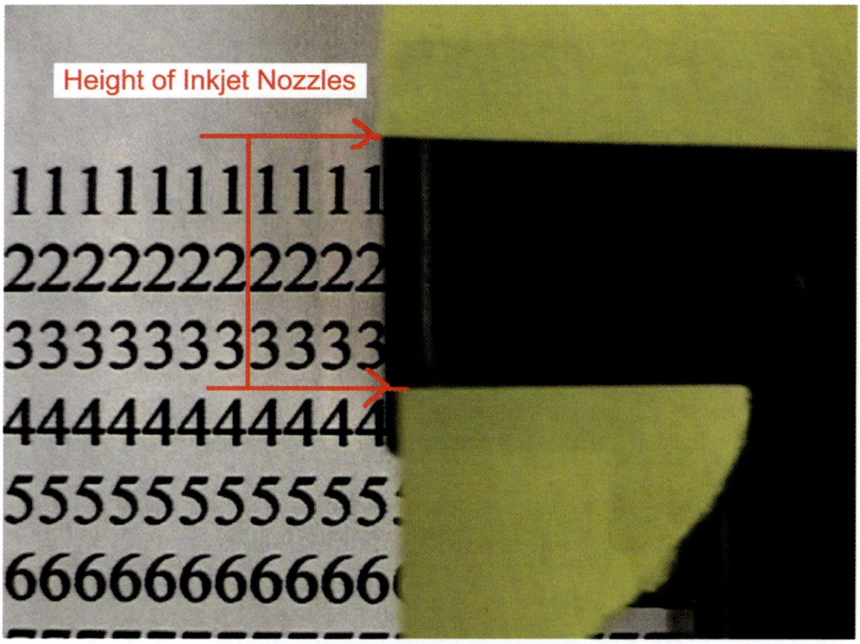

Figure 16.2 The vertical size of the nozzle array determines how many lines can be printed in one pass. In the illustration, yellow tape marks the uppermost and lowermost nozzles on the black-ink print cartridge from an HP inkjet printer. The lines of 10-point Times New Roman numbers are printed three lines to the left, then three lines to the right in a repeating pattern.

Figure 16.3 Photomicrograph of a modern inkjet printhead array.

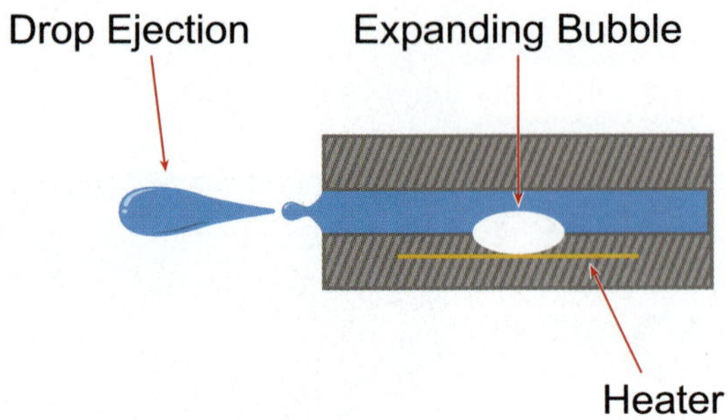

Figure 16.4 Schematic of a thermal drop-on-demand inkjet mechanism.

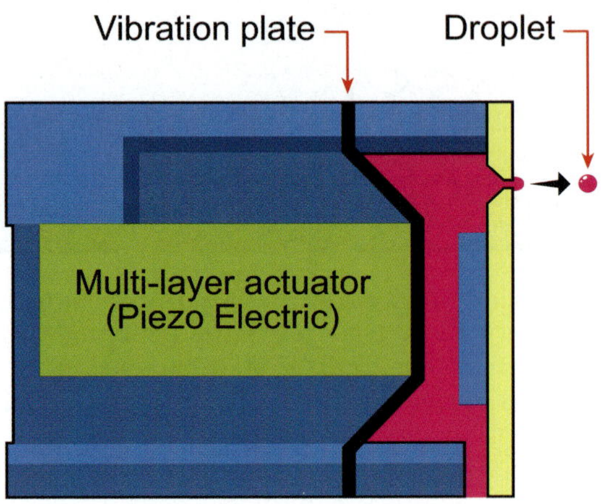

Figure 16.5 Schematic of piezoelectric inkjet mechanism.

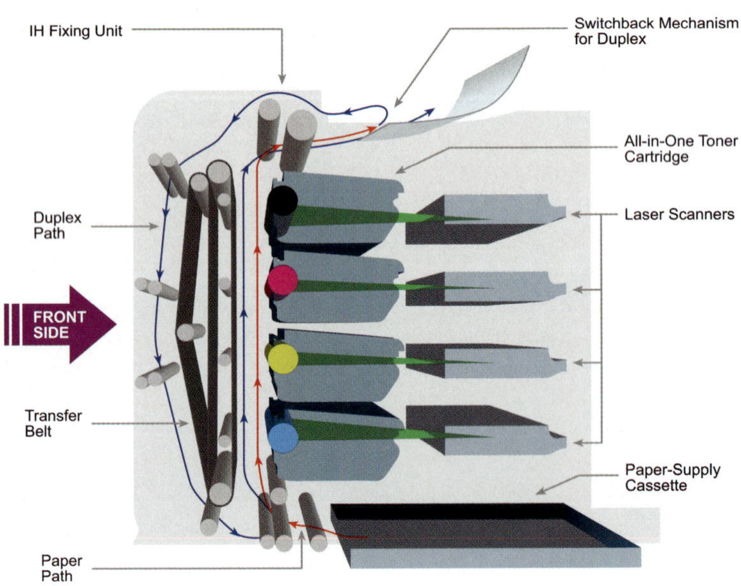

Figure 16.7 Schematic of a Canon four-color laser printer, in-line, vertical print engine.

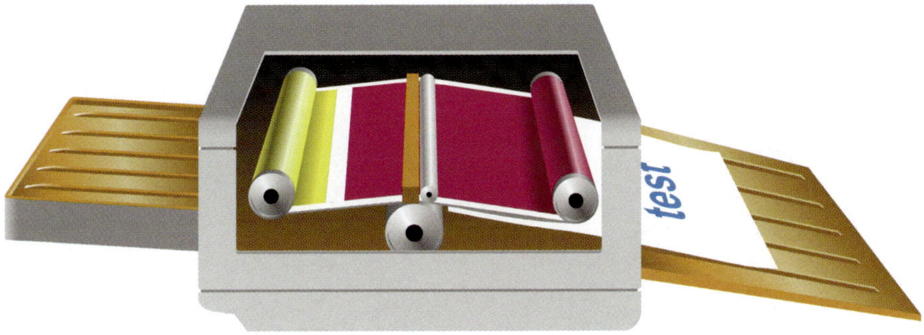

Figure 16.8 Cutaway view of a thermal wax printer.

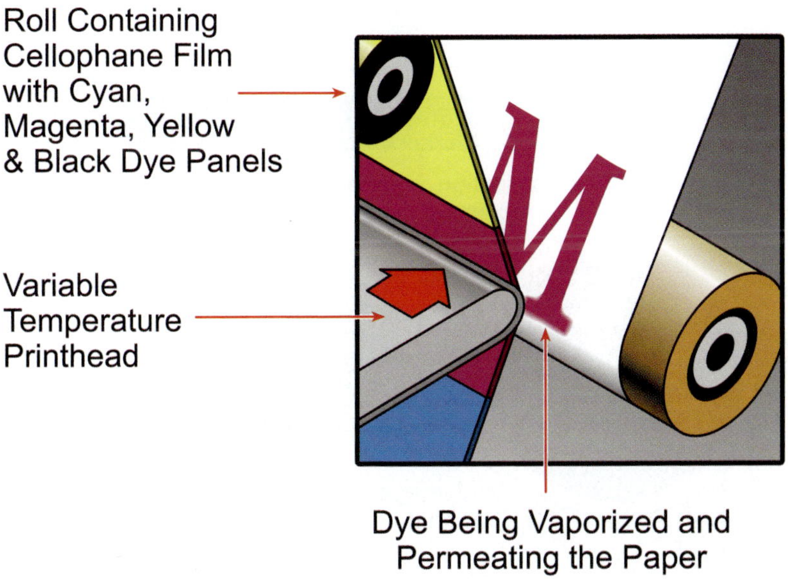

Figure 16.9 Dye sublimation print mechanism.

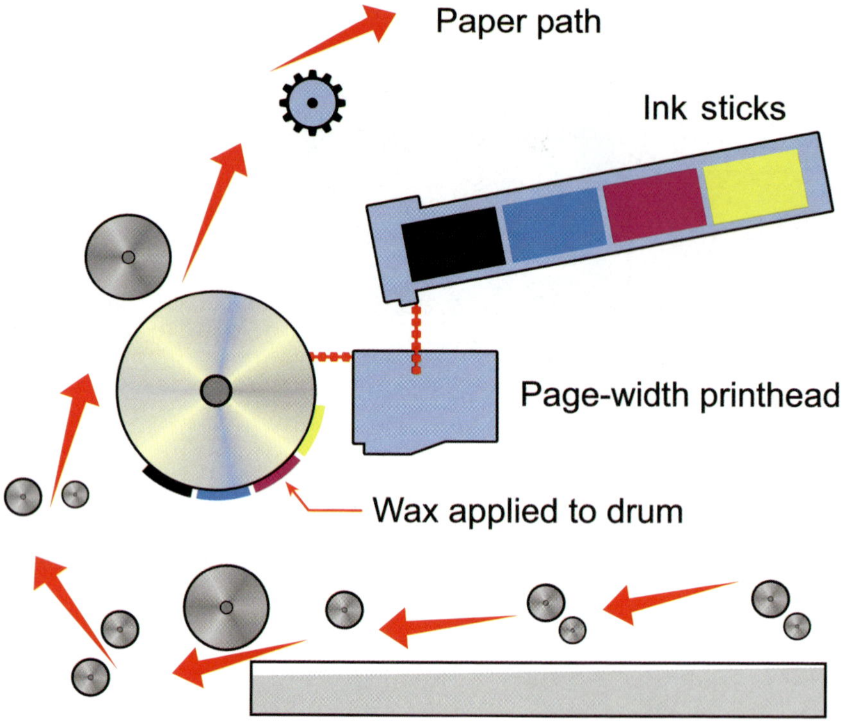

Figure 16.11 Schematic of a Xerox 8400 dry-ink stick-type printer.

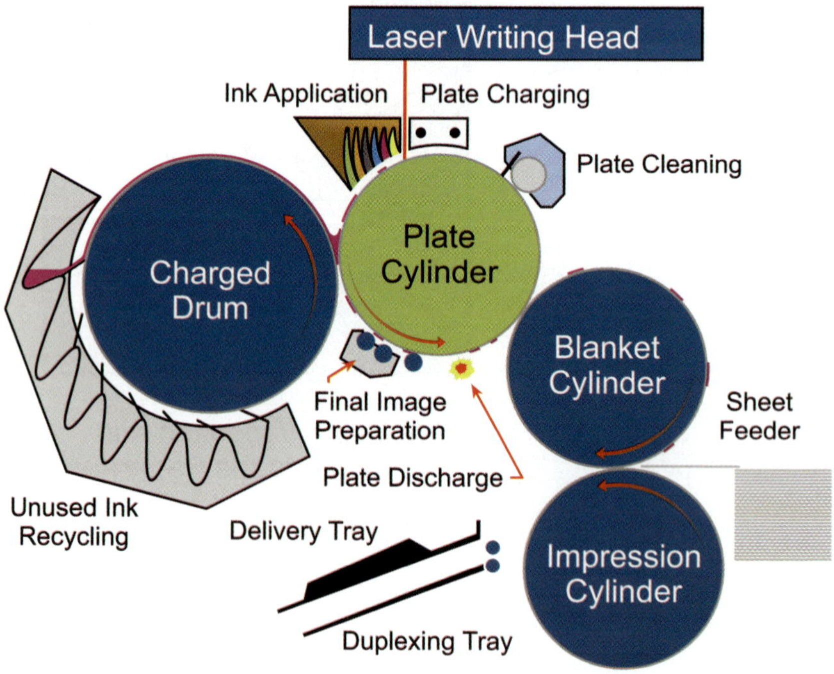

Figure 16.12 Diagram of a digital offset printer.

(a)

(b)

Figure 16.17 (a) Inkjet print sample with printhead moving in the leftward direction. Satellite drops that appear on one side of the character are typical indications of inkjet printing and can be used to determine the direction of the printhead travel. (b) This image shows a typical four-color (cyan, magenta, yellow, and black) dithering pattern associated with an inkjet-printed color graphic.

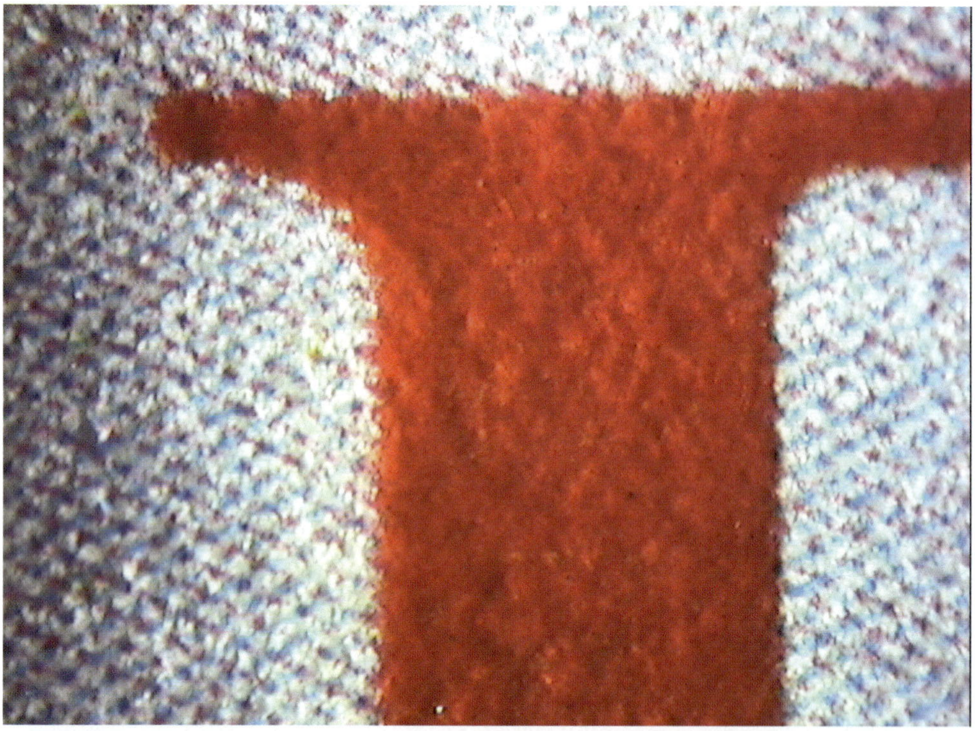

Figure 16.19 Four-color (CMYK) laser print sample.

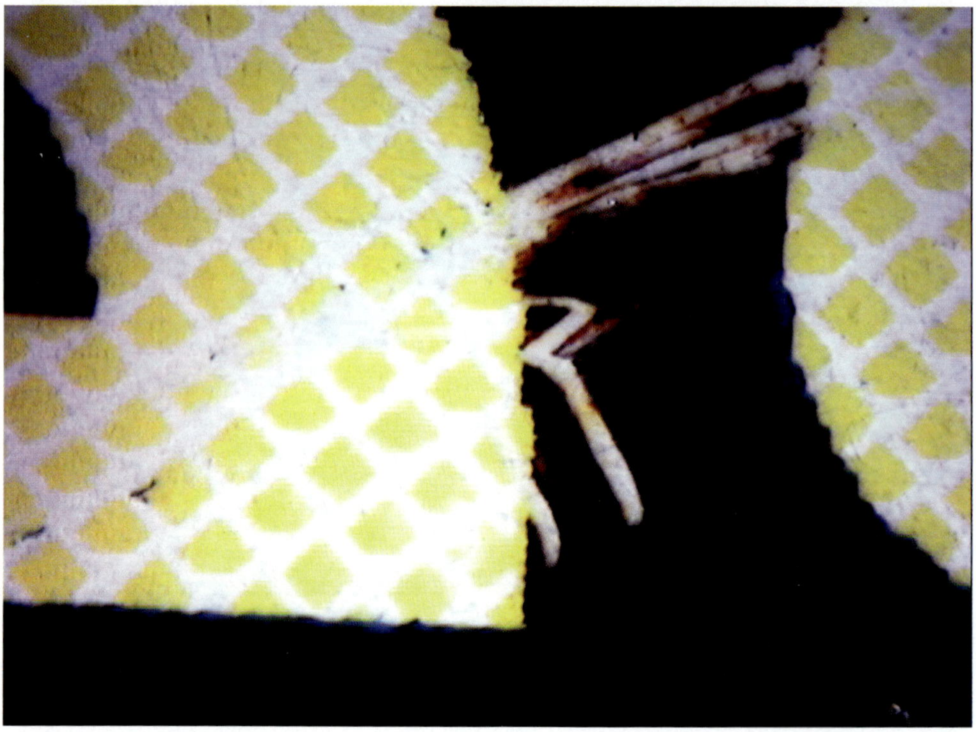

Figure 16.20 Sample from a thermal wax printer. Note that the waxy material can be easily removed with a sharp object.

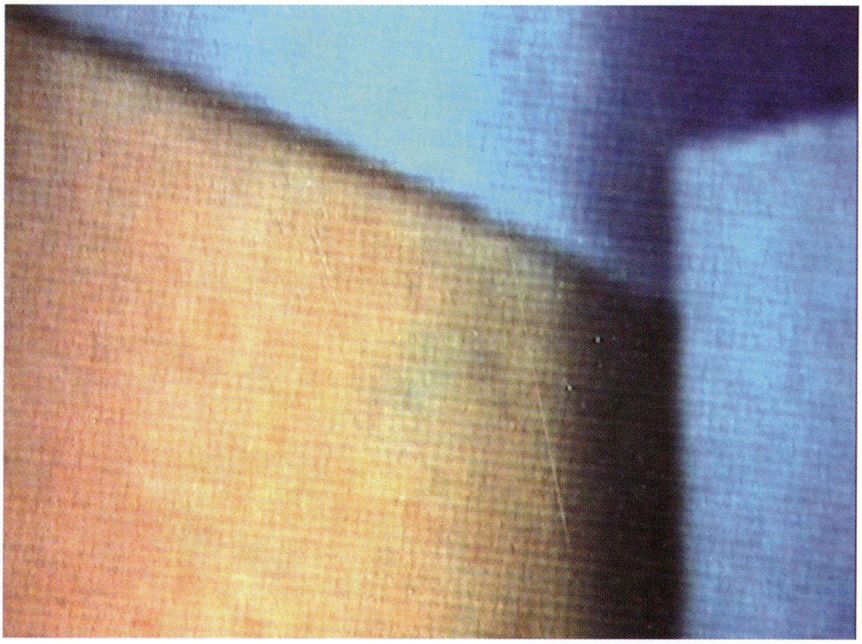

Figure 16.21 Sample output from a dye sublimation printer. This image is not out of focus. The dyes are transferred to the paper as a gas, thereby producing a typical fuzzy image under high magnification.

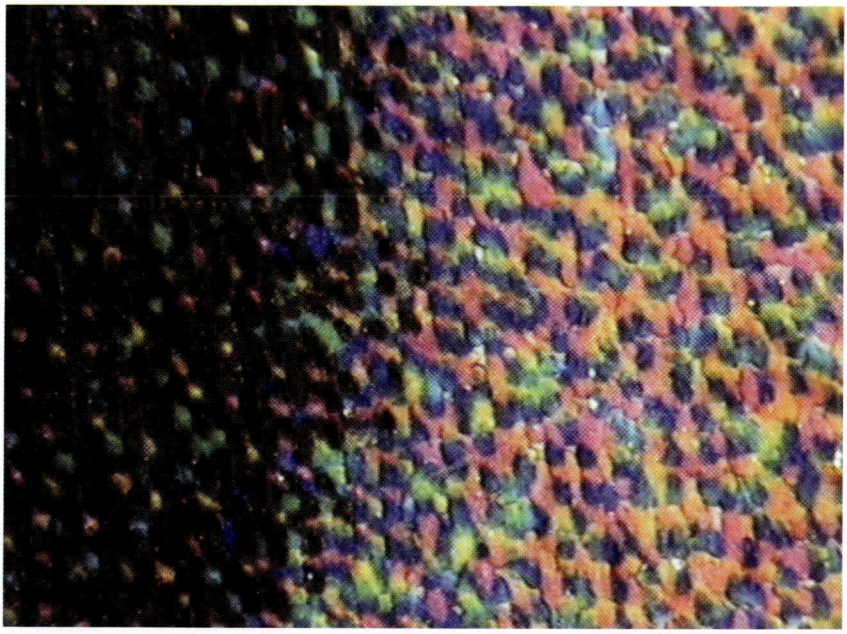

Figure 16.23 Photomicrograph of the output from a solid-ink stick-type printer. The CMYK ink droplets from a solid-ink printer are pressed into the paper after being sprayed onto a transfer drum. On the older models these drops tended to be teardrop shaped or oval along the drum rotation axis. It is reported that the newer models now produce more rounded drops.

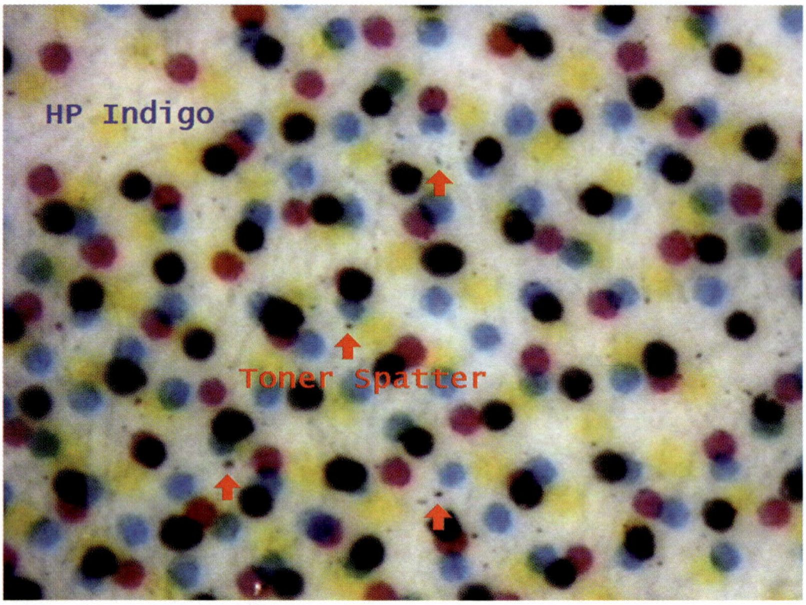

Figure 17.2 Shown here is the halftoning of the HP Indigo print; observe the minute toner particles spattered between the halftoning lines. (Courtesy of William J. Flynn.)

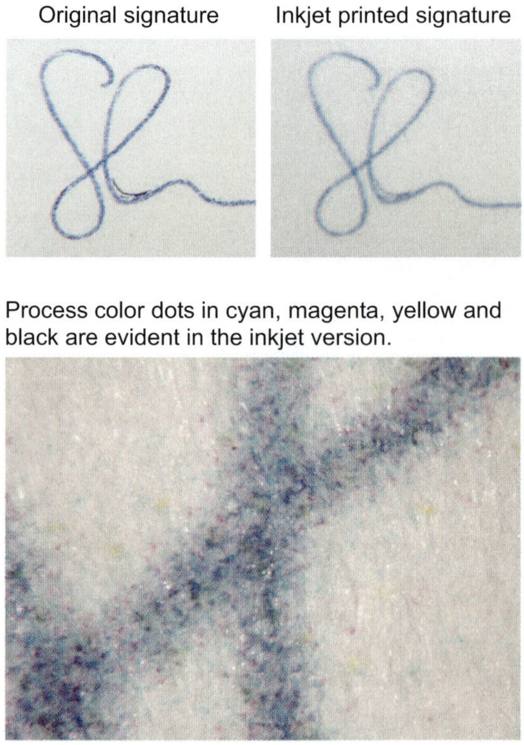

Figure 19.1 When a document is presented as original, it should be examined microscopically to determine whether there is an actual ink line or whether there is evidence of reproduction.

Figure 22.3 Remnants of a document that was removed from a stapled set of documents. One might also find trace evidence or blood attached to a staple.

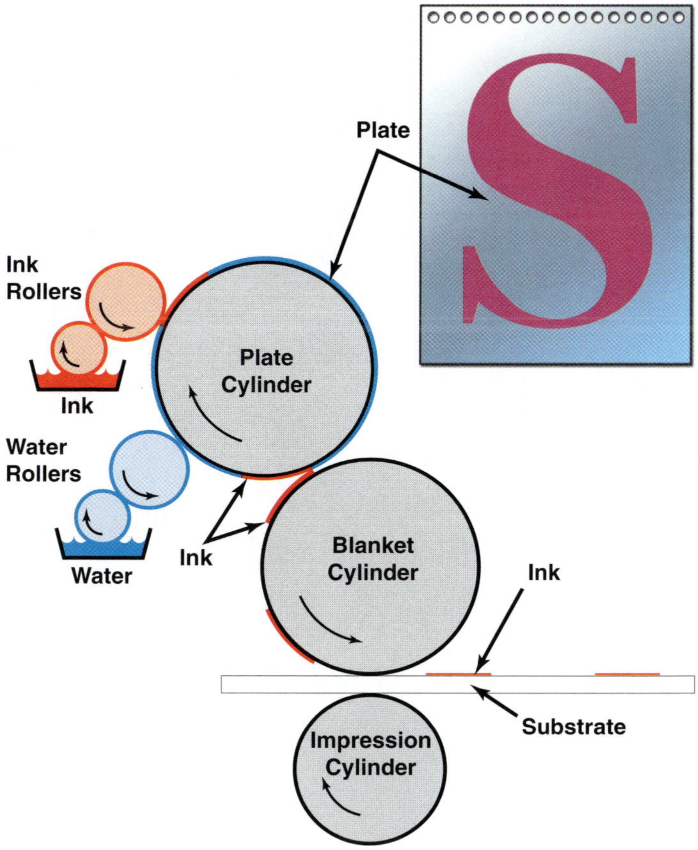

Figure 23.1 A graphic representation of the offset lithographic printing process.

Figure 23.2 Single-color offset lithography. Note the clear lines and even printing.

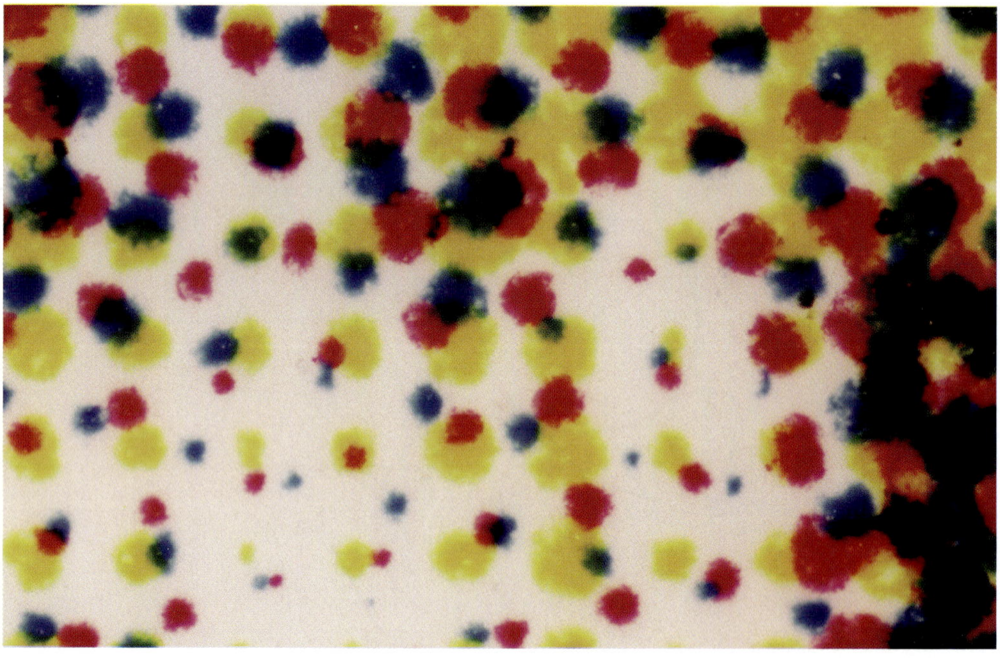

Figure 23.3 Four-color offset lithography. Note the use of halftone screens to vary the dot size and the highly organized nature of the dot pattern.

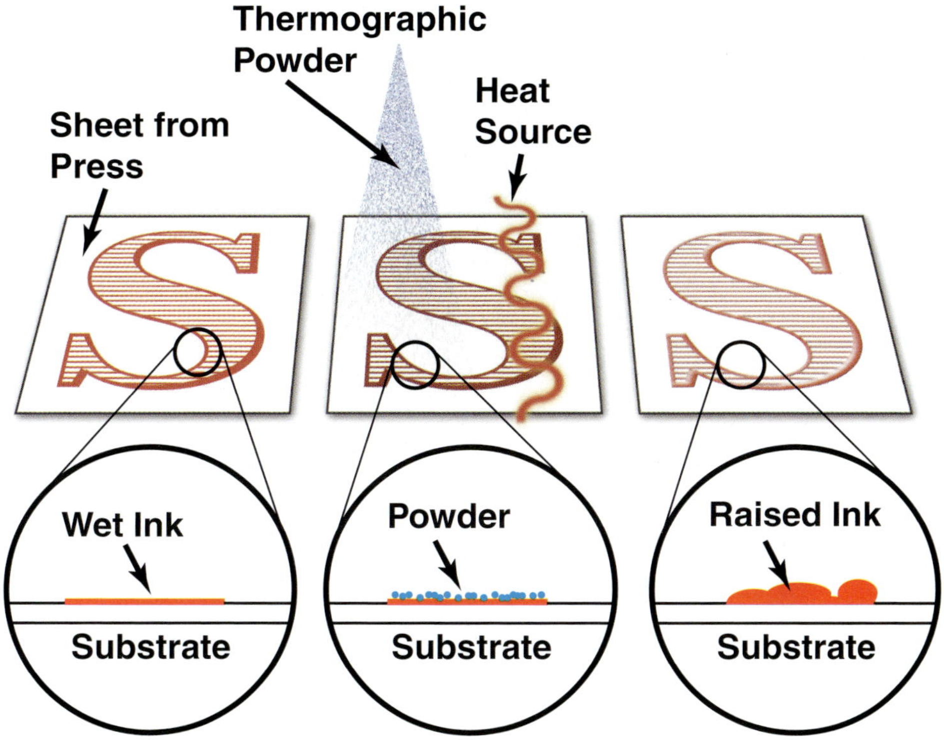

Figure 23.4 A graphic representation of thermography.

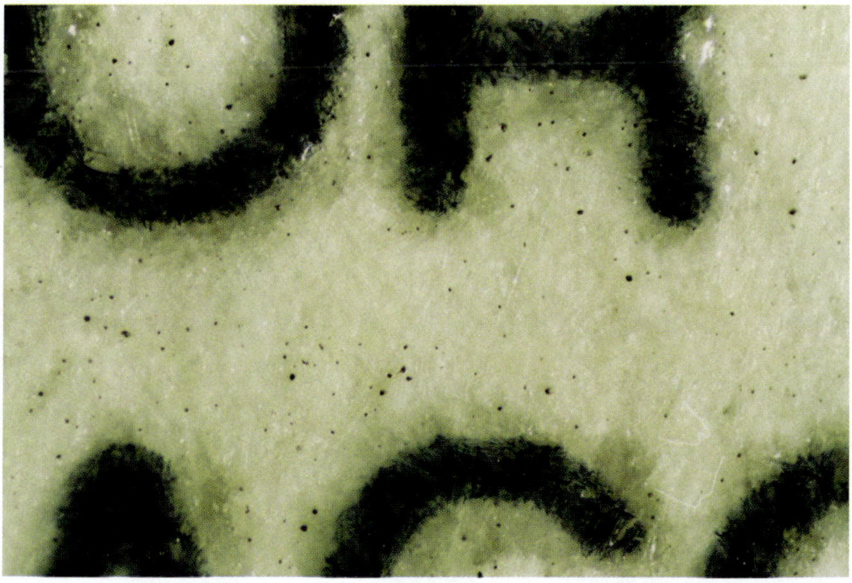

Figure 23.5 Note the presence of some bubbles in the thermographic resin.

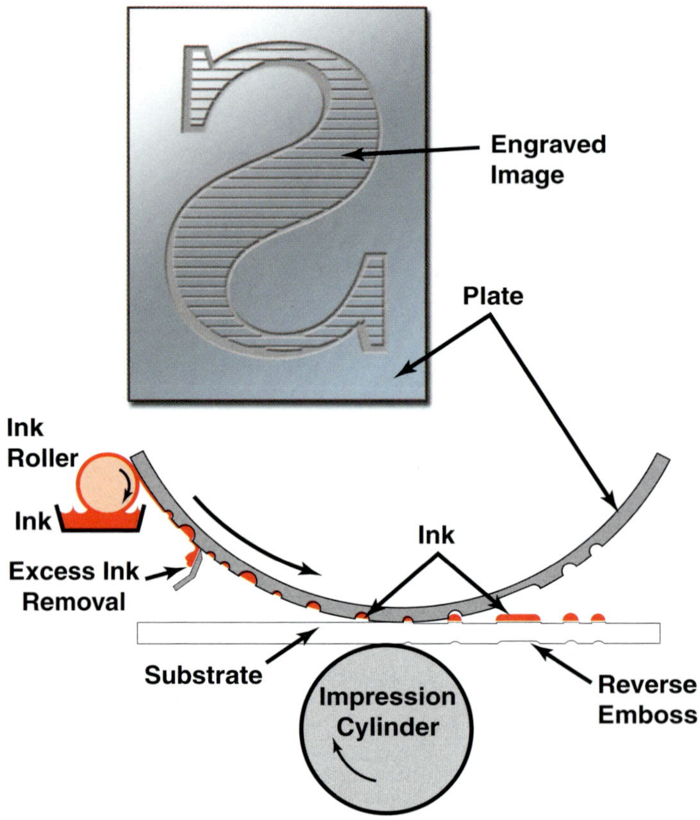

Figure 23.6 A graphic representation of intaglio printing.

Figure 23.7 Note the thickness of the ink and the characteristic feathering in intaglio printing.

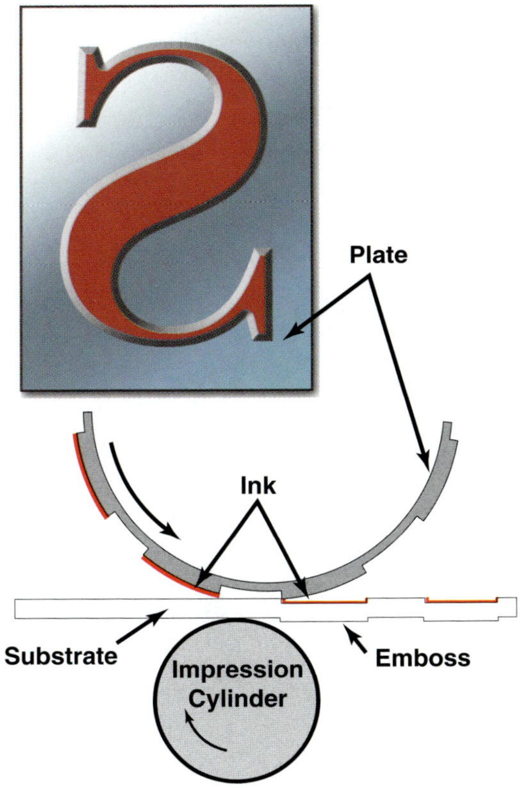

Figure 23.8 A graphic representation of letterpress printing.

Figure 23.9 Note the halo of ink around the edges and the slight embossing around the letter in letterpress printing.

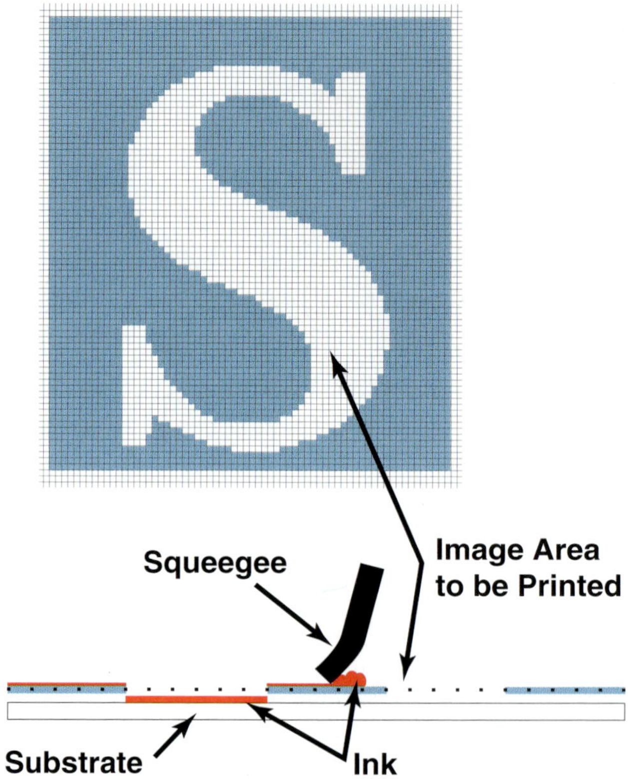

Figure 23.10 A graphic representation of screen printing.

Figure 23.11 Note the jagged appearance at the edge of the printed areas in screen printing.

Figure 23.12 In screen printing, the appearance of the screen is readily apparent throughout these images.

Figure 25.1 A typical carbonless paper document. This invoice form has three copies of different colors. It is common that the top copy is white, the middle copy is yellow, and the bottom copy is pink. However, this standard need not always apply.

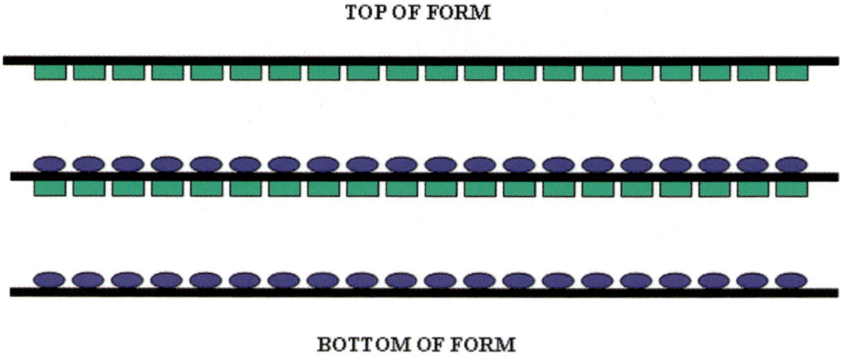

Figure 25.2 The chemical-mated system of carbonless paper consisting of colorless dye microcapsules (rectangles) and a colorless coreactant (ovals).

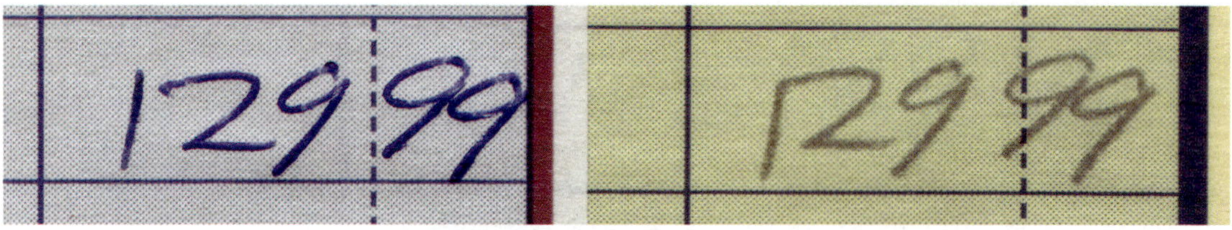

Figure 25.4 The numeral 1 was added after the layers were separated and repositioned. The alignment variation on the yellow page provides evidence of this change.

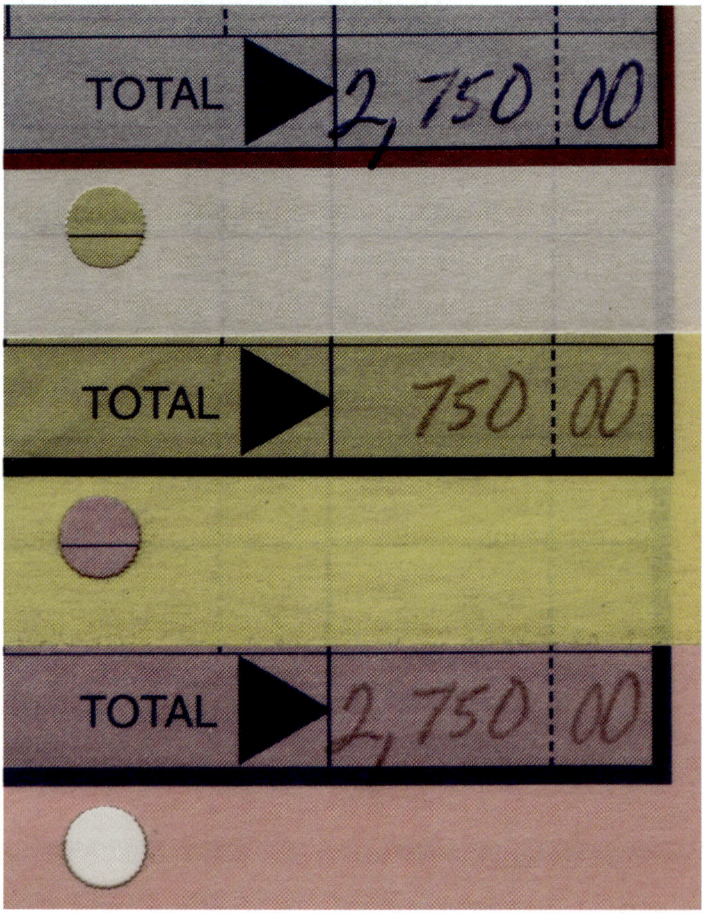

Figure 25.5 A customer's complaint that an invoice was altered was supported when the bottom (pink) copy of a disputed invoice was discovered to reflect a significant difference in the "Total" entry from the yellow copy, but the same amount as the white (top) copy. The claim by the store that an object must have inadvertently wedged itself in the form is not supported by the physical evidence, as any object interfering with the pressure needed to burst the microcapsules would do so for all subsequent pages.

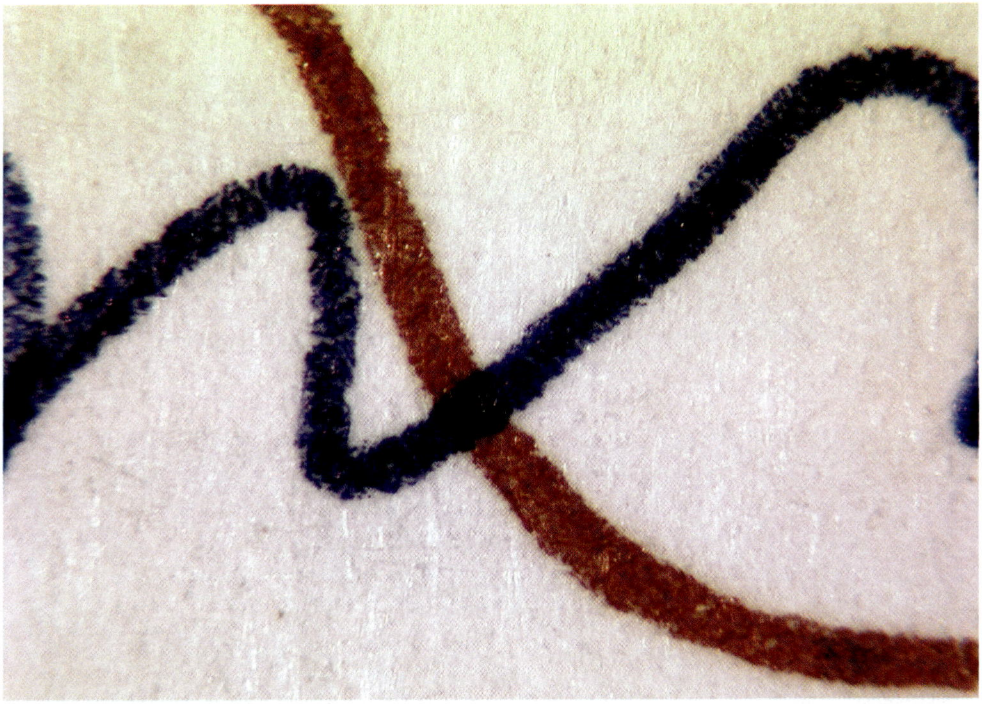

Figure 27.8 An optical illusion is created at the intersection of a darker and lighter ink. The blue ink appears to be on top, when actually the red ink was written last.

Figure 27.10 Fluid ink intersects toner lines. Spectral reflectance is observed at the left, where the red fluid ink intersects the printed *t*. This demonstrates that the red ink is on top of the toner. The opposite is seen at the intersection of the *n* arch and diagonal red line, where no spectral reflectance is seen, the result of the ink being beneath the toner.

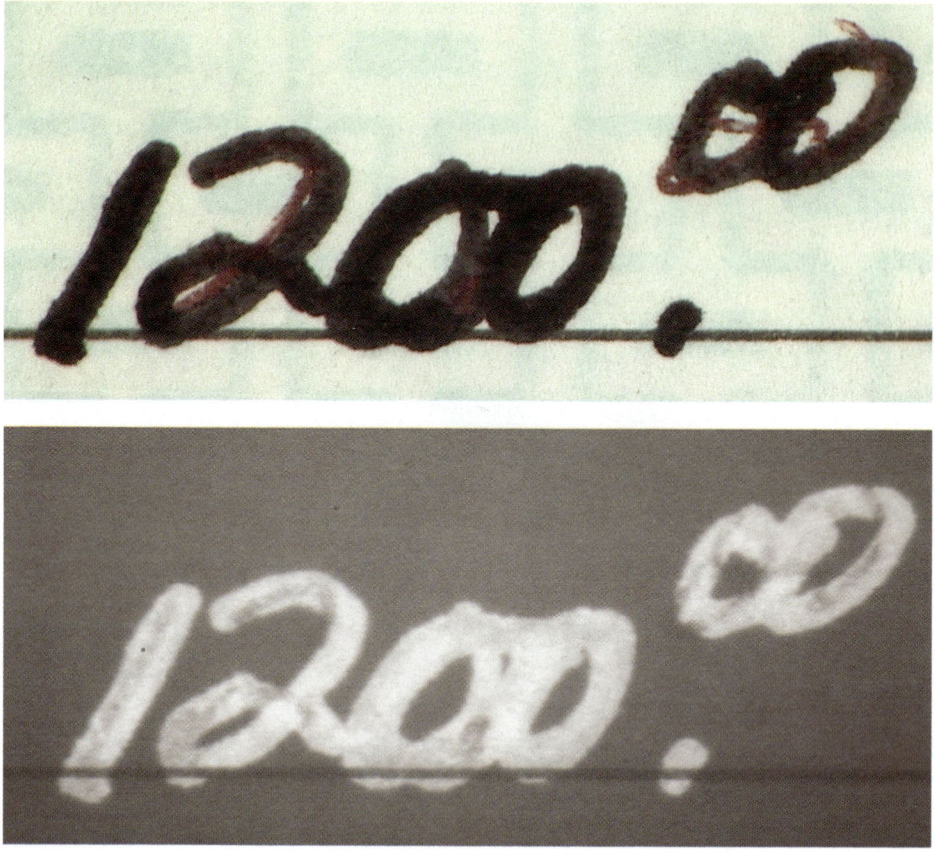

Figure 27.15 Two views of the numerical amount on a check. The top image was prepared with normal lighting, while the bottom frame shows the reaction under IR luminescence. The original amount was $12.00, raised to $1200.00. The overwriting luminesces, while the original amount, in red ink, is absorbent. Note the original writing is only visible where it intersects with the luminescent black fluid ink.

Figure 27.16 The same check as in Figure 27.15, but showing the written amount. As with the numeric amount, the black fluid ink is luminescent, while the underlying red ink is absorbent. There is no indication of an absorbing ink in the word *HUNDRED*, only in *TWELVE*.

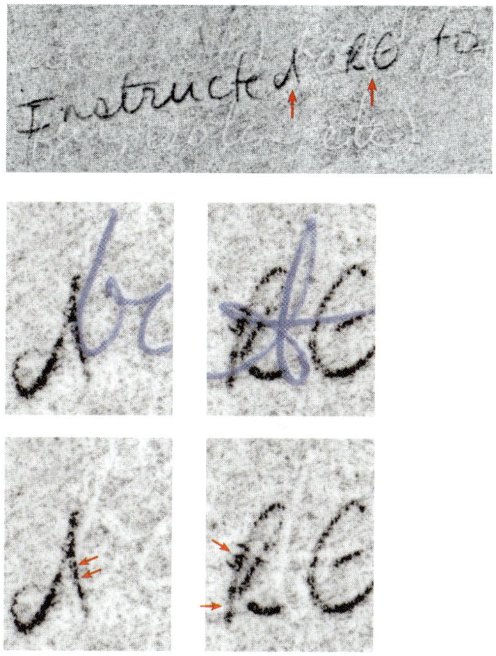

Figure 29.3 The first panel illustrates the intersection of indented writing impressions with the negative image of the visible handwriting on the page tested. The arrows indicate two areas that are enlarged in the panels below. In the second row the transparency sheet is registered with the page tested, showing the locations at which the visible writing and indented impressions overlap. The complete ink line at each intersection point makes these locations suitable for sequencing interpretation. The final row shows a number of points, indicated by arrows, at which the solid indentation line (black) breaks the negative (clear) image of the visible writing. Because the indentation line is solid, it can be stated that the handwriting came first, followed by the indented impressions.

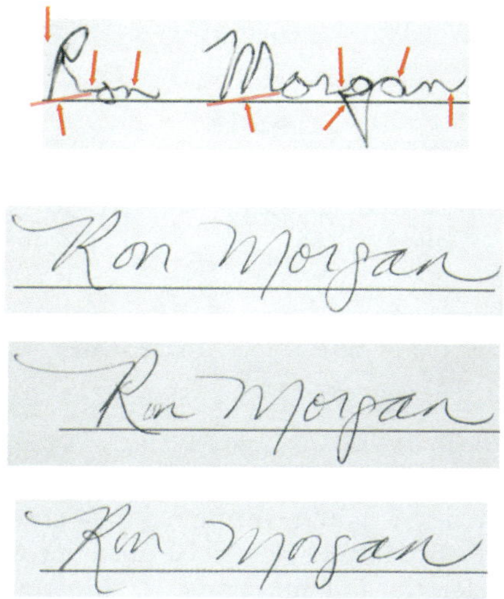

Figure 34.1 Morgan denied his signature on a contract. He furnished a large quantity of known course of business signatures. This chart shows three genuine signatures under the questioned signature. The arrows with letters at the end refer to text in the report describing some of the differences noted.

Hand Printing Comparison

Q1 @ 200%

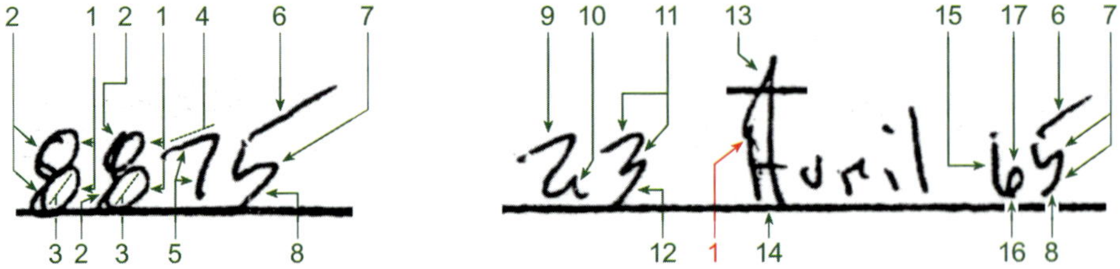

Legend: Similarites are shown in green; features not fully accounted for in red.

1. Two separate eyelets in the *8*'s
2. eyelets in the *8*'s begin and conclude at the left
3. diagonal orientation of the lower eyelet in the *8*'s
4. downward orientation of the horizontal stroke of the *7*
5. relative lengths of the horizontal and vertical strokes of the *7*
6. long, upward sloping and disconnected horizontal bar of the *5*
7. very angular lower body of the *5*
8. downward orientation of the final movement in the *5*
9. length and angle of the horizontal movement of the *2*
10. movement in the termination of the *2*
11. slope of the horizontal strokes in the *3*
12. strong downward termination of the *3*
13. angularity across the top of the *A*
14. narrow, elongated *A* body
15. orientation of the *6* staff, with slight curvature
16. extent of curl in the *6* bulb
17. orientation of the *6* bulb

1. Possible disconnection on the left staff of the *A*, or a break resulting from the photocopying process.

Questioned - Q1

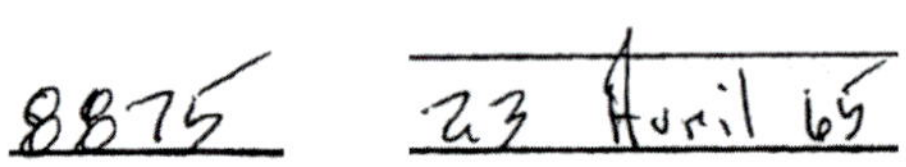

Specimens

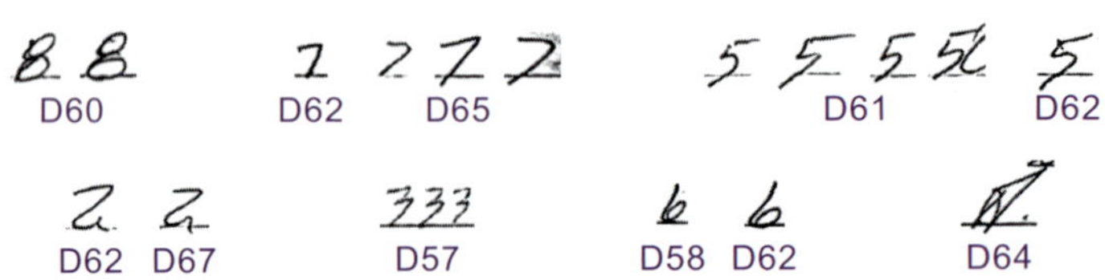

Images are shown at approximately actual size.

Figure 34.2 In this illustration the hand printing features are identified on an enlargement with arrows and numbers, with the features described in an accompanying legend. Toward the bottom of the chart, similar letters and numerals are illustrated. Enlargement differences between the upper and lower portions of the chart are indicated.

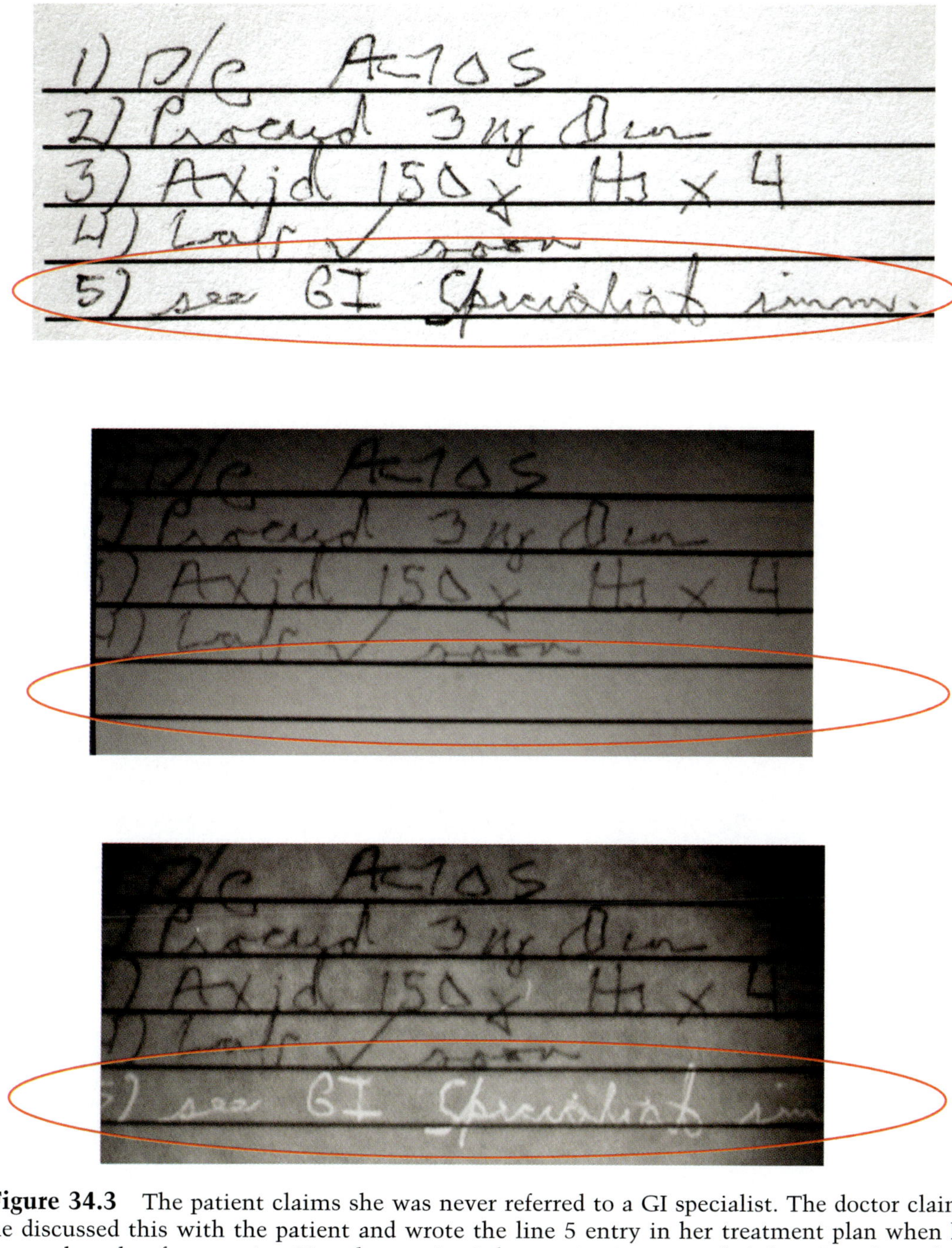

Figure 34.3 The patient claims she was never referred to a GI specialist. The doctor claims he discussed this with the patient and wrote the line 5 entry in her treatment plan when he wrote the other four entries. Non-destructive ink examinations revealed that the ink used on line 5 was different from that used in the other entries.

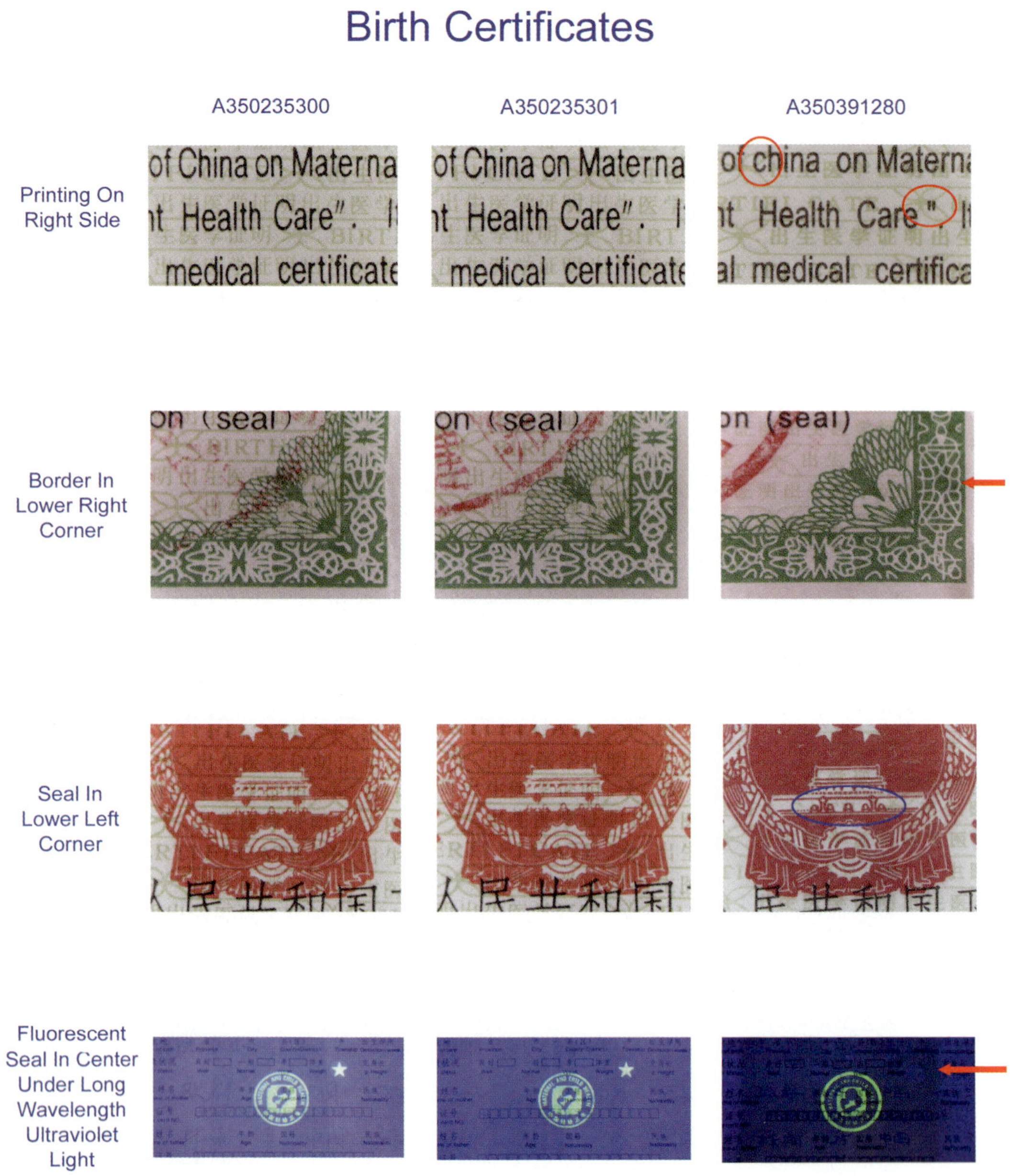

Figure 34.4 Three birth certificates from the People's Republic of China were submitted for examination. The question was whether any of them might be counterfeit. Because no confirmed genuine certificates were submitted, it was not possible to answer that question, but by isolating four areas from each certificate, it was clear that the certificate on the right was different from the other two. The arrows and circles highlight some of the more obvious differences.

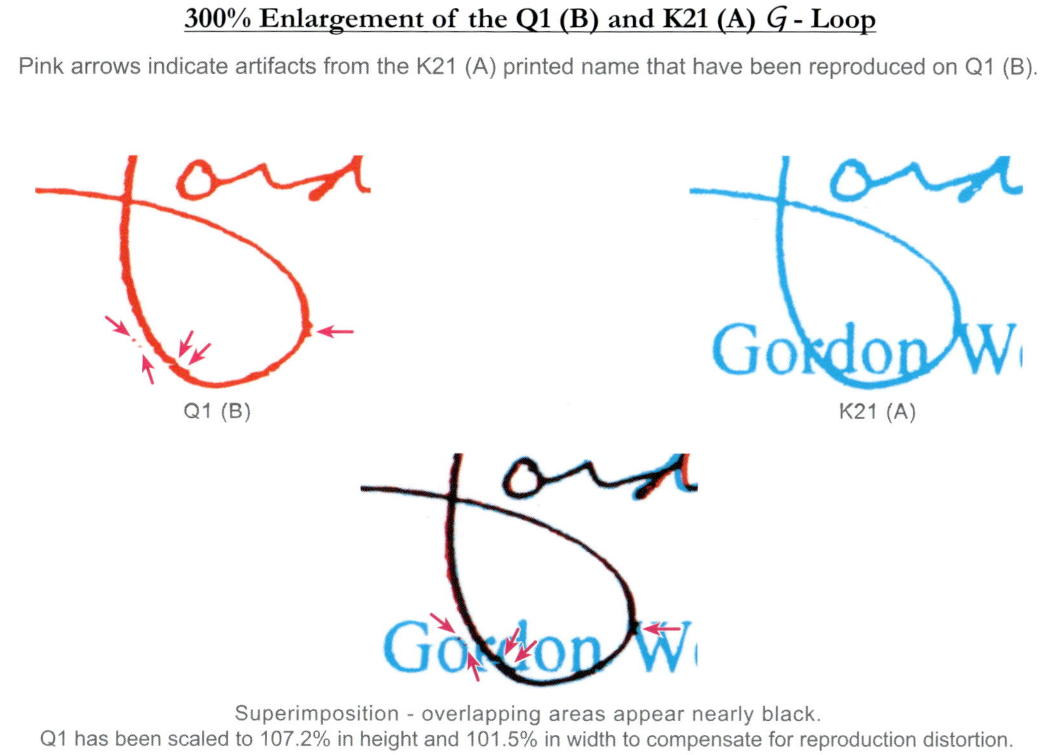

Figure 34.5 This chart makes use of image colorizing and superimposition to facilitate comparison of related items. The signature appearing on K21 (A) was copied onto Q1 (B), and in the process, portions of overlapping text remain in the transferred signature. Scaling was necessary to compensate for reproduction distortion and has been clearly described in the chart.

Figure 34.6(b) In an attempt to evaluate whether the position of the endorsement block was consistent with the remainder of the document, line spacing characteristics were considered. To illustrate the findings, the questioned text was superimposed by a layer in which a progressive typing was completed. The type font and point size on the original document were replicated in the overlay and the "show all" function of the word processing program was activated. A progression of windows was then created, each with successively more of the text typed. The chart demonstrates that there are no spacing anomalies.

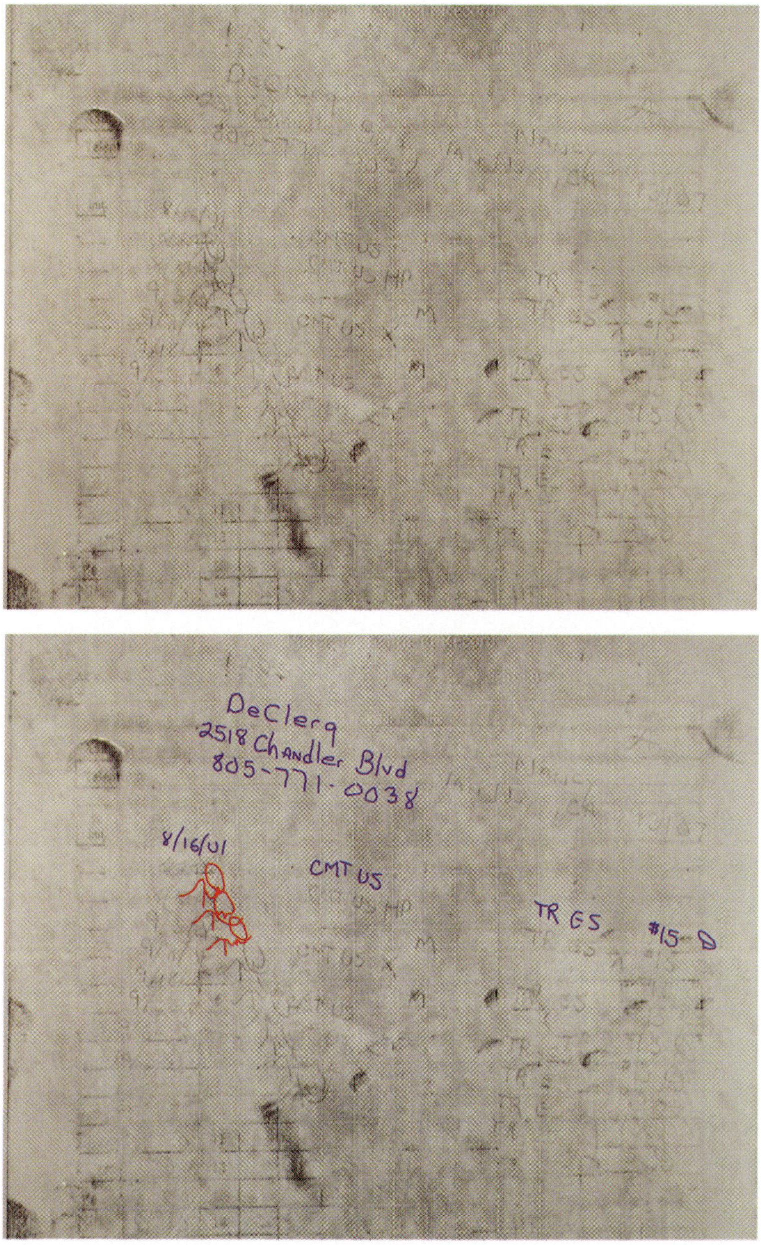

Figure 34.7(b) The electrostatic detection apparatus (ESDA) lift image at the top shows the indentations developed from the lower page in Figure 34.7a. These impressions originate from the writing on the second sheet. In the lower image, some of the indentations have been colored to highlight their content and location. Two registration groups were identified. One includes all of the doctor's entries, examples of which are highlighted in blue. A second group, partially highlighted in red, consists of all of the patient's initials and is found to be indented at a different angle relative to the other set of indentations. Only a select number of the impressions have been highlighted, so as to allow the reader an unhindered view of the remainder. Evidence of the alignment groups demonstrates that the entries were not done progressively over the dates recorded, but rather each group was executed in a single session.

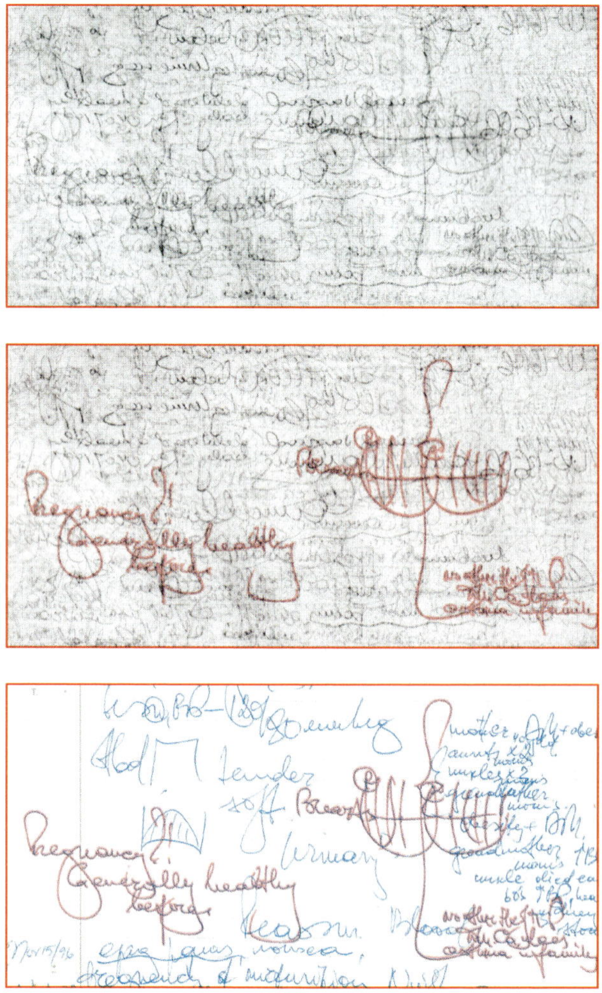

a)

b)

c)

Figure 34.8 The top image (a) shows a portion of an indented impression lift taken from a page in a medical file. Because the image contains multiple overlapping indented impressions, a second, highlighted version (b) is also provided in order to focus the viewer on the relevant content. Finally, the document bearing the source handwriting is shown with the corresponding entries, also highlighted in red (c). In this case, the date of the source handwriting significantly preceded that of the document bearing the indented impressions and provided proof of back-dating. Once again, the raw image of the impression transparency was presented first. The use of color in the final image helps to distinguish the entries on the original document from indented impressions seen in other panels of the chart.

Conclusions and Guidelines 32

CARL R. MCCLARY

In 1997, a group of government and private document examiners, realizing a need for standard guidelines for document examinations, formed the Technical Working Group for Document Examiners (TWGDOC). Later to be called the Scientific Working Group for Document Examiners (SWGDOC), this group of examiners was initially composed of representatives of numerous federal agencies and was one of the first TWGs to be formed. Until 1999, SWGDOC met at various locations in the Washington, D.C. area. SWGDOC then joined other Scientific Working Groups in other forensic disciplines at the FBI Academy in Quantico, VA, where meeting space was offered.

After 1997, the attendees were expanded to include examiners from the Coalition of Private Practice Examiners (COPPE), various state forensic laboratory representatives, and national and regional organizational representatives. Membership of SWGDOC then evolved, in 1999, from representatives of organizations and agencies to individuals who had particular experience or expertise in the examination standard guide or terminology being drafted at the time. The group meets, usually on a biannual basis, to draft standard guides for different types of document examinations and terminologies. These standardized guides serve the purpose of not only providing nationally recognized consensus methods of forensic document examination, but also satisfying admissibility, reliability, and *Daubert* criteria set forth by the federal court system.

Although standards produced by SWGDOC were considered consensus standards, a need to place them in a recognized and respected publication was realized. There already existed four standards on document examination published by the American Society for Testing and Materials International® (ASTM): E1658, *Standard Terminology for Expressing Conclusions of Forensic Document Examiners**; E444, *Standard Descriptions of Scope of Work Relating to Forensic Document Examiners**; E1789, *Standard Guide for Writing Ink Identification**; and E1422, *Standard Guide for Test Methods for Forensic Writing Ink Comparison**. Thus, ASTM was chosen as the publisher of documents drafted by SWGDOC. Additional

*Reprinted with permission. Copyright ASTM International, 100 Barr Harbor Drive, West Conshohocken, PA, 19428.

factors in this choice were the critical considerations that are given to votes cast by members, review by other professionals in the forensic arena, and mandatory 5-year revisions.

ASTM is one of the largest standards development organizations in the world. Having been in existence for over 100 years, its members come from private manufacturing, government, and academia backgrounds to draft and publish voluntary consensus standards. The Forensics Committee, E30, is one of the newer committees and has published numerous standard guides, terminologies, and practices in its short history. E30 is comprised of various subcommittees, including Interdisciplinary Forensic Science Standards, Criminalistics, Odontology, and Psychiatry and Behavioral Science. ASTM procedures provide for the drafting of standards, usually by task groups; the balloting of those drafts to a subcommittee that then considers the votes of those ballots at an annual meeting; a balloting of the draft document to all the members of a main committee; and a final consideration and vote by the subcommittee and main committees given at the next meeting. Approval for publication is then sought from the Committee on Standards.

The E30.02 standards published to date are:

E444 *Standard Descriptions Relating to the Scope of Work of Forensic Document Examiners**
E1422 *Standard Guide for Test Methods for Forensic Writing Ink Comparison**
E1658 *Standard Terminology for Expressing Conclusions of Forensic Document Examiners**
E1789 *Standard Guide for Writing Ink Identification**
E2195 *Standard Terminology Relating to the Examination of Questioned Documents**
E2285 *Standard Guide for Examination of Mechanical Checkwriter Impressions**
E2286 *Standard Guide for Examination of Dry Seal Impressions**
E2287 *Standard Guide for Examination of Fracture Patterns and Paper Fiber Impressions on Single-Strike Film Ribbons and Typed Text**
E2288 *Standard Guide for Physical Match of Paper Cuts, Tears, and Perforations in Forensic Document Examinations**
E2289 *Standard Guide for Examination of Rubber Stamp Impressions**
E2290 *Standard Guide for the Examination of Handwritten Items**
E2291 *Standard Guide for Indentation Examinations**
E2325 *Standard Guide for the Non-Destructive Examination of Paper**
E2331 *Standard Guide for Examination of Altered Documents**
E2388 *Standard Guide for the Minimum Training Requirements for Forensic Document Examiners**

Standardization in the discipline has long been a topic of discussion and the passing of ASTM E1658 and E444 were vanguard accomplishments in the field. Considering suggested criteria for U.S. federal court admissibility of expert testimony pursuant to the 1993 *Daubert v. Merrell Dow Pharmaceuticals, Inc.*, decision, the reliability and admissibility of qualified document examiners' testimony is significantly bolstered by the achievements of SWGDOC and ASTM. The *Daubert* decision specifically addresses standards as follows: "the court should ordinarily consider the known or potential rate of error, … and the existence and maintenance of standards controlling the technique's operation."[1] The consensus standards now available most certainly help to satisfy this consideration.

*Reprinted with permission. Copyright ASTM International, 100 Barr Harbor Drive, West Conshohocken, PA, 19428.

Although ASTM standards are being developed and used primarily in the U.S., the European community also has developed several standard procedures through the European Document Experts Working Group (EDEWG) and the European Network of Forensic Handwriting Experts (ENFHEX). EDEWG was formed in 1998 and is a European Network of Forensic Science Institutes (ENFSI) working group. EDEWG standards include *Examination of Alterations*, *Examination of Intersecting Lines*, and *Examination of Indented Impressions*, among others. The format of these standards are much the same as that used by ASTM and include sections, such as Glossary, Health and Safety, Equipment/Instrumentation/Operating Conditions, and Recording and Interpretation of Results. In addition to organizing a European conference every two years, including workshops, EDEWG also develops quality assurance trials and joint research and development.

ENFHEX was established in 1996 and is also an ENFSI working group. Its goals include building communication, organizing workshops, promoting expertise, and developing and revising methods and procedures.

Like the European community, examiners in Australia and New Zealand have developed methods and terminologies through the Australian and New Zealand Document Examination Special Advisory Group (DocSAG). One of six forensic special advisory groups, DocSAG is the combined Criminalistics and Document Examination Special Advisory Group. These groups participate in research, workshop organization, assessment of quality issues, and the promotion and exchange of knowledge. They also provide advice on these topics to the Senior Managers of Australian and New Zealand Forensic Laboratories (SMANZFL).

References

1. U.S. Supreme Court ruling, *Daubert v. Merrell Dow Pharmaceuticals*, 509 U.S. 579, 1993.

Section XII

Preparing for Court Testimony

Consideration in Preparing for Court

33

FRANK HICKS

The culmination of any document examiner's work on a case can be the presentation of findings at a deposition, at a preliminary hearing, in a courtroom, or before a quasi-judicial board. This requires a thorough review of all the documents submitted for examination, plus any paperwork generated during the actual examination and the preparation of the report. A review ensures that the FDE is prepared to suitably explain his or her work, the nature of the testing conducted, and the bases for the opinions to be offered.

While there are no carved-in-stone rules concerning how to conduct a case review in preparation for court, there are some logical steps that should be taken. First, a review of the initial contact on the case should be conducted. This may have been a phone call, letter, or e-mail from the submitting agent. The date of this initial contact can sometimes become important during the trial for reasons that have nothing to do with the FDE's examination. This contact information also should include the nature of the original request made of the document examiner and, in civil matters, the results of any conflict check.

Next, a review of the evidence submitted and the examinations conducted on that evidence should be completed. If the initial examination was made many months prior, it may be necessary to undertake at least a brief reexamination of the evidence to be comfortable with the material. If instrumental examinations were performed, a review of these examinations and their results should be made. The document examiner should review the bases behind the instrumental techniques used and how to properly explain these techniques and the results to laypersons. Excellent work can all go for naught if the FDE is unable to explain the examinations in such a manner that the audience (judge, jury, hearing officers, etc.) can understand. When testifying to a technique that is not a frequent topic of testimony for the document examiner, some extensive review of the technique is called for. The witness stand is a poor place to realize that you cannot remember why or how an instrumental technique works.

Since every detail of the FDE's report can be a source for questions during his or her testimony, any reports issued in the case should be reviewed in detail. All observations reported, plus the opinions expressed, should be very familiar to the document examiner so he or she can quickly and competently answer any questions posed about them.

This is also a good time to organize all of the papers in the case folder. To avoid having to shuffle through many sheets of paper on the witness stand, place like documents together in the folder. For instance, paper clipping together all letters from the submitter makes them easier to find if needed. The same is true of all the notes prepared by the document examiner during the examination. Finding specific types of documents in the folder can be made easier if the FDE uses colored paper for notes, instead of white paper. For instance, bench notes could be in one color, telephone records in another, and so forth. Another approach is to put the documents into a three-ring binder with dividers between different categories of documents. This has a somewhat more professional appearance than a file folder full of loose sheets of paper and avoids the embarrassment of dropping those sheets on the floor.

Finally, if it has not already been done as a part of the report preparation, the case should be evaluated to determine if a demonstrative exhibit, like a chart or an electronic presentation, should be made to help the court understand the examiner's opinion. If the FDE decides to proceed with the preparation of this demonstrative exhibit, it should be prepared prior to the pretrial conference so that it can be reviewed by the attorney.

Demonstrative Charts 34

FRANK HICKS
BRIAN S. LINDBLOM
ROBERT GERVAIS

Contents

Although the topics typically covered by forensic document examinations deal with subjects familiar to most people, the specific factors involved in the opinions expressed by the FDE can be difficult to convey without some type of demonstrative charts. These charts find a place in both reports and courtroom presentations.

A well-prepared and properly labeled chart can make a document report much easier for the recipient to comprehend. In some instances, the courts mandate what must be included in forensic reports. For example, U.S. federal courts require that an expert's findings cannot simply state the ultimate opinions, but rather must include all the details that form their basis. A good chart will ensure that the relevant characteristics are clearly illustrated and understood.

Typically, charts that accompany a report will be photographic in nature. For many decades, these illustrations were prepared with traditional wet photographic techniques. After preparing photographs of the questioned and known writings, specific words or letters would be physically cut from these photographs, arranged in the necessary format, and then photographed again. This same procedure can still be used today, but modern technology allows other approaches to be used. Much chart composition is now done with the use of a computer and digital photo processing software. There are several methods to input the images into the computer, including digital cameras and flatbed scanners. It is also possible to demonstrate the findings using electronic slides and a computer projector.

34.1 Choosing the Presentation Method

The first step in chart preparation is to choose the presentation method that will be used. This can take the form of printed charts, overhead projector transparencies, 35-mm slides, or computer presentation programs and videos. The classic 35-mm slides have effectively been replaced by the use of electronic computer presentation methods. There are some factors that will help the FDE to decide which method will be the most effective:

- What needs to be demonstrated? If the issue is a photocopy manipulation and the source document is available, then transparencies and an overhead projector may be the simplest presentation method. If you need to show the court how an instrumental technique works, a video of the equipment as it is being used is very helpful.
- What is the courtroom layout like? If you want to use video, is there enough space in the courtroom to do this? If this cannot be determined ahead of time, it is best to either plan on another method or have a backup plan available.
- How much preparation time is available? No matter how effective a photographic chart might be, it does require some time to prepare. If there is not adequate time, then an alternative must be considered.
- To whom is the presentation to be made? A one-on-one presentation to a judge or hearing officer may call for one form of presentation, while another may be better when presenting the material to a jury.
- What equipment and skills are available to the document examiner? While a video may be desirable, the FDE may lack the required equipment or skills, making it necessary to contract out this assignment. A computer presentation method requires a computer, digital projector, and screen. The court may not be equipped with the necessary projector and screen, making it necessary to rent and arrange setup prior to testifying.

34.2 Choosing the Content

The next step in the chart preparation process is selecting the content to be included. There are two approaches to chart content. One that was used in the early years of the profession involved incorporating a large number of images in a single chart. This style of chart provided a great deal of information, but also made it difficult for the judge and jury to focus their attention on specific characteristics, given the amount of detail present. A second style, which is more in favor today, involves simpler illustrations with less detail. In some cases, multiple charts may be used, each clearly demonstrating one specific point or a small number of characteristics.

In a signature case, the obvious decision is to include the questioned signature as one of the images on the chart. If there is more than one questioned signature, a decision has to be made about which signature to include if all of them are not to be presented. Where only a sampling can be included, it is important to select those that are the most legible and, at the same time, representative of the group. This can be important in cases where only copies of the documents were submitted. Some copies may be clearer than others. The FDE should be sure that any questioned signature selected for the chart is from a document that is actually going to be a part of the trial. Just because six questioned documents were submitted to the document examiner does not necessarily mean that all six are going to be at issue in the trial. This is one of the issues that should be discussed in the pretrial conference.

When selecting the known signatures to include in the chart, it is wise to ensure that the samples you anticipate using will, in fact, be introduced in the trial as genuine signatures. A chart containing a signature that has not been accepted into evidence will not be allowed. A portion of the known signatures to be included should be those that are closest in date to the questioned signature. These will reflect the writer's habits on or about the date at issue. It is sometimes necessary to display a pattern of writing deterioration over a course of months or perhaps years. This may require a separate chart in which the genuine signatures are presented chronologically. In some cases it may be discovered that the questioned signature displays a specific, unusual characteristic infrequently found in the known signatures. In such a case it is good practice to include at least one such example.

If the issue involves a typewriter or computer font, examples of the appropriate letters must be chosen, and it may be necessary to enlarge them considerably to show the relevant defect or characteristic. Likewise, it may be necessary to enlarge "trash marks" from photocopiers and computer printers. Often, the location of the trash mark relative to a page edge, a pre-printed form element, or a segment of text is important in the formation of the opinion, and care must be taken to capture the surrounding area in any image placed into a chart.

34.3 Selecting the Input Device

If conventional film photography is being used to create the chart, a two-step process will likely be involved. The first is to photograph the pertinent areas of the document. These images are then cut from the prints and mounted on chart board. This process involves properly aligning the individual clips with both the edges and centerline of the chart, as well as properly reflecting their orientation on the actual document examined. The second step requires photographing the chart and enlarging it to the desirable size. A large-format camera is preferable for this method, since less enlarging will be required of the final image.

The majority of illustrative charts today are produced through the use of a computer. Typically, one of two input devices is used to capture an image and import it into a computer: flatbed scanners and digital cameras. Both of these devices are quite capable of capturing images at resolutions suitable for the construction of charts. In addition, both of these methods produce electronic files that can be reviewed and evaluated almost instantly.

Flatbed scanners offer one advantage — the certainty that the document being imaged is on a flat plane. The weight of the scanner's lid on the document ensures that it is flattened and that the effect of folds and creases is minimized. The light source and optics in the scanner work on the same plane as the document platen. Most scanners offer the ability to adjust the resolution of the scan, with settings of 150, 300, and 600 dots per inch (dpi), and so forth, being typical. A decision must be made prior to the scanning about the resolution necessary. This is important for two reasons. First, if the image is to be significantly enlarged, either for inclusion on a poster-size chart or to demonstrate microfeatures that may require considerable enlargement, a higher resolution may be necessary in order to limit the pixilation. Second, higher-resolution images result in larger file sizes in the computer. Utilizing a higher-than-necessary resolution will result in wasted space in the computer's memory and also will slow any processing done to the image in the computer. For charts that will be enlarged to no more than handheld size, a resolution of 300 dpi is almost always sufficient. This is the recommendation of the Scientific Working Group for Imaging Technology.

In the past, carrying a flatbed scanner into the field to image documents was at best unwieldy. Newer scanners and notebook computers have removed most of the limitations in this area. The scanner and computer can now be placed in one computer bag and transported with ease.

Digital cameras have the advantage of being even more portable than the best flatbed scanners. They also do not require any outside power source, which can be a problem in some locations. A camera may be more useful than a scanner in imaging certain bulky documents, like a page in a book or a document that cannot be removed from a medical file. Digital cameras can be used either with or without a camera stand. If handheld, there is the potential problem of not having the lens parallel to the document being photographed. Evidence of this will be two straight, parallel lines from the document that are no longer parallel in the resulting image. Mounting the camera on a camera stand may alleviate this problem, but negatively impacts on the portability of the camera. A further potential problem of the digital camera is the need for proper lighting. Most cameras have built-in flash units that are certainly adequate at the distances for most document photography, but these flash units are usually not directly adjacent to the lens. This will result in uneven lighting, particularly if the camera is very close to the document being photographed. The lights available on many camera stands may overcome this uneven lighting problem. Also, macro ring lights, which surround the camera lens itself, are useful for macro-photography.

Whether using conventional or digital photography, when possible the document should be flattened using either a glass panel or a vacuum box. Where the former is used, proper consideration must be given to lighting to ensure that reflection from the glass is not captured in the image.

Finally, before any photography is done, a decision needs to be made concerning whether or not to include a ruler or some other scale so that the degree of enlargement can be calculated. This is certainly necessary in instances in which the examination is comparative in nature, such as those involving handwriting, computer printer and typewriter impressions, trash marks, counterfeits, and cut-and-paste manipulations. As a general practice, it is not advisable to enlarge individual components of a chart to different percentages, as this may give the reader the false impression that the actual characters vary in size when that may not be the case. Where it is necessary to employ differing degrees of enlargement, this should be clearly indicated in the chart by stating the percent enlargement for each item.

34.4 Processing the Images

It is sometimes necessary to process an image either before or after it is captured. Different computer software packages offer various methods for editing images, but there are some general techniques that are fairly universal. Two basic procedures are cropping and rotating. Cropping allows areas of the captured image that are of interest to be cut from the larger image. This does away with the surrounding portions of the document that serve no purpose in illustrating the findings. Rotating the image allows the baseline of the image to be adjusted so that it reflects the actual orientation on the document from which it was captured.

Somewhat more advanced adjustments include brightness, contrast, levels, and curves. The brightness and contrast commands can be applied selectively or uniformly over the entire image. In some situations these commands may result in an excessively harsh change to the image; however, there are circumstances that require just such an adjustment to be made. More versatile changes can be made to the brightness and contrast of an image by using the levels and curves commands. An argument can be made that by utilizing these commands the operator is changing the appearance of the image and, thus, altering the evidence. However, changes with similar effect can, and have been, achieved through film processing and development techniques, the results of which have been entered as exhibits in thousands of testimonies. Furthermore, it is certainly possible that due to lighting or camera capabilities, the image captured does not accurately represent the original image. When applying any enhancement to an image, it is important that changes be made to a sufficiently large area so that a minute portion or section of an image is not given undo emphasis, resulting in an end product that no longer correctly depicts the original information. Of course, all adjustments made to the image should be recorded so that the FDE can answer questions about what was done. Some of the imaging software now available allow for the careful tracking of all actions as they occur. Specialized software, designed to track changes, is available as well.

For many decades it has been the convention to use gray-scale images in charts. The justification for this is twofold. First, gray scale allows for the capture and representation of very subtle differences in shades. Second, the viewer is not distracted by the many colors that may be present in a given image. The FDE must always be cognizant of the purpose of the illustration and consider whether color aids in meeting this purpose. For instance, in a signature chart, color would add little to appreciating the similarities and differences in letter structures. Furthermore, color images may sometimes fail to accurately reflect the subtleties in hues and shades that are present in the original document and may lead to challenges that the chart is not an acceptable representation of the original. For this reason, unless color is necessary, it is frequently a good idea to print gray-scale images and add color through the labeling placed on the chart. Color Figure 34.1 (following p. 366) provides an example of the effective use of gray scale and color labeling.

34.5 Formatting and Labeling the Chart

During the FDE's testimony, the opportunity is presented to educate and inform the judge and jury. Through the use of effective, well-designed charts the audience's attention can be directed to specific areas of importance as they relate to the opinion being expressed. Certainly it is possible to give testimony without the use of charts by verbally describing the relevant characteristics, but this can be both taxing and tedious for the audience and the FDE. There is potential for misunderstandings on the part of the audience and there is a risk that the judge and jury will lose interest. Our society has become accustomed to visual displays in all walks of life. A properly structured and labeled chart can make the observed facts more apparent and easily understood by the viewers.

The format chosen will depend on a number of factors, not the least of which is the method of presentation. Testimony given with the aid of a computer, presentation software, and a projector will allow much more flexibility than one given with photographic prints. However, this flexibility can be exploited to such an extent that it overshadows the purpose of the display, which is to illustrate to the court the basis for the opinion(s) rendered. Gimmicks such as sound effects, unnecessary animations, and flashing lights should be avoided in the legal setting, as they add nothing to the facts and just detract from the point being made. Any text used with slides in this type of presentation should be large enough to be read by all parties. If the room size is not known prior to the creation of the slides, always use larger fonts than you think may be necessary. As discussed earlier, simpler, less cluttered slides are preferable, as they are easier for the audience to comprehend.

Another consideration when constructing photographic or electronic charts is whether to prepare the chart in landscape or portrait mode. The nature of the images to be included will be a contributing factor, as will the screen's dimensions when using projected slides. The landscape mode obviously provides more lateral spacing for putting images in columns.

If a chart was included with the report, it is often satisfactory to use copies of the same illustration in the courtroom. Multiple copies may be made and distributed. This has the advantage of providing the judge or jury members with an illustration they can refer to later during deliberation. To complement the report chart, an enlarged version can be made and mounted on an easel so that the FDE can point to the characteristics more readily during testimony.

With a signature chart, various approaches exist in arranging the images. One is to place a questioned signature at the top of the chart and several known signatures below. Another technique is to position the questioned signature in the middle of the chart with known signatures both above and below. This latter method is useful in placing the questioned signature in the chronological midst of known signatures, which may be an important part of the examination process.

Modern image editing software allows the user to label a chart in many useful ways. Keep in mind that the chart's purpose is to be used as a teaching aid in explaining the basis of the opinion(s) rendered. If the viewer does not know where the images originate, it will be difficult to follow the testimony. Thus, every image should be clearly labeled with its source and, in most instances, the date of the document.

Color Figure 34.2 to color Figure 34.8 (following p. 366), Figure 34.6(a), and Figure 34.7(a) are examples of chart styles used to illustrate a variety of subject matter. In some cases images have been colorized to emphasize similarities between different elements. In others, arrows, circles, and other highlighting methods are employed to emphasize important characteristics. Some use a series of charts to present the relevant details.

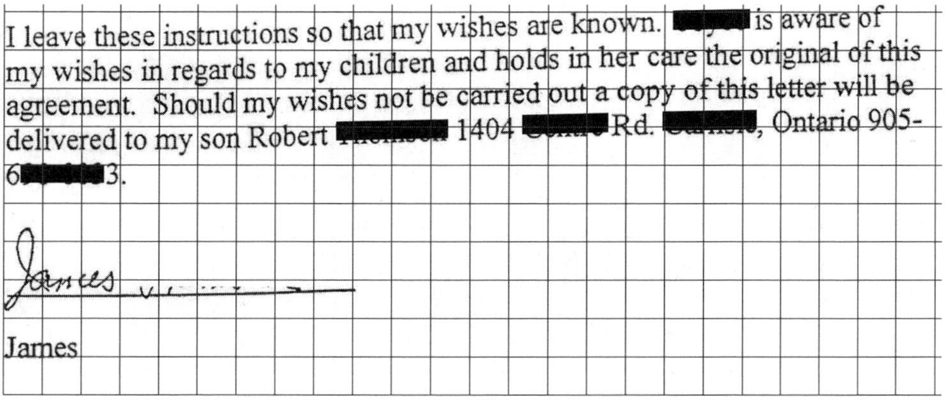

I leave these instructions so that my wishes are known. [redacted] is aware of my wishes in regards to my children and holds in her care the original of this agreement. Should my wishes not be carried out a copy of this letter will be delivered to my son Robert [redacted] 1404 [redacted] Rd. [redacted], Ontario 905-6[redacted]3.

James

Figure 34.6 (a) In a case involving the authenticity of a poor quality photocopy, a two-part chart has been used to demonstrate copying distortion and alignment considerations. In chart a, an electronic grid is overlaid on a portion of the text. It serves to highlight the non-linear nature of the distorted image.

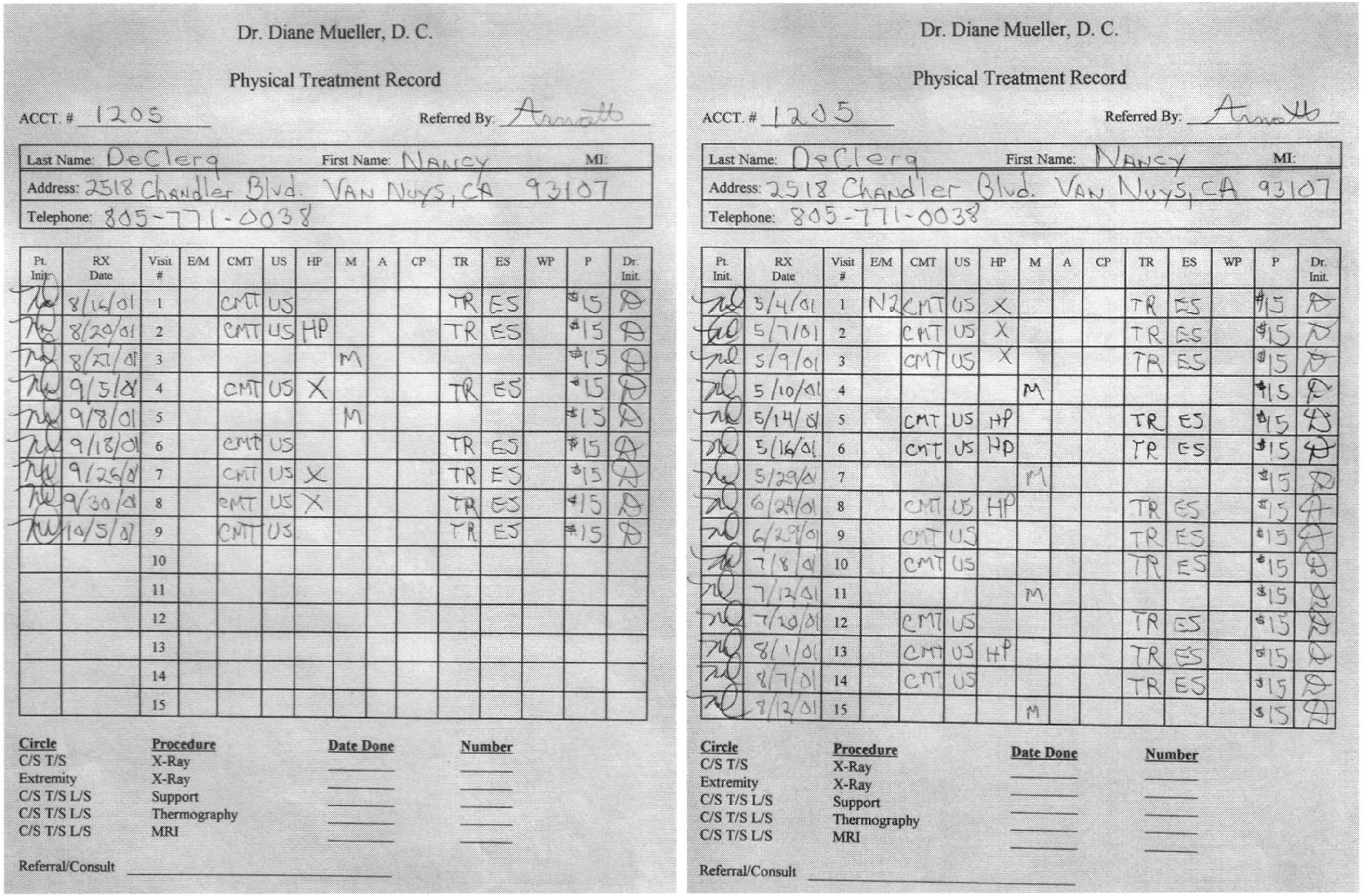

Dr. Diane Mueller, D. C.

Physical Treatment Record

ACCT. # 1205 Referred By: Arnott

Last Name: DeClerq First Name: Nancy MI:

Address: 2518 Chandler Blvd. Van Nuys, CA 93107

Telephone: 805-771-0038

Pt. Init.	RX Date	Visit #	E/M	CMT	US	HP	M	A	CP	TR	ES	WP	P	Dr. Init.
	8/16/01	1		CMT	US					TR	ES		$15	
	8/20/01	2		CMT	US	HP				TR	ES		$15	
	8/22/01	3					M						$15	
	9/5/01	4		CMT	US	X				TR	ES		$15	
	9/8/01	5					M						$15	
	9/18/01	6		CMT	US					TR	ES		$15	
	9/26/01	7		CMT	US	X				TR	ES		$15	
	9/30/01	8		CMT	US	X				TR	ES		$15	
	10/5/01	9		CMT	US					TR	ES		$15	
		10												
		11												
		12												
		13												
		14												
		15												

Circle	Procedure	Date Done	Number
C/S T/S	X-Ray		
Extremity	X-Ray		
C/S T/S L/S	Support		
C/S T/S L/S	Thermography		
C/S T/S L/S	MRI		

Referral/Consult

Dr. Diane Mueller, D. C.

Physical Treatment Record

ACCT. # 1205 Referred By: Arnott

Last Name: DeClerq First Name: Nancy MI:

Address: 2518 Chandler Blvd. Van Nuys, CA 93107

Telephone: 805-771-0038

Pt. Init.	RX Date	Visit #	E/M	CMT	US	HP	M	A	CP	TR	ES	WP	P	Dr. Init.
	5/4/01	1	N2	CMT	US	X				TR	ES		$15	
	5/7/01	2		CMT	US	X				TR	ES		$15	
	5/9/01	3		CMT	US	X				TR	ES		$15	
	5/10/01	4					M						$15	
	5/14/01	5		CMT	US	HP				TR	ES		$15	
	5/16/01	6		CMT	US	HP				TR	ES		$15	
	5/29/01	7					M						$15	
	6/24/01	8		CMT	US	HP				TR	ES		$15	
	6/29/01	9		CMT	US					TR	ES		$15	
	7/8/01	10		CMT	US					TR	ES		$15	
	7/12/01	11					M						$15	
	7/20/01	12		CMT	US					TR	ES		$15	
	8/1/01	13		CMT	US	HP				TR	ES		$15	
	8/7/01	14		CMT	US					TR	ES		$15	
	8/12/01	15					M						$15	

Circle	Procedure	Date Done	Number
C/S T/S	X-Ray		
Extremity	X-Ray		
C/S T/S L/S	Support		
C/S T/S L/S	Thermography		
C/S T/S L/S	MRI		

Referral/Consult

Figure 34.7(a) This two-part chart is used to present findings in a medical case. The documents at issue are two pages of physical treatment records. The left form contains the most contemporary date entries and was positioned on top of the right form that has earlier dated entries. The writing from the left page should be indented into the underlying page.

35 Pretrial Conference

FRANK HICKS

A pretrial conference between the document examiner and the attorney is an excellent way to prepare both parties for the upcoming testimony. In more complex cases, this conference can mean the difference between a powerful, effective testimony and a disjointed, weak testimony. In most cases, the retaining counsel will have had few, if any, experiences with an FDE and will welcome any suggestions that he or she may have.

The document examiner should present information about the documents he or she examined and the opinions expressed. If there is more than one opinion expressed by the FDE, the attorney may decide to present only some of the opinions in the trial. The FDE can discuss this with the attorney and cover any difficulties this may cause. If the opinion(s) to be offered in court is limited in its strength, the factors that result in this limitation should be thoroughly discussed. This will give the attorney the opportunity to prepare direct questions to explain the nature of these restrictions.

Since most trial attorneys have had little or no experience with a document examiner in the courtroom, it is entirely proper to offer a list of questions that he or she may use to qualify the witness. Sources for these types of questions are available in various publications; however, it should be stressed that the questions being offered by the FDE are tailored to his or her specific qualifications. It may put the attorney's mind at rest somewhat if the document examiner delivers his or her answers to each of these qualifying questions during the pretrial conference. This should assure both parties that at least the initial phase of the testimony will go smoothly.

If there are any issues about the FDE's past testimonies, these should be disclosed to the attorney. This way, counsel can prepare an effective strategy in which to respond to these issues if they are brought up while the document examiner is testifying. Attorneys do not appreciate being surprised by such information while their own witness is undergoing cross-examination.

If the document examiner's testimony is to include very technical issues, these should be discussed in detail with counsel. The FDE can suggest questions that will elicit concise answers that will make the testimony clearer to the court. Both parties should agree that the goal is to present the facts in a manner that the court can easily understand.

The use of visual aids during the FDE's testimony can be a very powerful tool. This too needs to be considered, with specific regard to the amount of detail to be presented, the order in which the charts will be used, and the method of delivery — poster, video, PowerPoint, or slides. The advantages/disadvantages of each should be outlined. The logistics for introducing the photographs and making the presentation need to be determined. The types of illustrations and methods for preparing them are discussed in detail in Chapter 34.

The use of computer-aided presentations is becoming more prevalent in courtrooms today. Many modern courtrooms are being designed with this in mind. For other courtrooms, due to either their size or their physical layout, this style of presentation is not only awkward, but may be impossible. If the document examiner is contemplating using such a visual aid, it is wise to ascertain any limitations the courtroom may impose.

If the opposing side of the case plans to introduce the testimony of someone on the same topic about which the FDE is testifying, the pretrial conference is the time to address this person's qualifications. This same proffered expert's curriculum vitae should be obtained so that any deficiencies in training or background can be discussed. If the witness's training does not meet the requirements of the mainstream international FDE organizations, such as the American Society of Questioned Document Examiners (ASQDE) and the American Board of Forensic Document Examiners (ABFDE), a copy of these requirements can be furnished to the attorney to use in cross-examination. If the responding witness's report is available, this should be reviewed in detail. Any instances of failures to follow accepted methodology or conclusions that are not supported by the evidence must be noted and an appropriate strategy of cross-examination devised. Chapter 37 provides numerous recommendations for dealing with these types of witnesses.

Preparation for a *Daubert* Hearing

36

JAN SEAMAN KELLY
KIRSTEN JACKSON

Contents

Daubert challenges have become a recurring phenomenon in U.S. courts since the Supreme Court's seminal decision in 1993, and forensic document examination has been one of the first and most active of the forensic sciences to face such challenges. Prior to the *Daubert* decision, testimony of the FDE typically consisted of a brief description of professional credentials, followed by an explanation of the examination conducted and conclusions. Post-*Daubert* testimony requires this and now often a great deal more. FDEs, like all experts, must now be prepared to answer questions regarding the science of their expertise.

36.1 *Daubert* Factors

In the *Daubert* decision, the Supreme Court provided criteria for the determination of the reliability of an expertise. The criteria, which have come to be known as the *Daubert* factors, are

- General acceptance
- Whether the theory or technique can be or has been tested
- Whether the technique has been subject to peer review and publication
- The existence and maintenance of standards controlling the technique's operation
- The known or potential error rate

36.1.1 How the Forensic Document Examination Profession Meets Each Factor

Even though the Supreme Court wrote in its decision that a science did not have to meet all five criteria, explanation of how the profession satisfactorily meets each factor has been the approach taken by the forensic document community. Testimony and documentation submitted to the court are the only sources of information the presiding judge can consider. Therefore, the FDE should be prepared to address each factor thoroughly.

The Supreme Court wrote that "general acceptance" refers to the acceptance of a technique by the relevant scientific community. The acceptance of the expertise of forensic document examination is evinced through several means: inclusion in professional, multi-discipline forensic science organizations worldwide; presence of the expertise in hundreds of federal, state, local, and international forensic science laboratories; courses and degrees offered in undergraduate and graduate university and college forensic programs; and evaluation of the expertise by accrediting bodies and standards organizations. Specifically, general acceptance of the theory and the field is met by the following:

- The American Academy of Forensic Sciences (AAFS) is a national multi-discipline organization (e.g., criminalistics, pathology/biology, toxicology, etc.) established in 1948. The Questioned Documents section was one of the first disciplines to be part of this national organization.
- The International Association of Identification (IAI), the Mid-Western Association of Forensic Scientists (MAFS), the Mid-Atlantic Association of Forensic Scientists (MAAFS), the Northeastern Association of Forensic Sciences (NEAFS), and the British, Australian, German, and Canadian Forensic Science Societies each have a Questioned Documents section.
- Forensic document examination courses are included in forensic science programs at George Washington University, (Washington, D.C.) Michigan State University, John Jay College (New York City), National University (San Diego), University of Alabama at Birmingham, University of New Haven, University of Central Oklahoma, University of Illinois at Chicago, and Oklahoma State University.
- The American Society for Testing and Materials International (ASTM) has a committee dedicated to forensic science (E30) and a subcommittee dedicated to questioned documents (E30.02).
- Methodology guidelines are published by ASTM. The guidelines are reviewed not only by the forensic document community, but also by scientists in other forensic disciplines (e.g., physical anthropology, engineering, pathology/biology, criminalistics, etc.) prior to their publication.
- The American Society of Crime Laboratory Directors–Laboratory Accreditation Board (ASCLD-LAB) evaluates forensic science laboratories to determine whether each discipline meets specific standards. Questioned documents is one of the disciplines evaluated.
- Collaborative Testing Services (CTS) provides proficiency tests to various forensic disciplines, including forensic document examination.

The second *Daubert* factor is whether the technique has been or can be tested. The basic premises of forensic handwriting examination are (1) no two writers share the same combination of handwriting characteristics, given a sufficient quantity of writing, and (2) a writer cannot duplicate the same letterforms exactly the same way twice (variation), but will continue to follow habitual patterns of movement when writing. These basic premises have been supported through the following:

- Publications and paper presentations relevant to the specific forensic document examination task that has been brought before the court. For example, to satisfy this second *Daubert* factor, testimony regarding the task of handwriting examination should include, but not be limited to, the listing of published articles on handwriting of twins, research on class characteristics, and published guidelines for examination methodology.
- The Forensic Information System for Handwriting (FISH) is a database maintained by the U.S. Secret Service Forensic Laboratory since 1991. The German Federal Police developed FISH to assist in investigations of crimes involving handwritten documents. The U.S. Secret Service database contains handwriting samples from more than 10,000 different individuals, and the German database contains handwriting samples of more than 100,000 individual writers. No two individuals have been found to have the same combination of handwriting characteristics in either the U.S. or German databases.
- The Center of Excellence for Document Analysis and Recognition (CEDAR) at the State University of New York in Buffalo conducted a study that offers scientific support for one of the basic premises of handwriting identification: that handwriting is individualistic. Published in the *Journal of Forensic Sciences* in 2002, the CEDAR study used computer software to measure selected handwriting features of over 1500 writers. Based on a combination of these measurements, the computer system was able to identify the writer with a 95% confidence level.

The third *Daubert* factor, regarding peer review of the theory or technique, is demonstrated through the numerous peer-reviewed journals containing articles written by FDEs. The purpose of peer review is to ensure that the relevant scientific community has had the opportunity to review the theory to detect any fallacies. Publishing research or methodology guidelines in an engineering journal, for example, would be meaningless, as engineering science is not a relevant scientific community for FDEs. Meetings of professional organizations and societies, research, training seminars, proficiency testing, and publications are all venues for peer review.

Research presentations before the regional, national, and international forensic science organizations, such as the International Association of Forensic Sciences (IAFS), ASQDE, AAFS, MAAFS, MAFS, the Canadian Society of Forensic Science (CSFS), the Southwestern Association of Forensic Document Examiners (SWAFDE), and the Southeastern Association of Forensic Document Examiners (SAFDE) provide one method of peer review. Even though only some of the research presented at the above-listed organizations is published, those papers that remain unpublished are still subjected to peer review by the FDE community.

Peer review also is achieved through publication in peer review journals, such as the *Journal of Forensic Sciences*, *Science and Justice*, the *Journal of Forensic Identification*, the *Journal of the American Society of Questioned Document Examiners*, the *International Journal of Forensic Document Examiners*, the *Canadian Society of Forensic Science Journal*, *Forensic*

Science International, the *Journal of Police Science and Administration*, and *the Journal of Criminal Law and Criminology.*

The peer review process in forensic document examination is ongoing. An FDE who is researching a subject for either a research project or casework can locate copies of the published and unpublished works at the libraries of professional organizations, such as ASQDE and SWAFDE. Published articles can also be ordered through a local library. In addition to the local library, two sources of reference databases are available to the FDE. ABFDE lists 475 published articles (by category) in its syllabus. The Questioned Document Reference Article Collection Index (QDRAC) has, as of 1999, over 7393 articles.

Peer review may include the evaluation of an individual's technical proficiency. ABFDE is an independent certification body in North America. Its certification is a form of peer review. In order to become an ABFDE diplomate, the test candidate must successfully complete the three-tier certification test. Once certified, the diplomate must meet recertification requirements every 5 years. Recertification is achieved through continuing education and presenting research to the FDE community. The certification process and testing standards of the ABFDE are recognized by several forensic societies in both the U.S. and Canada.

A fourth *Daubert* factor involves the existence and maintenance of standards controlling the technique's operation. The voluntary, consensus standards organization, ASTM, has established a subcommittee devoted to questioned documents (E30.02), including a main committee for the forensic sciences (E30). Guidelines published by ASTM provide objective documentation that standards of examination methodology are in place. Publication in ASTM ensures that the guideline reflects the accepted methodology that has been approved by both the FDE and other forensic science communities. Published standards now exist for the following:

- ASTM E2290 *Standard Guide for Examination of Handwritten Items**
- ASTM E1658 *Standard Terminology for Expressing Conclusions of Forensic Document Examiners**
- ASTM E444 *Standard Guide for Scope Relating to Forensic Document Examiners**
- ASTM E1422 *Standard Guide for Test Methods for Forensic Writing Ink Comparison**
- ASTM E1789 *Standard Guide for Writing Ink Identification**
- ASTM E1732 *Standard Terminology Relating to Forensic Science**
- ASTM E1492 *Standard Practice for Receiving, Documenting, Storing, and Retrieving Evidence in a Forensic Science Laboratory**
- ASTM E2195 *Terminology Relating to the Examination of Questioned Documents**
- ASTM E2388 *Standard Guide for the Minimum Training Requirements for Forensic Document Examiners**

ASCLD-LAB requires written policies regarding examination techniques, case notes, and handling of evidence. The written policy of most law enforcement laboratories on examination techniques parallel the methodologies published in ASTM. In law enforcement agencies, the forensic document section policy manual will typically reference the guidelines published by ASTM as well as published articles. For those laboratories that are

ASCLD-LAB accredited, this provides another avenue to satisfy the *Daubert* criteria of a science as having established standards.

The fifth *Daubert* factor seeks a known or potential error rate for a given scientific technique. FDEs were the first forensic scientists to submit to a battery of blind proficiency tests conducted by a member of the academia to establish a group error rate. Testing began in 1994 and was proctored at FDE organizational seminars/meetings. Tests were given in a group setting, timed, with limited examination equipment. Such controlled conditions ensured consistency in testing, but may have adversely affected the results, as the testing environment did not reflect the laboratory conditions under which FDEs normally conduct examinations. Nevertheless, the error rate of professional document examiners was found to be far less than that of laypersons in handwriting comparison tasks. Testing by Dr. Kam continues today and will be the subject of future publications.

An understanding of the performance percentages by FDEs and laypersons is necessary in order to deliver articulate testimony. The tests are as follows:

Kam 1994 — Kam pilot study[3] (both handwriting and hand printing): This study consisted of 7 FDEs (from the FBI) and 10 laypersons (control group). The results were as follows:
False ID (incorrect identification): FDEs, 0.14285%; laypersons, 8.0%
False elimination (incorrect elimination): FDEs, 0.45857%; laypersons, 16.7%
In this 1994 study, the results indicate that laypersons were 56 times more likely to wrongly associate handwriting than FDEs. Laypersons were also 36 times more likely to wrongly differentiate handwriting than the document examiners.

Kam 1997 — large-scale handwriting study[4] (both handwriting and hand printing): One hundred five FDEs and 41 laypersons (control group) participated. In this larger study, 38.3% of the laypersons and 6.5% of the FDEs falsely identified the writer to a set of writing. This result indicates that the laypersons were six times more likely to identify the wrong writer than FDEs.

Kam 1998 — monetary incentives study: This study was conducted in response to criticisms that of the two groups, FDEs were motivated to perform their best on the tests because they had the greatest risk. According to the criticism, poor test results would be an end to the profession's existence in the legal arena. Kam responded by offering various monetary incentives to laypersons for reaching the correct answer and deducting a set monetary amount for incorrect answers. Variable monetary incentives did not make a statistically significant difference in layperson performance. Layperson performance remained the same as in the two earlier tests.

Kam 2001 — signature study: Sixty-nine FDEs and 50 laypersons (control group) participated in this signature authentication/simulation detection test. Results of this study provided further documentation that laypersons were far more prone to identify the wrong writer than were document examiners. For example, the results demonstrate that laypersons were 3 ½ times more likely than FDEs to misidentify a genuine signature and state that it is a simulation. Laypersons were also about 13 times more likely than document examiners to identify a simulation as a genuine signature.

Kam 2003 — hand printing: (re-analysis of the Kam 1997 study, hand printing portion): The results of 90 FDEs and 34 laypersons (control group) were examined to determine if FDEs outperformed laypersons in the identification of hand printing. The tests were categorized into two sets. The first set, only hand printed documents, consisted

> of documents containing only hand printing and no cursive writing. The second set, only non-hand printed documents, consisted of cursive writing and hand printing that was a mix of cursive and hand printed letters in a word. In the set comprised of only hand printing, laypersons were 4.35% more likely than FDEs to identify the wrong writer to a set of hand printing. Laypersons also were 7 times more likely than FDEs to identify the wrong writer in the set of non-hand printed documents.

Many document examiners participate in proficiency testing programs. CTS is one company that provides proficiency tests to several of the forensic sciences. Participation in a proficiency program provides the FDE with an individual error rate for a specific task.

Each of the five *Daubert* factors represents one link in the reliability chain of a science. Each factor actually reinforces the remaining four. For example, publication of articles in peer-reviewed journals ensures that the relevant community has had the opportunity to scrutinize the work. Articles are published in general forensic science journals as well as journals specifically published for FDEs. Some courts have questioned the publication of articles in FDE-specific journals. The fact the article is published in this type of journal does not lessen its significance. Where else should an FDE publish? How much credence should be given to the results of a drug study being published in *People Magazine* instead of the *New England Journal of Medicine*? *Daubert* specifies peer review should take place in the relevant community in order to provide assurance that the technique being presented before the court has received scrutiny by the professionals who know and understand the methodology of that expertise. Testing of the technique, published standards, and a known error rate join the general acceptance and peer review links to reinforce the intent of *Daubert*: the methodology used in the examination presented before the court is sound and reliable science.

36.1.2 How the Forensic Document Examiner Meets Each Factor

The previous discussion focused on how both the forensic document examination profession and the applied techniques met *Daubert*. Even though *Daubert* addresses the reliability of the technique, the FDE's testimony should include an explanation as to how he or she as an individual meets each of the *Daubert* factors. In establishing the individual's reliability as an expert in forensic document examination, the FDE's testimony will need to be thorough. Professional certification, participation in external or internal proficiency testing, and peer review of casework all provide support to the reliability of the individual document examiner. In a *Daubert* hearing, the FDE must keep in mind that the only facts the presiding judge can consider are those that have been admitted into the court through testimony or as exhibits.

To complete the testimony in a *Daubert* hearing, the FDE's qualifications should be described in detail. Testimony should include a detailed description of training, including formal training as an apprentice with a qualified document examiner, formal education, seminars, and workshops. Mainstream FDEs (also called Osbornians) believe it is crucial for a trainee to complete a minimum 2-year full-time training program with a qualified and experienced examiner. Training based solely on correspondence courses or a self-study literature review of FDE reference books is not an accepted avenue of training. The importance of the formal training program under the tutelage of an experienced FDE must be made known to the presiding judge. If the document examiner's training included research projects, periodic testing, or moot courts, these must be included in the testimony.

The document examiner's testimony should include a listing of the following: memberships in professional organizations, committee service, offices held in the relevant FDE organizations, dates of attendance at professional meetings, and papers and research the FDE has presented. A listing of published articles should also be a part of the testimony. Presentation and publication of research are evidence that the FDE is involved in the field. Attendance at forensic meetings and workshops that include the presentation of papers demonstrates that the document examiner is exposed to both sides (researcher and reviewer) of the peer review process.

The qualification testimony should include information about whether the document examiner is the primary trainer of a trainee or was an instructor in a workshop. Detailed testimony should be given on what is involved in being a primary trainer or an instructor.

Use of published standards needs to be discussed from the perspective of how the document examiner meets this *Daubert* factor. If he or she works in an ASCLD-LAB-accredited forensic laboratory, testimony regarding the policies and procedures that must be adhered to should be given. Evidence handling, case notes, safety, and peer review are established procedures to be followed as a member of an accredited laboratory. By describing these procedures, the FDE is letting the judge know that he or she follows a standard or protocol in every facet of the examination process.

Certification testing is one method of meeting the peer review factor of *Daubert*. There are a few law enforcement agencies that certify their own document examiners. In the U.S., the only recognized independent forensic document examination certification body is the American Board of Forensic Document Examiners. Whether certification stems from an employer who requires in-house certification or the ABFDE, testimony should include a detailed description of the testing process.

36.2 Conclusion

From the content of this chapter, it is obvious that a great deal is required to adequately prepare for a *Daubert* hearing. Even though the Supreme Court wrote in its decision that the expert did not have to meet all five of the factors, there are courts that require the proffered witness to testify how he or she meets each factor. Therefore, testimony and documentation should be presented to the presiding judge describing how the field, the technique employed, and the FDE meet each one of the *Daubert* factors. Preparing a file containing the information necessary to satisfy each *Daubert* factor is a wise investment of time.

References

1. *Daubert v. Merrill Dow Pharmaceuticals, Inc.*, U.S. Supreme Court, 509 U.S. 579, 1993.
2. Srihari, S.N., Cha, S.H., Arora, H., and Lee, S., Individuality of handwriting, *J. Forensic Sci.*, 47, 856–872, 2002.
3. Kam, M., Westein, J., and Conn, R., Proficiency of professional document examiners in writer identification, *J. Forensic Sci.*, 39, 5–14, 1994.

4. Kam, M., Fielding, G., and Conn, R., Writer identification by professional document examiners, *J. Forensic Sci.*, 42, 778–786, 1997.
5. Kam, M., Fielding, G., and Conn, R., The effect of monetary incentives on document examination by non-professionals, *J. Forensic Sci.*, 43, 1000–1004, 1998.
6. Kam, M., Gummadidala, K., Fielding, G., and Conn, R., Signature authentication by forensic document examiners, *J. Forensic Sci.*, 46, 884–888, 2001.
7. Kam, M. and Lin, E., Writer identification using handprinted and non-handprinted questioned documents, *J. Forensic Sci.*, 48 (6), 1391–1395, 2003.

37 The Untrained Document Expert Witness

JAN SEAMAN KELLY

Contents

According to Rule 702 in the Federal Rules of Evidence a witness may testify as an expert if that person possesses knowledge, skill, experience, training, or education on a specific topic.[1] This is a judgment call by the court that does not necessarily follow the standards of qualifications held by the profession's organizations or certifying bodies. Untrained document expert witnesses provide a unique, and at times frustrating, experience for qualified FDEs. This untrained expert witness is not necessarily one who is an advocate for the party who retained him or her, but is one who lacks the requisite training in forensic document examination. Graphologists/graphoanalysts, fraud/forgery detectives, criminalists, and critics of the forensic document profession are considered untrained document experts because they lack the two-year formal training period in forensic document examination. This chapter will discuss how the two groups differ and areas of inquiry during cross-examination.

37.1 Qualified Forensic Document Examiners

As with all professions (including doctors, nurses, and lawyers), early practitioners were self-taught. As the science of forensic document examination evolved, a formalized training program consisting of studying and training under the tutelage of a qualified FDE became an accepted standard; this program has been in place for the past 60 years. Ordway Hilton, as the founding father of the Questioned Documents Section of the American Academy of Forensic Sciences (AAFS), endorsed this standard as it is a requirement for membership in this section. Other forensic document organizations such as the American Society of Questioned Document Examiners (ASQDE), the Southeastern Association of Forensic Document Examiners (SAFDE), the Southwestern Association of Forensic Document Examiners (SWAFDE), and the Questioned Document Section of the Midwest Association of Forensic Scientists (MAFS) also require completion of a formal training program. A diplomate of the American Board of Forensic Document Examiners (ABFDE) must prove that he or she had completed the requisite two-year formal training program under the tutelage of a recognized FDE in order to qualify as a testing candidate.

Training programs based solely on correspondence courses or literature review of FDE reference books are not accepted avenues of study. Working side by side with an experienced examiner provides the training for correct interpretation of the significant characteristics observed in questioned and known material. In addition to the FDE apprenticeship, a well-rounded training program consists of studying numerous texts and articles, and attending workshops and seminars sponsored by the organizations mentioned earlier in this section. The FDE's direct testimony must be thorough in its discussion of training, certification, and organization memberships.

37.2 Graphoanalysts

Even though document examiners and graphoanalysts/graphologists both examine handwriting, their focus of study and objectives differ. Past texts have discussed the primary differences between the two professions: graphoanalysts, also called graphologists, examine features of handwriting to determine personality traits; FDEs examine handwriting and signatures to determine identification or elimination of a writer. In addition to the specialized training in handwriting identification, FDEs have special training to examine documents for alterations, determination of printing processes, sequence of writing problems, indented writing, etc.[2–5]

Sometimes the layman is confused as to the differences between a graphologist whose training is solely in graphoanalysis and a qualified forensic document examiner. The two professions differ greatly in training, professional memberships, and objectives.[2–5] An analogy of how the two differ can be demonstrated by comparing a neurosurgeon to a cardiologist. Both are medical doctors and each possesses a specialized skill that allows for diagnosis and treatment of an illness affecting a particular area of the body such as the heart. Few would argue that if one has a heart attack, the cardiologist, not the neurosurgeon, should be the treating physician.

A number of graphologists receive training through correspondence courses from graphology organizations. Their curriculum vitae may list the correspondence school that issued the certificate or master graphologist certification. Correspondence courses offered by the International Graphoanalysis Society (IGAS) focus on graphology, not forensic document examination. In a letter from officials of IGAS, it clearly states that it offers a one-week lesson consisting of an overview of the questioned document profession. IGAS recognizes that forensic document examination is a separate field with its own technical and professional requirements. According to IGAS, the study of graphoanalysis does not qualify one as an FDE.[6]

37.3 Fraud/Forgery Investigators

A law enforcement agency or financial institution may employ fraud or forgery investigators whose careers are spent investigating fraudulent documents. Due to their exposure to spurious signatures or writings, they are often under the incorrect assumption that this experience qualifies them to testify as FDEs. Law enforcement personnel can attend the two-week Questioned Document Course offered by the U.S. Secret Service. Upon returning from this training, it is not uncommon for a detective to "set up shop" and claim to be a qualified FDE. In some instances, detectives become the police department's FDE because the agency feels that two weeks of training is adequate. Unfortunately, their assignment as the agency's expert lends credibility to their examinations and conclusions. The abuse by a few of those who attended the Secret Service course caused officials of that agency to issue a letter stating the school is not intended to be the complete training program.[7] This letter detailing the course's objective, along with additional disclaimers in the correspondence sent to the applicants and their supervisors provide a clear commentary on the program's intent. The Secret Service makes it very clear that the investigator is far from being a qualified forensic document examination expert.

The comment, "I'm no expert, but the writing looks the same to me" is heard more often from fraud/forgery detectives than any other investigator. This comment serves more to annoy the FDE than to give him or her a hint as to the "right" answer. Because certain fraud/forgery detectives fail to recognize the complexity of a forensic document examination, the detective may not seek the services of an FDE and may move forward to file charges on a suspect based upon his or her own handwriting examination. This preconceived notion prevents the investigator from recognizing his or her own limitations, as well as the limitations that are inherent in some of the evidence.

Fraud or forgery investigators are usually the lead detectives on their case. As the investigator, he or she contacts the victim and develops a suspect. His or her dual role as investigator and handwriting expert prevents him or her from being an advocate for the evidence. As is common with laypeople, the detectives place a great deal of weight to gross similarities in a writing or signature while disregarding or failing to recognize the subtle, inconspicuous characteristics. And finally, since they have not completed a minimum formal two-year training program, investigators lack an understanding of the methodologies applied in document cases.

37.4 Criminalists

Just as with fraud/forgery investigators, criminalists are sometimes accepted as qualified FDEs because the courts fail to recognize their lack of formal training in document examination. Criminalists who testify as FDEs can be problematic if the proffered expert is not questioned extensively in *voire dire* about his or her training in document examination. In forensic sciences, training in one specific discipline is not sufficient for one to conduct examinations in other areas of the comparative sciences. Each comparative discipline is a complex science comprised of its own methodologies that are unique to that specific expertise. Without the proper and specific training, it is improper for an FDE to offer expert testimony in firearms or latent prints. The same is true for the criminalist who has not received the two-year formal training program in forensic document examination. A minimum training period is required for each discipline in order that the trainee is able to correctly interpret the evidence.

37.5 Critics of the Forensic Document Profession

Professional critics of the forensic document discipline are the final group of untrained individuals to be discussed. Critics have challenged several forensic disciplines, including documents, in the courts of the United States. These critics are usually from the academia field — Ph.Ds in areas of study separate from the forensic sciences. Based upon a cursory literature review, the critic believes that the forensic sciences lack reliability.

As a rule, the critic does not possess an understanding of terminology and examination methodologies of the forensic discipline. A limited literature review results in a narrow scope of understanding, causing the critic to reach a conclusion that obviously reflects an inability to discern information in its proper context. For example, members of this group have testified that FDEs cannot identify hand printing, basing their testimony on articles and books published prior to 1955. Literature from this time period describes hand printing as a new writing style that was introduced into the United States from Great Britain. Even though none of the pre-1955 literature states hand printing lacks identifying features, some of the observations discussed in these writings are understandable when taken in the context that hand printing was a new writing style for U.S. writers in the 1920–1940 era. During that time in history, most adults had not been taught hand printing in grammar school, as it was not an established subject. Additionally, the critics overlook current literature consisting of numerous research studies on hand printing, including class characteristics of certain letterforms. The contemporary research is contrary to their position and advocacy. Therefore, the FDE must include these studies in direct or redirect testimony so the judge will know that forensic document examination practitioners use contemporary methodology that is not based on antiquated information, but rather on ongoing research that has a historical progression.

37.6 Cross-Examination of Credentials

Revealing a lack of proper credentials in forensic document examination can only be done through vigorous cross-examination with the assistance of a qualified FDE. It is the attorney's responsibility to investigate the claims of training, education, professional organization memberships, awards, and references on any proffered expert's curriculum vitae. The attorney should contact the organizations to make sure that the expert is a member. The FDE can be of assistance by having an understanding of the membership requirements of the mainstream organizations.

Not all certifications are equal. If certification is claimed, the certifying body should be contacted to ascertain the certification process, recertification requirements, and whether the listed expert currently possesses certification through that organization. As discussed earlier, some graphoanalysts earned certification as a master graphologist, which does not make them certified as a forensic document examiner. The Association of Certified Fraud Examiners offers certification to those involved in fraud/forgery investigations. Again, this particular certification does not make the investigator a certified FDE. Since there are several organizations that offer certification, the attorney's cross-examination should include questions seeking clarification as to who issued the certification, detailed description of the testing process, whether the certifying body is a separate (stand alone) organization (such as ABFDE), or instead, is through a professional membership organization or government agency. For example, there are some certification programs offered on the Internet. One such program offers certification upon successful completion of a written test consisting of ethics questions. Forensic scientists who claim certification from one of these programs are referred to as having "checkbook credentials."[8]

It cannot be over-emphasized that the attorney should verify training and references listed on an expert's curriculum vitae. There have been instances where an untrained document expert claimed training by Ordway Hilton. This became such a frequent occurrence that Hilton wrote a letter in 1996 explaining he has never trained anyone and that his presentation at seminars should not be considered as one-on-one training.[9] This demonstrates that the listing of training references should not be taken at face value.

37.7 Conclusion

It is unfortunate that a number of individuals mistakenly believe they can conduct document examinations based on their ability to recognize superficial similarities. Few other disciplines face this problem. A latent print examiner, for example, will not be told by a detective that he can tell that the latent print looks the same as the print on the known fingerprint card. This lack of appreciation for the complexity of forensic document examination leads to a proliferation of untrained document experts who lack the requisite training to recognize they are not qualified to make conclusions regarding the authenticity of a signature or extended writing.

To properly expose the untrained document expert in court, the FDE should make sure the attorney is aware of the mainstream organizations, including membership requirements and requirements of certifying bodies. He or she should also have an understanding of the difference between graphoanalysis and forensic document examination. The attorney should investigate listings on the curriculum vitae of training, references, memberships, and certifications in order to expose those who have embellished their credentials or falsely claimed one-on-one training. Although it is the attorney's responsibility to investigate an expert's credentials, the FDE should be willing to aid the attorney with his or her inquiries.

The document examiner should be knowledgeable about the background and past testimonies of the critics who try to exclude the FDE from testifying. As with all of the untrained witnesses, vigorous cross-examination of the critic will reveal the lack of proper credentials, and his or her advocacy to exclude the forensic document examination profession from the courtroom.

All of the untrained document expert witnesses have one thing in common: the lack of a structured formal two-year training program. This lack of training will factor quite heavily in contemporary U.S. courtrooms as a result of *Daubert*. In the United States, the requirements of *Daubert* and FRE 702 demand an expert's use of a methodology that has been tested, generally accepted, and properly applied. This is rather difficult to do if the proffered expert lacks the training to understand the significance and application of the methodologies used in forensic document examination.

References

1. Rules of Evidence (2004). Available at www.law.cornell.edu/rules/fre/rules.htm
2. Morris, Ron N. (2000). *Forensic Handwriting Identification: Fundamental Concepts and Principles*, Academic Press, San Diego, p. 224.
3. Levinson, Jay (2001). *Questioned Documents: A Lawyer's Handbook*, Academic Press, San Diego, p. 56.
4. Harrison, Wilson R. (1958). *Suspect Documents*, Nelson-Hall Publishers, Chicago, pp. 518–519.
5. Hilton, Ordway (1982). *Scientific Examination of Questioned Documents*, Elsevier Science Publishing, New York, p. 5
6. ABFDE Resource Kit, letter to Brian Lindblom from Corrine A. Lamb, Director, Office of Public Affairs of the International Graphoanalysis Society, November 9, 1995.
7. ABFDE Resource Kit, correspondence from Secret Service Chief Document Examiner John W. Hargett, December 2, 1993.
8. Hansen, M. (2000). *Expertise to Go*. Available at http://www.abanet.org/journal
9. ABFDE Resource Kit, correspondence from Ordway Hilton to Floyd Whiting, August 13, 1996.

The Document Examiner's Testimony in Court

38

JAN SEAMAN KELLY

Contents

The document examiner's testimony reflects the culmination of the examination process and the findings issued in the particular case. Demonstration exhibits and the pretrial conference are two tasks that have been completed. Court preparation, *Daubert* hearing notification, demonstrative charts, and pretrial conferences have been discussed in earlier chapters in this section. This chapter will focus on the knowledge an FDE should articulate before the judge and/or jury to effectively meet the objective of speaking for the evidence.

The court systems of the U.S., Canada, Great Britain, and the European Union will also be discussed. All free countries have established court systems designed to provide equal justice for their citizens. Successful delivery of testimony begins with having a basic understanding of that country's court system. Understanding the judicial procedure will greatly assist the FDE in being an effective advocate for the evidence.

38.1 Knowledge of Judicial Process for Court Hearing/Testimony

38.1.1 United States

The judicial system in the U.S. is comprised of local, state, and federal courts. Military and bankruptcy courts are included in the federal court system. Court testimony is conducted in an adversarial climate; i.e., there are two sides in every legal dispute — a plaintiff (or prosecutor) and a defense. The plaintiff and defense attorneys each have the responsibility of representing their client to the best of their ability. Each one is an advocate for his or her position. The presiding judge is the referee who makes sure the laws and court procedures are properly interpreted and followed. This adversarial court system is contrary to the position of forensic scientists. A forensic scientist's role, no matter the discipline, is to be an advocate for the evidence. His or her testimony is to reflect the accuracy of the evidence no matter whose side it will help or hinder. For this reason, the forensic scientist can expect vigorous cross-examination from opposing counsel in court jurisdictions based in the adversarial process.

The Federal Rules of Evidence (FRE) are the rules that govern the court proceedings in the U.S. Article VII in the Federal Rules of Evidence focuses on opinions and expert testimony. FRE Rule 702[1] specifically addresses the admittance of experts and their testimony. Revised in 2004, this rule states that if the scientific, technical, or other specialized knowledge will assist the trier of fact to understand the evidence, the witness who is qualified as an expert by knowledge, skill, experience, training, or education may testify to his opinion providing the testimony meets the following three criteria:

1. Testimony is based upon sufficient facts.
2. Testimony is the product of reliable principles and methods.
3. The principles and methods have been applied reliably to the facts of the case.

Rule 702 is a close companion to the *Daubert* factors. The objective of Rule 702 is to ensure that the expert's testimony is reliable and assists the jury (trier of fact) by providing information outside of the knowledge base of the jury. Determining if the testimony assists the trier of fact lies solely with the presiding judge. For this reason, it is extremely important for the FDE to articulate that his or her conclusion was based on an objective examination using accepted and established methodology.

38.1.2 Canada

The federal and provincial governments form the judicial system in Canada. The federal government appoints the judges of the superior courts in the provinces. Parliament created the Supreme Court of Canada, the Federal Court, and the Tax Court. Authority for criminal law and procedure lies with Parliament.[2]

The provincial government divides its court system into two categories: provincial courts and superior courts.[2] Provincial courts hear cases involving criminal violations and, in some provinces, civil cases involving small amounts of money. Trial and appellate courts comprise the superior courts. Serious criminal and civil cases are heard at the trial court. The appellate court in the superior courts' system hears the case appeals from the trial court.

In *R vs. Morin*,[3] the Supreme Court of Canada outlined a number of limitations on the use of forensic experts in trial proceedings. Admission of expert evidence must meet the below listed criteria:[3]

1. **Relevance** is the threshold requirement determined by the Supreme Court of Canada. The judge determines this first requirement, as it is a question of law. The

issue of relevancy is a complex one, as the judge has to balance the expert's testimony between its probative and prejudicial values. The judge is concerned that his decision may allow the jury to give more weight to the expert's findings, thus endangering the fact-finding process. The Supreme Court of Canada instructed the trial judge to determine the following:[4]
   a. The probative value of the evidence
   b. The danger of misleading or confusing the jury by allowing the expert to testify
   c. Achieve equilibrium between the probative value and the prejudicial effect of the evidence
2. **Necessity in assisting the trier of fact** is the second criteria in determining if the expert evidence should be admitted. To meet this criteria, the information to be given to the jury by the expert must be outside of the jury's experience and knowledge.
3. **A properly qualified expert** is one who has acquired specialized knowledge through study or experience of a specific field. The judge makes this determination in the *voir dire* portion of the expert witness's testimony.
4. **The absence of any exclusionary rule** is the final criteria. The trial judge determines whether the testimony violates any other exclusionary rule of evidence. If it does, then the expert's testimony cannot be admitted, even if it meets the other three criteria. For example, the expert's testimony is not admissible if its only purpose is to bolster the credibility of another witness's testimony.

Even though the court system operates within a parliamentary government, testimony occurs in an adversarial atmosphere. As of 2004, *Daubert* hearings have not taken place in Canada. However, *Daubert*-style questions have been asked during the cross-examination phase of an FDE's testimony. Direct testimony should include an explanation of the accepted methodology used in order to satisfy *Daubert*-style questions and to assist the trial judge in his or her decision as to whether the FDE's expertise will assist the trier of fact.

38.1.3 Great Britain, Northern Ireland, and Wales

Parliamentary sovereignty reigns over all government institutions, including the monarch and the courts. Legislative and judicial functions are overseen by Parliament. Since 1948, criminal cases have been heard before regular juries.[5]

In civil cases, permission from the court must be obtained before the expert can be called or his or her report can be admitted into evidence. The rules in civil cases pertaining to the requirements for expert evidence and the form and content of the expert's reports can be found in the *Practice Direction: Experts and Assessors*, Supplement CPR Part 35.[6] An interesting requirement listed in this supplement states that it is the duty of the expert to help the court on matters within his own expertise. The requirement goes on to state this duty is paramount and overrides any obligation to the person from whom the expert has received instructions or by whom he is paid.[6] This requirement certainly is contrary to the systems in the U.S. and Canada.

The presiding magistrate determines if the expert's opinion testimony is outside the experience or knowledge of the judge or jury. If the information is within his knowledge, most likely the expert will not be allowed to testify.[7]

38.1.4 European Union

In 1992, the Treaty on the European Union was signed. Its objective was to recognize an individual's rights as a fundamental of community law. Twenty-seven legal systems (countries referred to as member states) are members of the European Union. The objective of this treaty was to set procedural safeguards for suspects in criminal proceedings.[8] In addition to the United Kingdom and Ireland, Austria, Belgium, Cyprus, the Czech Republic, Denmark, Estonia, Finland, France, Germany, Greece, Hungary, Italy, Latvia, Lithuania, Luxembourg, Malta, the Netherlands, Poland, Portugal, Slovakia, Slovenia, Spain, and Sweden are member states.

Even though the member states are unified in their goal of setting standards for judicial procedures, the use of expert testimony varies. In a few countries, such as Austria, Denmark, Estonia, and Malta, the court appoints the expert. In Finland, France, Italy, and Poland, communications between the expert and the retaining party are only privileged if the party is the defense. If the retaining party is the prosecution or the court, the remaining member states do not recognize privileged communications; therefore, all communications are disclosed.[9]

The discussion regarding the rules for experts in Canada, England, and the European Union is not an in-depth one. The purpose of this brief outline is to stress the importance of understanding the judicial system where the testimony is to be delivered. Knowing what is and is not allowed will ensure that the FDE delivers his or her testimony according to the requirements of the particular court.

38.2 Direct Testimony

Upon entering the courtroom, the document examiner will be asked to hold up his or her right hand and take an oath to tell the truth. Once seated, the attorney who retained the FDE's services will begin the testimony phase by asking questions detailing his or her credentials. This initial phase of questioning is called *voir dire*. The opposition can offer to stipulate to the qualifications of the FDE. The attorney who will be conducting the direct *voir dire* can agree to accept the stipulation or tell the presiding judge that he or she prefers that the judge or jury hear the expert's qualifications. Questions to elicit information regarding the FDE's academic education, training specific to forensic document examination, memberships in recognized organizations, offices held in these organizations, presentations or publications of research, and continuing education or training are discussed in the direct *voir dire* testimony. If the document examiner is certified, testimony should include a detailed description of the testing process, including the requirements for recertification. Once the direct *voir dire* is complete, opposing counsel can elect to cross-examine the FDE or choose not to challenge the qualifications. Upon acceptance by the court that the FDE has the qualifications to testify before the judge or jury as a forensic document examiner, direct testimony will follow.

The FDE's testimony should always follow the rule WYSIWYM (what you say is what you mean). A savvy litigator may try to convince the document examiner to modify statements in his or her evidence or credentials to avoid what the attorney perceives to be a weakness. A change in verbiage, or the level of certainty, not only impacts on the accuracy of the evidence, but also can affect the judge or jury's perception of the FDE's integrity. For example, a report or testimony stating "The results of the examination are inconclusive" could be somewhat damaging to the prosecution, but beneficial to the defense in a criminal trial. "I cannot eliminate Mr. Jones" is another way of stating an inconclusive examination. Since the second statement emphasizes the inability to eliminate Mr. Jones, the testimony now implies the FDE is issuing some degree of identification instead of a neutral conclusion. Another example is to eliminate qualifying words during the *voir dire*. For example, the FDE testifies, "I have attended the 2-week document examination school at the FBI Academy." This statement has a qualifier: the training is 2 weeks in length. Some attorneys feel the length of the school minimizes the significance of the training. They may encourage the FDE to state, "I have attended the document examination school at the FBI Training Academy." This latter statement sounds much more impressive than the first. However, if during cross-examination the FDE is asked the length of the FBI class, he or she has given the jury a reason to not trust the rest of the testimony due to the omission of the training class's length.

Direct testimony focuses on the examination and the results rendered in the report. As a rule, testimony regarding the examination methodology and issued conclusion is straightforward. However, there are instances where the examination leads to an inconclusive or qualified opinion. The significance of a mixed bag of similarities and dissimilarities resulting in a qualified or limited conclusion must be explained to the judge or jury by the FDE in his or her direct testimony. Limitations may be due to the following: the condition of the questioned text, for example, the disputed signature is a tracing of a genuine; or an inadequate amount of comparable known exemplars is preventing a definitive conclusion. Whatever the limitation, the FDE must articulate its cause as well as explain what the qualified conclusion means. By taking the time to give thorough testimony discussing the strengths and weaknesses that led to the issuance of a less than definitive conclusion, the FDE is fulfilling his or her role as an advocate for the evidence.

During direct testimony, the attorney asks the document examiner to identify the documents that have been admitted to court as exhibits. The FDE should take his or her time to study each exhibit to determine if it is one of the documents submitted for examination. Careful review of each document will allow the FDE to identify any documents that were not submitted in the examination, but have been included (intentionally or not) into the group of exhibits. The document examiner should not hesitate to state that an exhibit does not look familiar.

The basic premises of identification/elimination and the methodology used in reaching the conclusion are important portions of direct testimony. For instance, testimony regarding a handwriting case should include why handwriting is identifiable by explaining the two primary premises of this discipline: no two people write exactly alike, and no one writes the same letters exactly the same way twice. Even though the primary premises are basic to seasoned FDEs, direct testimony should explain the foundations of forensic document examination since this information is outside the expected knowledge of the trier of fact. Once the foundation has been laid, direct testimony narrows the focus to the examination and conclusion that is relevant to the case being heard before the court.

Use of demonstrative charts is an integral part of direct testimony. Prior to his or her court appearance, the FDE must make sure his or her choice of media presentation (PowerPoint, ELMO, slide projector, or an enlarged chart) can be used in the assigned courtroom. If, for example, the courtroom cannot accommodate a PowerPoint or slide presentation, then an enlarged chart may be the only option. No matter how the demonstrative chart is presented, the examples must accurately reflect the original evidence. If the exhibit contains any detail that underwent enhancement techniques, the FDE should state this while using the exhibit. It might be wise to have a demonstrative exhibit to show the condition of the non-enhanced image and then to follow with the enhanced exhibit accompanied with the explanation of the steps taken to enhance the text. The FDE should state that the enhancement only increased the clarity of the image and that it was not used as a tool to alter the evidence. For example, where multiple indented impressions have been discovered on a tested sheet, it may be advantageous to improve the legibility of these impressions through the use of autoadjust settings in a graphics software program. The resulting image will be clearer and more readily appreciated by the judge or jury. An explanation of how the visibility of the indented impressions was improved should accompany the presentation of the charts.

Cross-examination follows completion of the FDE's direct testimony. Opposing counsel is given great latitude in questioning the document examiner on any of the information discussed in his or her direct testimony. During all phases of testimony, the FDE must listen carefully to the questions in order to answer them completely. A favorite ploy of the opposition is to insist that the FDE be allowed to answer only yes or no to his or her questions. If the question cannot be answered honestly with this type of answer, the document examiner needs to advise the court that a narrative answer is required. Prior to *Daubert* (in the U.S. courts), the tactic of one-word answers was successfully used because the presiding judge ordered the witness to answer the question in the manner directed by opposing counsel. The *Daubert* era has mostly neutralized the strategy of one-word answers, since the presiding judge knows the expert must testify about a great deal more information. If this type of situation occurs, the attorney for whom the expert is appearing should ask for an explanation of the same questions asked during the cross-examination. By asking for further elaboration, the re-direct testimony will provide an accurate explanation of the information for the judge or jury. Cross-examination can last a few minutes or a few days. Anticipating an extensive cross-examination should be considered when preparing for court testimony. Investing the time to review the work notes, demonstrative court exhibits, and foundational information will provide the confidence the FDE needs to endure both the direct and cross-examination portions of testimony.

Re-direct questioning from the retaining attorney is the final phase in the testimony process. It is worthy to note that opposing counsel will have an opportunity to cross-examine the FDE on any information discussed in the redirect phase of the testimony. Re-direct questions are limited to the information discussed during cross-examination. As a rule, re-direct testimony is used to clarify a question asked of the FDE during cross-examination. This provides the opportunity for the FDE to elaborate more completely on questions asked during that phase of the testimony.

38.3 Conclusion

Direct testimony is the document examiner's opportunity to speak as an advocate for the evidence. Preparation includes reviewing the basic principles of forensic document examination, the case work notes, and report. The FDE also should take the time to familiarize himself or herself with the courtroom procedures. This is important especially if the FDE is going to testify in a jurisdiction he or she rarely testifies in, or before a judge he or she has never testified in front of. Embellished testimony or implying something is more than what it actually is produces tainted testimony that does not serve justice. Even though the FDE must give testimony in an adversarial atmosphere, communicating the scientist's objective position of representing the evidence, not the plaintiff or defense, supplies the information the trier of fact needs in order to render a just verdict.

References

1. Federal Rules of Evidence, LII 2004 ed.
2. Canada Department of Justice government website, http://www.canada.justice.gc.ca
3. *R v. Morin*, 89 C.C.C (3d) 402 at p. 411, 114 D.L.R. (4th) 419 (S.C.C.), 1994.
4. Chayko, G.M. and Gulliver, E.D., Forensic evidence in Canada, in *Canada Law Book*, Aurora, Ontario, 1999.
5. Wikipedia, the Free Encyclopedia, Parliament of the United Kingdom, http://wikipedia.org/wiki/Parliament_of_the_United_Kingdom
6. Department for Constitutional Affairs, Civil Procedure Rules, *Practice Direction: Experts and Assessors* (this Practice Direction supplements CPR Part 35).
7. Radley, R., E-mail correspondence, 2005.
8. Treaty on the European Union, February 7, 1992.
9. Law Society of England and Wales, Study of the laws of evidence in criminal proceedings throughout the European Union, 2004, www.lawsociety.org.uk

Glossary

Accidental feature An unusual feature that is not found in a representative sampling of an individual's handwriting or signature. This accidental feature may be a one-time-only occurrence and the result of an unknown condition or set of conditions. Whether this anomaly, if found on a document in question, is significant or not is a subjective determination.

Alteration A change (revision or modification) in a document that changes its legal effect either in the obligation it imparts or its force as legal evidence.

Altered document A document that contains some change, either as an addition or a deletion.

American Board of Forensic Document Examiners Certifying body for forensic document examiners, sponsored by the American Academy of Forensic Sciences, the American Society of Questioned Document Examiners, and the Canadian Society of Forensic Sciences.

American Society of Questioned Document Examiners Professional association of forensic document examiners.

Analog copier A photocopier utilizing an imaging system based on xerographic technology in which the image of the original is photographically transferred to the drum.

Apex The uppermost point of a character (ASTM 2195).*

Arcade Stroke used to form the arcs of the cursive letters *m* and *n*; also a cursive writing style where upward arc forms predominate.

Ascender A stroke that rises above the height of the body of the letter formation (ASTM 2195).*

ASCLD-LAB American Society of Crime Laboratory Directors–Laboratory Accreditation Board. Independent accrediting body for government and private laboratories. ASCLD-LAB laboratories are primarily found in the U.S.

Assisted hand signature A signature that is executed while the writer's hand or arm is steadied, or stabilized, by another individual (ASTM 2195).*

Awkward hand *See* opposite-hand writing.

Ball element An element used in an impact printing device in which the fully formed characters are located on the outer surface of a sphere-like device. Frequently called *golf ball element* (ASTM F 909).*

Ballpoint pen A writing instrument having as its marking tip a small, freely rotating ball bearing that rolls the ink onto the paper. The ink is typically highly viscous and non-aqueous.

Baseline The ruled or imaginary line upon which the writing or typewriting appears to rest (ASTM 2195).*

Bitmap A representation, consisting of rows and columns of dots, of a graphics image in computer memory. The value of each dot (whether it is filled in or not) is stored in one or more bits of data. For simple monochrome images, one bit is sufficient to represent each dot, but for colors and shades of gray, each dot requires more than one bit of data. The more bits used to represent a dot, the more colors and shades of gray that can be represented. The density of the dots, known as the resolution, determines how sharply the image is represented. This is often expressed in dots per inch (dpi) or simply by the number of rows and columns, such as 640 × 480. There are many bitmapped file formats, including, but not limited to, .bmp, .tif, .jpg, .gif, and .png. Bitmap images are also referred to as raster or paint images.

Blemish A small extraneous spot found near inked regions of checkwriter impressions that is characteristic of machines that use ribbons as their ink source (ASTM 2285).*

Bowl A curved stroke joining a stem that completely or partially encloses an area forming a bowl-like shape.

Brightness (1) A measure of how light or dark an image is. (2) In paper, a characteristic of white paper measured in terms of reflectance in the blue and violet portions of the spectrum (ASTM F149).*

Capital letters Large, uppercase letters; also called majuscules or caps.

Carbonless paper A paper stock that is coated with a pressure-activated inking system. Commonly referred to as NCR (no carbon required) paper.

Case notes Written record of the FDE's activity as it relates to a specific case. Case notes are comprised of copies of the submitted documents and notes, whether written, drawn, highlighted, or electronically recorded, that describe the FDE's observations in the examination process as well as the methodology used. Also known as work notes.

Case review To assess the contents of a case folder in preparation for testimony.

CB An abbreviation for coated back, referring to a carbonless paper of the two-coat transfer type (ASTM F549).*

CCD *See* charge-coupled device.

CF An abbreviation for coated front, referring to a carbonless paper of the two-coat transfer type (ASTM F549).*

CFB An abbreviation for coated front and back, referring to a carbonless paper of the two-coat transfer type (ASTM F549).*

Character Any language symbol (e.g., letter, numeral, punctuation mark, or other sign), other symbol, or ornament (ASTM 2195).*

Charge-coupled device A computer chip that converts light into electric signals. Commonly referred to using the acronym CCD.

Chart A visual display of information specifically created to illustrate the findings of the FDE.

Checkwriter A device manually or electrically powered or computer generated, designed to ink, emboss, print, perforate, or shred a monetary value, along with other peripheral information, onto a document (ASTM 2285).*

Chemical toner Toner that is polymerized into very tiny, uniform spheres.

Class characteristic A characteristic common to a group or class.

CMYK These letters stand for cyan, magenta, yellow, and black, and are the four colors used in most four-color printing technologies. Using halftoning, the eye can be tricked into seeing millions of colors, when just these four are combined properly. Cyan, magenta, and yellow are the three primaries in this subtractive color model.

Collected specimens Handwriting, machine-generated, paper, or other material samples taken from personal or business files. *See also* course of business exemplar.

Compact flash A form of portable digital memory storage.

Connecting stroke A line joining two adjacent characters (ASTM 2195).*

Contrast A measure of the tonal range of an image.

Copier fabrication The use of a photocopier to construct a document for the purpose of deception, often through the transfer of a signature or other text from an existing document into a new document.

Copybook form Letter designs that are peculiar to a particular system of writing and are provided for students to imitate.

Copying The act of producing an image on a document or other receptor media that is a duplication of the image of another document, such as by a photographic, xerographic, or facsimile process or by carbon or carbonless papers (ASTM F335).*

Counterfeit document A document created or altered without authority. In many instances it is an attempt to imitate a genuine negotiable instrument or some other document with monetary or identification value, such as currency, transit passes, identification cards, and passports.

Course of business exemplar In handwriting or machine-generated documents, samples executed during the course of everyday business. These samples were prepared without any thought that they might be used in a forensic document examination. Examples of course of business exemplars are cancelled checks, contracts, lease or rental agreements, and personal correspondence.

Cropping The process of removing portions from the periphery of an image.

Cross-examination Following direct testimony, this is the opposing counsel's opportunity to question a witness.

Cross mark Historically, a simple mark or design, many times an X or a cross that is used by an individual who is either uneducated or unable to write out a more individualizing signature. Legally, this simple mark or symbol can and will be accepted as a legitimate signature of an individual, if properly witnessed.

Curriculum vitae A summary of one's education, professional history, and job experience.

Cursive writing A type of writing in which the letters are joined and the writing instrument is not lifted after most strokes (ASTM 2195).*

Curves A procedure in a digital photo processing software program that allows any brightness value in an image to be mapped to any other brightness value.

Daisy wheel *See* typewheel.

Dandy roll A cylinder that is covered with woven wire that may ride on the mat of fibers in the papermaking process. It may have a raised or recessed design affixed to the wire cover to mark the passing sheet and create a watermark.

***Daubert* hearing** Hearing to determine if a scientific discipline meets some or all of the following five factors: (1) whether the method or theory has been tested, (2) whether the technique or theory has been subjected to peer review, (3) the existence and maintenance of standards controlling the technique's operation, (4) general acceptance of the technique or theory, and (5) the known or potential error rate. Supreme Court decision in *Daubert v. Merrell Dow Pharmaceuticals, Inc.* (92-102), 509 U.S. 579 (1993).

Decipherment The process of making out what is illegible or what has been effaced. In this book decipherment refers to the process of reading or interpreting the erased or obliterated material that is illegible.

Defendant In civil litigation the person against whom the action has been filed; in criminal matters it is the person charged with the crime.

Descender A stroke that extends below the baseline of the body of the letter formation (ASTM 2195).*

Diacritic A sign, such as an accent or cedilla, used to indicate different sounds or values of a letter or to distinguish words that are otherwise graphically identical.

* Reprinted with permission. Copyright ASTM International, 100 Barr Harbor Drive, West Conshohocken, PA, 19428.

Difference A characteristic that appears in the questioned but does not appear in the known. In handwriting it may be repetitive and falls outside the person's range of writing and may lead to a non-identification.

Digital Operating on data represented as a series of binary digits or in a similar discreet form.

Digital copier A copier whereby the image of the original document goes through a digital processor before the laser charges the photoconductor drum.

Digital enhancement A change made to an image with the aid of computer software intended to improve the visibility of information contained within it.

Digital offset printing A process that merges electrostatic imaging, liquid toners, and an offset blanket and impression roller. Most of the new digital offset printers can utilize more than the four traditional offset colors (CMYK).

Digital photo processing software Computer programs that are capable of editing digitally stored images.

Direct testimony Evidence given by a witness who is testifying for the side that subpoenaed him or her.

Disguised writing Deliberately altered writing intended to hide the identity of the writer by changing his or her habits.

Disputed document Document in argument or controversy; interchangeable with questioned document.

Distorted writing Writing that does not appear to be, but may be, natural. This appearance can be due to either voluntary factors (e.g., disguise, simulation) or involuntary factors (e.g., physical condition of the writer, writing conditions) (ASTM 2290).*

Dithering Refers to any of several models of halftoning in which varying patterns of dots are used to simulate shades of gray. This technique is often extended to color models in which dithering is applied to individual channels to simulate a broad range of colors.

Document examiner One who scientifically studies the details and elements of documents in order to identify their source or determine their authenticity or integrity.

Documents Any materials containing marks, symbols, or signs visible, partially visible, or invisible that may ultimately convey meaning.

Dot matrix Impact printers wherein a column or multiple columns of electrically fired pins imprint characters through a ribbon.

Drag line A very thin ink stroke where the writer intended to raise the pen. Can be used to show the direction of the pen. Also called hairline stroke.

* Reprinted with permission. Copyright ASTM International, 100 Barr Harbor Drive, West Conshohocken, PA, 19428.

Dye sublimation A printing process typically using thermal dye ribbons. The selected organic dyes sublimate (change states directly from a solid to a gas) when heated, permeate the substrate surface, and return to a solid state. The substrate must be specifically designed to accept sublimation; thus, plain paper cannot be used.

Efface To rub out, to strike or scratch out, or to erase.

Electron beam printer Similar to a laser printer, except that electricity is used to create the image instead of light. This evolved from ion deposition and is used in very high speed page printers exceeding 800 ppm.

Electrostatic detection device (EDD) The generic name for various items of equipment used for the development of writing impressions. The electrostatic detection apparatus (ESDA) is the most common device.

Enlargement An image, photographic or electronic, that has been increased in its physical size.

Erasure The removal of writing, typewriting, or printing from a document. It may be accomplished by either of two means: a chemical eradication in which the writing is removed or bleached by chemical agents, or an abrasive erasure in which the writing is effaced by rubbing, such as with a rubber eraser.

Exemplar Samples of handwriting, printed text, paper, ink, etc., known to have been produced by a particular individual, machine, or manufacturer.

Eyelet A small oval or loop formation that is part of a cursive or hand printed letter.

FDE Abbreviation for forensic document examiner.

Federal Rules of Evidence Evidence code used in the U.S.

Fiber-tip pen *See* porous-tip pen.

Fixed-pitch font *See* monospaced font.

Flash memory A form of portable digital memory storage.

Flat-die stamps A stamp die that has a low or no relief and is manufactured using the thermal or light-burst technologies.

Flourish A stroke designed for ornamentation.

Fluency Line quality that is smooth, consistent, and has an element of rhythm in the writing line.

Font A design for a set of characters. A font is the combination of typeface and other qualities, such as size, pitch, and spacing. For example, Times Roman is a typeface that defines the shape of each character. Within Times Roman, however, there are many fonts to choose from, such as different sizes, italics, or bold.

Font family A group of fonts sharing a common basic design. One member of the family may be condensed and another use small caps.

Font size The height of a font, typically expressed in points (pts.), including internal leading, diacritic space, ascender and descender space, and cap height. As this measurement includes spacing for components that may not be visible, the set point size will almost always exceed the visible height of capital letters.

Font style The font style specifies whether the text is to be rendered using a normal, italic, bold, or oblique face.

Font variant A font variant is one of many possible alternatives to the normal glyphs that comprise a font. These can be in the form of alternative effects such as small caps, double underline, and shadow or fully rendered variants such as condensed, narrow, and demi.

Font weight Refers to the boldness or lightness of the glyphs used to render the text, relative to other fonts in the same font family. The weight of the font is usually proportional to the width of the strokes that comprise the characters.

Forensic document examiner One who studies scientifically the details and elements of documents in order to identify their source or determine their authenticity or integrity. Also known as an FDE or a document examiner.

Forgery A legal term that has as an element of its definition the intent of the writer. The term generally refers to writing that is not authentic. The determination that a signature or a document was created with the intent to defraud is a determination made by the trier of fact.

Formation Construction of a letter, numeral, or punctuation character.

Formatting The physical arrangement of characters, words, or sentences on a document.

Fourdrinier A paper machine developed by Louis Robert and financed by Henry and Sealy Fourdrinier that produces a continuous web of paper; also the term for the section of the paper machine that is a continuous wire or belt screen, through which the first removal of water occurs (creativepro.paperspecs.com/resources/glossary/f.htm).

Fraudulent signature A signature prepared by someone other than the actual individual and without the permission of the actual individual. A genuine signature may be considered to be a fraudulent signature if it is used for a purpose other than that for which it was prepared, e.g., photocopy manipulation.

Freehand imitation A fraudulent signature that was created purely by simulation rather than by tracing the outline of a genuine signature. Also called a simulated forgery.

Garland A rounded, trough-like stroke such as that in the *u*; an individual style of handwriting with a predominance of these types of strokes.

Gel pen Like ballpoint and roller ball pens, the ink is delivered using a housed metal ball. Initially this class of writing instrument used a high-viscosity, pigment-based, opaque ink, which does not tend to bleed into the paper fibers as much as is found with water-based roller ball or porous-tip pens. Modern black gel inks incorporate both dye-based and pigment coloring. A broad range of ink colors are available.

Glyphs In typography, a pictograph or other symbol that imparts information, especially referring to the design of individual characters in a typestyle.

Graphical User Interface (GUI) A graphical method of controlling how a user interacts with a computer to perform various tasks. Instead of issuing commands at a prompt, the user performs desired tasks by using a mouse to choose from a dashboard of options presented on the display screen. These are in the form of pictorial buttons (icons) and lists. Some GUI tools are dynamic, and the user must manipulate a graphical object on the screen to invoke a function, e.g., moving a slider bar to set a parameter value, as in setting the scale of a map (www.mcaggis.com/glossary.htm).

Graphoanalysis The art of attempting to interpret the character or personality of an individual through the study of his or her handwriting; also called graphology. Graphoanalysis has no relationship to handwriting identification.

Graphology *See* graphoanalysis.

Grayscale Composed of discrete shades of gray, especially as related to photography.

Guided signature *See* assisted hand signature.

Habit Any repeated element or detail that can, and many times does, individualize a person's writing.

Hand lettering *See* hand printing.

Hand printing A style of writing in which the letters are not joined and the writing instrument is lifted after most strokes (ASTM 2195).* Also called hand lettering.

Hand stamp Stamp die mounted on a wooden or plastic handle that requires the die to be impressed on a separate ink pad for inking.

Hesitation A pause or stop in the writing motion in which the writing instrument remains in contact with the writing surface.

Hiatus A void in the ink line caused by lifting the writing instrument at the end of a movement or character followed immediately by its reapplication to the writing surface.

High-relief dies Die characters that are raised above the die's background (non-print area). Stamp dies made of vulcanized rubber, photopolymer, pre-mix gel, salt-leached rubber, foam, and powder are examples of high-relief dies.

Holographic document A document written entirely by its author.

Image The optical counterpart of an object produced by a lens or mirror system or the graphic representation of an object (ASTM F335).*

Indentations Latent or visible impressions in paper or other media (ASTM 2195).*

Individual characteristic Any feature or element that is sufficiently uncommon and is repeated. These features, when found in combination, form the basis for identifying the writer of a signature, the machine used to produce printed text, etc.

Infrared examination The examination of documents employing invisible radiation beyond the red portion of the visible spectrum.

Infrared luminescence A phenomenon encountered with some dyes used in inks and colored pencils that, when illuminated with a narrow band of light in the blue-green portion of the spectrum, give off a luminescence in the far-red or near-infrared range.

Initial stroke The first stroke leading into a character, word, or signature.

Ink eradicator A chemical solution capable of bleaching ink.

Inkjet printer A non-impact printer in which the characters are formed by projecting droplet's of ink onto a substrate (ASTM F909).*

Insertion The addition of writing, text, or other material within a document, such as between lines or paragraphs, or the addition of whole pages.

Intaglio printing A security printing method in which the printing plate is engraved to various depths, resulting in a raised image on the document. Intaglio is used extensively to print bank notes, passports, and other identity documents.

Interlineation The act of inserting writing, typing, or printed words between two lines or paragraphs of information.

Internal consistency An element of handwriting where the same or similar strokes or sequences of strokes occur in different letters.

Interpolate To estimate a value of a function or series between two known values. Interpolation is used in the processing of a digital image.

JPEG A standard algorithm for the compression of digital images (Joint Photogaphic Experts Group).

Justification The typographic adjustment of letters and words to conform to a left margin, a right margin, or both margins. A paragraph is fully justified when each line of type begins and ends at the margins.

Kerning In typography, kerning refers to adjusting the space between characters, especially by placing two characters closer together than normal. Kerning is used for aesthetic purposes to make certain pairs of letters, such as *WA*, *TA*, and *VA*, look better.

Known *See* exemplar.

Latent prints Deposits or traces usually made by the friction ridge surfaces of the hand that are completely invisible. With proper treatment they can be developed or made visible.

Layering tool A feature in digital photo processing software that allows images to be overlaid.

LED An acronym for light-emitting diode. A semiconductor diode that emits monochromatic (single-color) light when forward biased. The emitted light can be red, yellow, orange, green, blue, or non-visible infrared. LEDs are sometimes used rather than a laser to form images on the light-sensitive coating of dry-toner printers.

Letterpress printing A traditional, widely used printing method in which the plate detail is raised off of the surface resulting in corresponding indentations of the characters on the printed document. Today it is primarily used to record serial numbers and other changeable information on invoices, applications, passports, tickets, stocks, and other documents.

Levels An adjustment that enables amendments to three tonal parts of an image: highlights, midtones, and shadows.

Light-burst® technology The use of high-intensity light bursts (such as those emitted by a Xenon flash) to seal the non-print areas on a flat-die stamp. This process allows ink to exit only through the text area on a flat-die pad. The rest of the pad (background) will be dry and free of ink.

Line quality An important characteristic of handwriting. It is the combination of writing speed, skill, freedom of movement, execution rhythm, and pen pressure.

Liquid toner Toner material composed of carbon particles or colorants suspended in a liquid carrier.

Lithography *See* offset lithography.

Lowercase A written or printed letter that is smaller than, and often different in design from, its corresponding uppercase, or capital, counterpart.

Low-relief dies Die characters are on an even or near-even plane with the die background. Opposite of a high-relief die, the die character does not rise above the die's background; it is flat. Also referred to as flat- or no-relief die.

Macro Large in scale. In photography, this term refers to the ability to capture images close-up without a loss of detail.

Manuscript writing *See* hand printing.

Master pattern The family-like resemblance in the general structure of a character in spite of some divergences in design details.

Mechanical toner Toner, produced by grinding resin, that is characterized by irregularly shaped particles.

Megapixel In digital imaging devices, the resolution range where the number of pixels is equal to or greater than 1,000,000.

Memory sticks A form of portable digital memory storage.

Micro-encapsulation As it relates to carbonless paper, the process by which inking chemicals are contained on the surface of the paper and broken open under the pressure.

Microphotography The capture of images through the eyepiece, or a phototube, of a microscope.

Model signature A signature that is used as a prototype for a simulation or copy, by manual, electronic, or other means (ASTM 2195).*

Monospaced font A typeface in which all characters, regardless of their shape or width, are assigned the same horizontal space. As a consequence, the characters fall into uniform columns across the width of the page. Courier is one such example. Also called a fixed-pitch font.

Mould watermark Watermarks produced on a cylinder paper machine. This type of machine makes paper by partially immersing rotating cylinders in vats of pulp stock. In this process, the paper forms into a sheet on wire-covered cylinders or moulds as the water drains through, leaving the fibers on the cylinder surface. Cylinder mould watermarks produce varying densities of the paper fibers and result in high-quality images and portraits that are used in security papers for currencies and other security documents throughout the world.

Movable type Individual letters and punctuation that can be set to form the text of a page. Usually refers to the letterpress printing process.

Movement An important element of handwriting. It embraces all the factors related to the motion of the writing instrument — skill, speed, freedom, hesitation, rhythm, emphasis, tremor, and the like. The manner in which the writing instrument is moved, that is, by finger, hand, or arm action, may influence each of these factors.

Multi-function devices A computer peripheral that can replace several separate devices. For instance, a printer, scanner, copier, and a facsimile machine are combined into one. Also referred to as all-in-one machines.

Natural writing Any specimen of writing executed without an attempt to control or alter its usual quality of execution (ASTM 2290).*

NCR paper *See* carbonless paper.

Oblique light examination An examination with the illumination controlled so that it grazes or strikes the surface of the document from one side at a very low angle; also referred to as a side light examination.

Obliteration Blocking out or covering up of writing, type text, or other information on a document to make the original entry invisible or indecipherable.

Offset lithography The most common form of graphic arts printing. The printing plate is flat in both the image and non-image areas. A latent image is transferred from the printing plate to an offset blanket and then onto the document. The printed image is neither raised nor indented.

Operating system An operating system (sometimes abbreviated OS) is software that, after being initially loaded into the computer during boot-up, manages all the other programs in a computer.

Opinions Judgments based on special knowledge and rendered by an expert.

Opposite-hand writing Any writing executed with the opposite hand from that normally used. Sometimes used as a means of disguising one's writing. Also unaccustomed hand or awkward hand.

Patching Retouching a portion of a written stroke (ASTM 2195).*

Payee perforation An optional device on a checkwriter that perforates or shreds a pattern above the numeric impression region for the purpose of protecting the payee entry from alteration (ASTM 2285).*

Pen emphasis The act of intermittently forcing the pen against the paper surface with increased pressure.

Pen lift An interruption in a stroke caused by removing the writing instrument from the writing surface (ASTM 2195).*

Pen nibs Two points that form the writing portion of a dip or fountain pen.

Pen position The manner of holding the pen relative to the paper or writing surface.

Pen pressure The force with which the writing instrument contacts the paper (ASTM 2195).*

Photocopy manipulation Use of a photocopy machine to fabricate or alter documents.

Pica (1) In typewriting, refers to a pitch in which 10 characters occupy an inch. (2) A typographic unit of measure equal to one sixth of an inch — 12 points.

Piezoelectric inkjet systems An inkjet firing system consisting of a chamber that holds the ink, an orifice where the ink is ejected, and a piezoelectric crystal. When an electric current is applied to the crystal, it flexes into the chamber, forcing ink to be ejected.

Pitch The horizontal spacing of typewriting.

Pixelation Rendering visible, usually through enlargement, the individual picture elements (pixels) in a digital image.

Pixels Short for picture element. Pixels are the smallest elements that can be resolved in a graphic.

Plaintiff In civil litigation, the person who has filed a complaint against another person or group (such as a business or corporation).

Point A typographic unit (abbreviated pt.) commonly used to measure fonts and leading. In North America the point is equivalent to 1/72.27 of an English inch (0.13837 inch), though it is often approximated to 1/72 inch. Twelve points is equal to one pica.

Porous-tip pen A writing instrument in which the marking element or point consists of porous material through which the ink can flow. These pens are commonly known as fiber- or felt-tip pens.

Prefix (1) The part of a checkwriter impression located immediately to the left of the left-most numeral. (2) The part of the checkwriter designed to create the aforementioned impression.

Pre-inked stamp One in which the ink and die are one unit. The die is saturated with ink, making it the ink source.

Printing element The parts of the total checkwriter impression that are not parts of the prefix, which may encompass the currency type, decimal points, and commas (ASTM 2285).*

Proportional type A typeface in which the horizontal space allotted to each character is proportional to its design. As a consequence, the characters do not fall into uniform columns across the width of the page. Also called a variable-pitch font.

Prosecutor In criminal litigation, the government attorney who is presenting the case against the defendant of the criminal charges.

Questioned document Document in argument or controversy.

Range of variation The observed parameters of variance around the master pattern of a letter, word, or numeral. The extent of variation can differ widely among individuals.

Raster image *See* bitmap.

RAW A file format for the storage of digital images in which the color information captured by the CCD has not yet been interpolated. Not all digital cameras allow access to this intermediate format.

Receive Terminal Identifier (RTI) This is usually a one-line string of data appearing across either the top or bottom of the received fax. The data are a function of the receiving fax device and can include identifying information, such as the sender's name, company, and telephone number; date; time; recipient's telephone number; transmission number; and page number.

Requested specimens Samples prepared specifically for the purpose of a forensic comparison.

Resolution The number of picture elements (pixels) per unit of length, usually specified in dots per inch (dpi).

Restoration Any process in which erased, faded, stained, or damaged writing or other information is recovered or made legible.

Retouching *See* patching.

Retracing A stroke written back over the preceding stroke in the reverse direction (ASTM 2195).*

RGB An abbreviation for the three primary additive colors: red, green, and blue.

Rhythm That element of the writing movement marked by regular or periodic recurrences. It may be classed as smooth, intermittent, or jerky in its quality.

Ridge and groove A type of checkwriter impression format resulting in a series of parallel embossed peaks and valleys.

Roller ball pen A writing instrument in which a ball bearing is used in combination with an aqueous fluid ink to transfer the ink to the paper.

RTI *See* receive terminal identifier.

Rubber stamp Marking device comprised of a low- or no-relief or a high-relief die made of materials such as vulcanized rubber, photopolymer, salt-leached rubber, or pre-mix gel attached to a stamp mount. When the die comes in contact with ink and is then pressed onto a surface, such as paper, an impression will be produced.

Rule 702 Section in the U.S. Federal Rules of Evidence that applies specifically to expert witnesses.

Scalable fonts Scalable fonts also are known as vector or outline fonts. The information needed to reproduce the characters is stored as a series of equations describing the curves or outlines that make up the characters. The most commonly available scalable fonts are TrueType and PostScript.

Screen printing A printing technique involving the passage of ink through a screen that has been stretched on a frame. Areas of the mesh are blocked, through the use of either a stencil or photoemulsion, thereby limiting the application of ink to the image area only.

Secure digital A form of portable digital memory storage.

Self-inking stamp The die and ink pad comprise the stamp. The die is in contact with the ink pad, and upon pressing down on the stamp, the die rotates and contacts the paper, resulting in an impression.

Sequence of strokes The order in which writing strokes are placed on the paper.

Shading As it applies to a handwritten stroke, a widening of the ink stroke caused by added pressure on a flexible pen point or through the use of a nib or fountain pen.

Significant difference An individualizing characteristic that is divergent between items, whether they be handwritten, typed, stamped, etc.

Significant similarity An individualizing characteristic in common between two or more items, whether they be handwritten, typed, stamped, etc.

Single-lens reflex Commonly known by the abbreviation SLR, the camera uses a mirror placed between the lens and the film to project the image seen through the lens to a matte focusing screen. Most SLRs use a pentaprism to observe the image via an eyepiece.

Skill Degree of ability or proficiency of someone to write, based on consistency, uniformity, writing instrument pressure, and general artistic qualities.

Slant The angle or inclination of the axis of letters relative to the baseline. Also referred to as slope.

Slope *See* slant.

Solid-ink printer Uses sticks of wax ink that are melted into a liquid. The ink is directed onto a drum, similar to a laser printer, and then transferred onto the paper to produce high-quality output.

Spacing The space between letters or words, whether they be typed or handwritten.

Speed of writing The swiftness with which a body of writing or signature is prepared. The consideration of writing speed may be a significant identifying element. It cannot be measured precisely from the finished product, but can be interpreted in broad terms as slow, moderate, or rapid.

Spurious signature A fraudulent signature in which there was no apparent attempt at simulation or imitation.

Standard Collected writing, instrument, or other types of specimens for which the authorship or source is established and acknowledged.

Stroke In a character, a single straight or curved line.

Sublimation The physical characteristic of a substance to go directly from a solid to a gas when heated, or from a gas to a solid when cooled. In dye sublimation printing, resinous materials are flash heated and transferred to the paper as a gas.

System (of writing) The combination of the basic letter and numeral designs and the associated writing movement as taught in school. Through practice, writing diverges from the system, but generally retains some influence from the basic training.

Technique The systematic procedure by which a complex or scientific task is accomplished.

Terminal stroke The final stroke leading out of a character, word, or signature.

Testimony A declaration by a witness under oath given before a court or deliberative body.

Thermal wax printer Printers that function by transferring a wax-based ink onto paper. As the paper and ribbon travel in unison beneath the thermal printhead, the wax-based ink on the ribbon is melted onto the paper. Monochrome printers use a black panel for each page to be printed, while color printers employ either three (CMY) or four (CMYK) panels.

Thin-layer chromatography (TLC) A technique for separating components in a mixture on the basis of their differing polarities. A spot of sample is placed on a flat plate coated with silica and then carried along by a solvent, or combination of solvents, that travels up the plate through capillary action. Different components will move different distances over the surface. TLC is a useful analytical tool for ink examinations.

TIFF An abbreviation for Tagged Image File Format, which is a file format used for still-image bitmaps stored in tagged fields.

Toner A material (either a dry powder, usually mixed with a binder, or particles suspended in a liquid) that, by means of an electrical charge, is transferred in a controlled pattern onto an imaging drum or belt and then to paper. Toner can be found in printers, fax machines, and copiers that utilize an electrophotographic (xerographic) process.

Traced signature The reproduction of a genuine signature prepared by actually following the outline of a genuine signature with a writing instrument. Tracing may be produced with the aid of carbon paper by first drawing an outline and then covering the outline with a suitable ink stroke. A second method involves using light transmitted through the document and actually drawing the signature over the model. This method many times will leave an indentation around the model signature.

Transmit Terminal Identifier (TTI) This is usually a one-line string of data appearing across the top of the received fax. The data are a function of the transmitting fax device and can include identifying information such as the sender's name, company, and telephone number; date; time; recipient's telephone number; and page number.

Trash marks Marks appearing on a photocopy, facsimile copy, or computer-printed document that do not originate from the original document or file, but rather result from either debris or a defect somewhere in the image processing or printing mechanism. These marks have identifying value.

Tremor Shaky, irregular handwriting movements that occur in handwriting or a signature that is slowly executed or written by persons suffering from advanced age, nervousness, or physical/mental infirmities. Also a symptom of forgery, manifested as shaky, irregular strokes caused by unnaturally slow pen movements due to conscious attention to letterforms in deliberate rather than spontaneous writing.

TTI *See* Transmit Terminal Identifier.

Typeball *See* ball element.

Typebar A block bearing letters and numerals attached to a metal arm. The typebars are engaged when a corresponding key is depressed.

Typeface (1) An interpretation of a character set (letters, numbers, punctuation, and symbols) that share a similar appearance and design, such as stroke width and the presence or absence of serifs. (2) The printing surfaces of the type block or type element.

Typewheel A typing element consisting of a hub surrounded by a series of arms, each of which contains a single character at the end. The wheel rotates and a plunger strikes the back of the type to produce the impression. Also called a daisy wheel.

Typography The arrangement, appearance, and style of printed matter.

Ultraviolet examination An examination utilizing ultraviolet radiation from the visible blue-violet end of the spectrum to react with some substances, which causes a visible light emission (UV fluorescence).

Unaccustomed hand *See* opposite-hand writing.

Uppercase *See* capital letters.

Variation Those deviations among repetitions of the same handwriting characteristics that are normally demonstrated in the habits of each writer (ASTM 2290).* The term can also be applied to deviations in printed matter, paper, and other materials that are the subject of forensic analysis.

* Reprinted with permission. Copyright ASTM International, 100 Barr Harbor Drive, West Conshohocken, PA, 19428.

Vector graphic Images that are comprised of outlines described using mathematical equations. A vector image can be enlarged and still maintain its integrity.

Virtual grids Electronic grids created using computer software allowing them to be overlaid on an image. Can be useful in evaluating alignment and spacing characteristics in a document.

Voir dire A preliminary examination of prospective jurors or witnesses under oath to determine their competence or suitability.

Watermark A translucent design impressed into certain papers during the course of their manufacture.

Witness stand The location in a courtroom where testimony is offered by a witness.

Work notes *See* Case notes.

Writing condition Both the circumstances under which the writing was prepared and the factors influencing the writer's ability. Circumstances pertaining to preparation involve the writer's position (e.g., sitting, standing, lying), the paper support and backing, and the writing instrument.

Writing habits Any characteristic of handwriting that is sufficiently unique and well established to serve as a fundamental feature of identification.

Writing impression (indentation) Impressions formed on a sheet of paper by the act of writing. They may be made on a sheet of paper immediately below the one where the original writing was prepared or on subsequent sheets. The impressions may be clearly visible indentations or latent impressions only detectable through specialized examination techniques. Also referred to as indented writing. *Primary indentations*: Impressions caused by the act of writing or other dynamic actions (ASTM 2291).* *Secondary impressions*: Fiber disturbances caused by contact with the embossed side of indentations and not caused by the act of writing (ASTM 2291).*

Writing offset The result of a paper coming in contact with fresh ink writing. It may be the mirror image of entire words or sentences, or merely fragments of words or letters. The offset may be visible to the naked eye or only observed through specialized examination techniques.

Wrong-handed writing *See* opposite-hand writing.

Xerography A process of making copies of written or printed material, pictures, etc., using a dry photographic or photocopying process in which an image, formed by a resinous powder on an electrically charged drum, is electrically transferred to and thermally fixed on the paper or other copying surface.

Index

A

B

C

D

G

H

P

Q

R

S

T

U

V

W

X